MATA REFERENCE MANUAL
VOLUME 2
[M-4]–[M-6]
RELEASE 11

A Stata Press Publication
StataCorp LP
College Station, Texas

Published by Stata Press, 4905 Lakeway Drive, College Station, Texas 77845
Typeset in TEX
Printed in the United States of America

10 9 8 7 6 5 4 3 2 1

ISBN-10: 1-59718-056-4 (volumes 1–2)
ISBN-10: 1-59718-057-2 (volume 1)
ISBN-10: 1-59718-058-0 (volume 2)
ISBN-13: 978-1-59718-056-6 (volumes 1–2)
ISBN-13: 978-1-59718-057-3 (volume 1)
ISBN-13: 978-1-59718-058-0 (volume 2)

The suggested citation for this software is

StataCorp. 2009. *Stata: Release 11*. Statistical Software. College Station, TX: StataCorp LP.

[M-4] Index and guide to functions

Title

[M-4] intro — Index and guide to functions

Contents

Description

The entries in this section provides an index to the functions, grouped according to purpose.

Remarks

The next section, section [M-5], presents all the Mata functions, in alphabetical order.

Also see

[M-0] **intro** — Introduction to the Mata manual

Title

[M-4] io — I/O functions

Contents

[M-5] Manual entry	Function	Purpose
Console output		
printf( )	printf()	display
	sprintf()	display into string
errprintf( )	errprintf()	display error message
display( )	display()	display text interpreting SMCL
displayas( )	displayas()	set whether output is displayed
displayflush( )	displayflush()	flush terminal output buffer
liststruct( )	liststruct()	list structure's contents
more( )	more()	create --more-- condition
	setmore()	query or set more on or off
	setmoreonexit()	set more on or off on exit
File directories		
direxists( )	direxists()	whether directory exists
dir( )	dir()	file list
chdir( )	pwd()	obtain current working directory
	chdir()	change current working directory
	mkdir()	make new directory
	rmdir()	remove directory
File management		
findfile( )	findfile()	find file
fileexists( )	fileexists()	whether file exists
cat( )	cat()	read file into string matrix
unlink( )	unlink()	erase file
adosubdir( )	adosubdir()	obtain ado-subdirectory for file

	File I/O		
fopen( )		`fopen()`	open file
		`fclose()`	close file
		`fget()`	read line of ASCII file
		`fgetnl()`	same, but include newline character
		`fread()`	read k bytes of binary file
		`fput()`	write line into ASCII file
		`fwrite()`	write k bytes into binary file
		`fgetmatrix()`	read matrix
		`fputmatrix()`	write matrix
		`fstatus()`	status of last I/O command
		`ftell()`	report location in file
		`fseek()`	seek to location in file
		`ftruncate()`	truncate file at current position
ferrortext( )		`ferrortext()`	error text of file error code
		`freturncode()`	return code of file error code
bufio( )		`bufio()`	initialize buffer
		`bufbyteorder()`	reset (specify) byte order
		`bufmissingvalue()`	reset (specify) missing-value encoding
		`bufput()`	copy into buffer
		`bufget()`	copy from buffer
		`fbufput()`	copy into and write buffer
		`fbufget()`	read and copy from buffer
		`bufbfmtlen()`	utility routine
		`bufbfmtisnum()`	utility routine

	Filename & path manipulation		
pathjoin( )		`pathjoin()`	join paths
		`pathsplit()`	split paths
		`pathbasename()`	path basename
		`pathsuffix()`	file suffix
		`pathrmsuffix()`	remove file suffix
		`pathisurl()`	whether path is URL
		`pathisabs()`	whether path is absolute
		`pathasciisuffix()`	whether file is ASCII
		`pathstatasuffix()`	whether file is Stata
		`pathlist()`	process path list
		`pathsubsysdir()`	substitute for system directories
		`pathsearchlist()`	path list to search for file

Description

The above functions have to do with

1. Displaying output at the terminal.

2. Reading and writing data in a file.

Remarks

To display the contents of a scalar, vector, or matrix, it is sufficient merely to code the identity of the scalar, vector, or matrix:

```
: x
                1              2              3              4
    1 | .1369840784    .643220668    .5578016951    .6047949435 |
```

You can follow this approach even in programs:

```
function example()
{
    ...
    "i am about to calculate the result"
    ...
    "the result is"
    b
}
```

On the other hand, display() and printf() (see [M-5] **display( )** and [M-5] **printf( )**) will allow you to exercise more control over how the output looks.

Changing the subject: you will find that many I/O functions come in two varieties: with and without an underscore in front of the name, such as _fopen() and fopen(). As always, functions that begin with an underscore are generally silent about their work and return flags indicating their success or failure. Functions that omit the underscore abort and issue the appropriate error message when things go wrong.

Also see

[M-4] **intro** — Index and guide to functions

Title

Contents

Editing		
_fillmissing( )	_fillmissing()	change matrix to contain missing values
editmissing( )	editmissing()	replace missing values in matrix
editvalue( )	editvalue()	replace values in matrix
edittozero( )	edittozero()	edit matrix for roundoff error (zeros)
	edittozerotol()	same, absolute tolerance
edittoint( )	edittoint()	edit matrix for roundoff error (integers)
	edittointtol()	same, absolute tolerance

Permutation vectors		
invorder( )	invorder()	inverse of permutation vector
	revorder()	reverse of permutation vector

Matrices into vectors & vice versa		
vec( )	vec()	convert matrix into column vector
	vech()	convert symmetric matrix into column vector
	invvech()	convert column vector into symmetric matrix
rowshape( )	rowshape()	reshape matrix to have r rows
	colshape()	reshape matrix to have c columns

Associative arrays		
asarray( )	asarray()	store or retrieve element in array
	asarray_*()	utility routines

Description

The above functions manipulate matrices, such as extracting the diagonal and sorting.

Remarks

There is a thin line between manipulation and utility; also see

[M-4] **utility**   Matrix utility functions

Also see

[M-4] **intro** — Index and guide to functions

Title

[M-4] mathematical — Important mathematical functions

Contents

Basics, *continued*

moptimize( )
`moptimize()`	function optimization
`moptimize_ado_cleanup()`	perform cleanup after ado
`moptimize_evaluate()`	evaluate function at initial values
`moptimize_init()`	begin setup of optimization problem
`moptimize_init_*()`	set details
`moptimize_result_*()`	access `maximize()` results
`moptimize_query()`	report settings
`moptimize_util_*()`	utility functions for writing evaluators and processing results

Fourier transform

fft( )
`fft()`	fast Fourier transform
`invfft()`	inverse fast Fourier transform
`convolve()`	convolution
`deconvolve()`	inverse of `convolve()`
`Corr()`	correlation
`ftperiodogram()`	power spectrum
`ftpad()`	pad to power-of-2 length
`ftwrap()`	convert to frequency-wraparound order
`ftunwrap()`	convert from frequency-wraparound order
`ftretime()`	change time scale of signal
`ftfreqs()`	frequencies of transform

Cubic splines

spline3( )
`spline3()`	fit cubic spline
`spline3eval()`	evaluate cubic spline

Polynomials

polyeval( )
`polyeval()`	evaluate polynomial
`polysolve()`	solve for polynomial
`polytrim()`	trim polynomial
`polyderiv()`	derivative of polynomial
`polyinteg()`	integral of polynomial
`polyadd()`	add polynomials
`polymult()`	multiply polynomials
`polydiv()`	divide polynomials
`polyroots()`	find roots of polynomial

	Number-theoretic point sets		

halton( )	halton()	generate a Halton or Hammersley set
	ghalton()	generate a generalized Halton sequence

	Base conversion		

inbase( )	inbase()	convert to specified base
	frombase()	convert from specified base

Description

The above functions are important mathematical functions that most people would not call either matrix functions or scalar functions, but that use matrices and scalars.

Remarks

For other mathematical functions, see

[M-4] **matrix**	Matrix mathematical functions
[M-4] **scalar**	Scalar mathematical functions
[M-4] **statistical**	Statistical functions

Also see

[M-4] **intro** — Index and guide to functions

Title

[M-4] matrix — Matrix functions

Contents

QR decomposition, solvers, & inverters		
qrd( )	qrd( )	QR decomposition $A = QR$
	qrdp( )	QR decomposition $A = QRP'$
	hqrd( )	QR decomposition $A = f(H)R_1$
	hqrdp( )	QR decomposition $A = f(H, tau)R_1 P'$
	hqrdmultq( )	return QX or $Q'X$, $Q = f(H, tau)$
	hqrdmultq1t( )	return $Q_1'X$, $Q_1 = f(H, tau)$
	hqrdq( )	return $Q = f(H, tau)$
	hqrdq1( )	return $Q_1 = f(H, tau)$
	hqrdr( )	return R
	hqrdr1( )	return R_1
qrsolve( )	qrsolve( )	solve $AX = B$ for X
qrinv( )	qrinv( )	generalized inverse of matrix

Hessenberg decomposition & generalized Hessenberg decomposition		
hessenbergd( )	hessenbergd( )	Hessenberg decomposition $T = Q'XQ$
ghessenbergd( )	ghessenbergd( )	gen. Hessenberg decomp. $T = Q'XQ$

Schur decomposition & generalized Schur decomposition		
schurd( )	schurd( )	Schur decomposition $T = U'AV$; $R = U'BA$
	schurdgroupby( )	Schur decomp. with grouping of results
gschurd( )	gschurd( )	gen. Schur decomposition $T = U'AV$; $R = U'BA$
	gschurdgroupby( )	gen. Schur decomp. with grouping of results

---

Singular value decomposition, solvers, & inverters		
svd( )	svd( )	singular value decomposition $A = UDV'$
	svdsv( )	singular values s
fullsvd( )	fullsvd( )	singular value decomposition $A = USV'$
	fullsdiag( )	convert s to S
svsolve( )	svsolve( )	solve $AX = B$ for X
pinv( )	pinv( )	Moore–Penrose pseudoinverse

Triangular solvers		
solvelower( )	solvelower()	solve $AX = B$ for X, A lower triangular
	solveupper()	solve $AX = B$ for X, A upper triangular

Eigensystems, powers, & transcendental		
eigensystem( )	eigensystem()	eigenvectors and eigenvalues
	eigenvalues()	eigenvalues
	lefteigensystem()	left eigenvectors and eigenvalues
	symeigensystem()	eigenvectors/eigenvalues of symmetric matrix
	symeigenvalues()	eigenvalues of symmetric matrix
eigensystemselect( )	eigensystemselect*() etc.	selected eigenvectors/eigenvalues
geigensystem( )	geigensystem() etc.	generalized eigenvectors/eigenvalues
matpowersym( )	matpowersym()	powers of symmetric matrix
matexpsym( )	matexpsym()	exponentiation of symmetric matrix
	matlogsym()	logarithm of symmetric matrix

Equilibration		
_equilrc( )	_equilrc()	row/column equilibration
	_equilr()	row equilibration
	_equilc()	column equilibration
	_perhapsequilrc()	row/column equilibration if necessary
	_perhapsequilr()	row equilibration if necessary
	_perhapsequilc()	column equilibration if necessary
	rowscalefactors()	row-scaling factors for equilibration
	colscalefactors()	column-scaling factors for equilibration

LAPACK		
lapack( )	LA_*()	LAPACK linear-algebra functions

Description

The above functions are what most people would call mathematical matrix functions.

Remarks

For other mathematical functions, see

      [M-4] **scalar**   Scalar mathematical functions

      [M-4] **mathematical**   Important mathematical functions

Also see

[M-4] **intro** — Index and guide to functions

Title

[M-4] programming — Programming functions

Contents

	Break key		
setbreakintr( )		`setbreakintr()`	turn off/on break-key interrupt
		`querybreakintr()`	whether break-key interrupt is off/on
		`breakkey()`	whether break key has been pressed
		`breakkeyreset()`	reset break key

	Associative arrays		
asarray( )		`asarray()`	store or retrieve element in array
		`asarray_*()`	utility routines
hash1( )		`hash1()`	Jenkins' one-at-a-time hash

	Miscellaneous		
assert( )		`assert()`	abort execution if not true
		`asserteq()`	abort execution if not equal
c( )		`c()`	access `c()` value
sizeof( )		`sizeof()`	number of bytes consumed by object
swap( )		`swap()`	interchange contents of variables

	System info		
byteorder( )		`byteorder()`	byte order used by computer
stataversion( )		`stataversion()`	version of Stata being used
		`statasetversion()`	version of Stata set

	Exiting		
exit( )		`exit()`	terminate execution
error( )		`error()`	issue standard Stata error message
		`_error()`	issue error message with traceback log

Also see

[M-4] **intro** — Index and guide to functions

Title

[M-4] scalar — Scalar mathematical functions

Contents

	Factorial & gamma		
factorial( )		`factorial()`	factorial
		`lnfactorial()`	natural logarithm of factorial
		`gamma()`	gamma function
		`lngamma()`	natural logarithm of gamma function
		`digamma()`	derivative of `lngamma()`
		`trigamma()`	second derivative of `lngamma()`

	Modulus & integer rounding		
mod( )		`mod()`	modulus
trunc( )		`trunc()`	truncate to integer
		`floor()`	round down to integer
		`ceil()`	round up to integer
		`round()`	round to closest integer or multiple

	Dates		
date( )		`clock()`	%tc of string
		`mdyhms()`	%tc of month, day, year, hour, minute, and second
		`dhms()`	%tc of %td, hour, minute, and second
		`hms()`	%tc of hour, minute, and second
		`hh()`	hour of %tc
		`mm()`	minute of %tc
		`ss()`	second of %tc
		`dofc()`	%td of %tc
		`Cofc()`	%tC of %tc
		`Clock()`	%tC of string
		`Cmdyhms()`	%tC of month, day, year, hour, minute, and second
		`Cdhms()`	%tC of %td, hour, minute, and second
		`Chms()`	%tC of hour, minute, and second
		`hhC()`	hour of %tC
		`mmC()`	minute of %tC
		`ssC()`	second of %tC
		`dofC()`	%td of %tC
		`date()`	%td of string
		`mdy()`	%td of month, day, and year
		`yw()`	%tw of year and week
		`ym()`	%tm of year and month
		`yq()`	%tq of year and quarter
		`yh()`	%th of year and half
		`cofd()`	%tc of %td
		`Cofd()`	%tC of %td

date( ), *continued*	month( )	month of %td
	day( )	day-of-month of %td
	year( )	year of %td
	dow( )	day-of-week of %td
	week( )	week of %td
	quarter( )	quarter of %td
	halfyear( )	half-of-year of %td
	doy( )	day-of-year of %td
	yearly( )	%ty of string
	yofd( )	%ty of %td
	dofy( )	%td of %ty
	halfyearly( )	%th of string
	hofd( )	%th of %td
	dofh( )	%td of %th
	quarterly( )	%tq of string
	qofd( )	%tq of %td
	dofq( )	%td of %tq
	monthly( )	%tm of string
	mofd( )	%tm of %td
	dofm( )	%td of %tm
	weekly( )	%tw of string
	wofd( )	%tw of %td
	dofw( )	%td of %tw
	hours( )	hours of milliseconds
	minutes( )	minutes of milliseconds
	seconds( )	seconds of milliseconds
	msofhours( )	milliseconds of hours
	msofminutes( )	milliseconds of minutes
	msofseconds( )	milliseconds of seconds

Description

With a few exceptions, the above functions are what most people would consider scalar functions, although in fact all will work with matrices, in an element-by-element fashion.

Remarks

For other mathematical functions, see

[M-4] **matrix**	Matrix functions
[M-4] **mathematical**	Important mathematical functions
[M-4] **statistical**	Statistical functions

Also see

[M-4] **intro** — Index and guide to functions

Title

[M-4] **solvers** — Functions to solve AX=B and to obtain A inverse

Contents

Description

The above functions solve $AX = B$ for X and solve for A^{-1}.

Remarks

Matrix solvers can be used to implement matrix inverters, and so the two nearly always come as a pair.

Solvers solve $AX = B$ for X. One way to obtain A^{-1} is to solve $AX = I$. If $f(A, B)$ solves AX=B, then $f(A, \text{I}(\text{rows}(A)))$ solves for the inverse. Some matrix inverters are in fact implemented this way, although usually custom code is written because memory savings are possible when it is known that $B = I$.

The pairings of inverter and solver are

inverter	solver
invsym()	(none)
cholinv()	cholsolve()
luinv()	lusolve()
qrinv()	qrsolve()
pinv()	svsolve()

Also see

[M-4] **intro** — Index and guide to functions

Title

[M-4] standard — Functions to create standard matrices

Contents

	vec() & vech() transform	
Dmatrix( )	Dmatrix()	duplication matrices
Kmatrix( )	Kmatrix()	commutation matrices
Lmatrix( )	Lmatrix()	elimination matrices

Description

The functions above create standard matrices such as the identity matrix, etc.

Remarks

For other mathematical functions, see

[M-4] **matrix**	Matrix mathematical functions
[M-4] **scalar**	Scalar mathematical functions
[M-4] **mathematical**	Important mathematical functions

Also see

[M-4] **intro** — Index and guide to functions

Title

[M-4] stata — Stata interface functions

Contents

Variable characteristics

st_varrename( )	st_varrename()	rename Stata variable
st_vartype( )	st_vartype()	storage type of Stata variable
	st_isnumvar()	whether variable is numeric
	st_isstrvar()	whether variable is string
st_varformat( )	st_varformat()	obtain/set format of Stata variable
	st_varlabel()	obtain/set variable label
	st_varvaluelabel()	obtain/set value label
st_vlexists( )	st_vlexists()	whether value label exists
	st_vldrop()	drop value
	st_vlmap()	map values
	st_vlsearch()	map text
	st_vlload()	load value label
	st_vlmodify()	create or modify value label

Temporary variables & time-series operators

st_tempname( )	st_tempname()	temporary variable name
	st_tempfilename()	temporary filename
st_tsrevar( )	st_tsrevar()	create time-series op.varname
	_st_tsrevar()	same

Adding & removing variables & observations

st_addobs( )	st_addobs()	add observations to Stata dataset
st_addvar( )	st_addvar()	add variable to Stata dataset
st_dropvar( )	st_dropvar()	drop variables
	st_dropobsin()	drop specified observations
	st_dropobsif()	drop selected observations
	st_keepvar()	keep variables
	st_keepobsin()	keep specified observations
	st_keepobsif()	keep selected observations
st_updata( )	st_updata()	query/set data-have-changed flag

Executing Stata commands

stata( )	stata()	execute Stata command
st_macroexpand( )	st_macroexpand()	expand Stata macros

Accessing e(), r(), s(), macros, matrices, etc.		
st_global( )	st_global()	obtain/set Stata global
st_local( )	st_local()	obtain/set local Stata macro
st_numscalar( )	st_numscalar()	obtain/set Stata numeric scalar
	st_strscalar()	obtain/set Stata string scalar
st_matrix( )	st_matrix()	obtain/set Stata matrix
	st_matrixrowstripe()	obtain/set row labels
	st_matrixcolstripe()	obtain/set column labels
	st_replacematrix()	replace existing Stata matrix
st_dir( )	st_dir()	obtain list of Stata objects
st_rclear( )	st_rclear()	clear r()
	st_eclear()	clear e()
	st_sclear()	clear s()

Parsing & verification		
st_isname( )	st_isname()	whether valid Stata name
	st_islmname()	whether valid local macro name
st_isfmt( )	st_isfmt()	whether valid %*fmt*
	st_isnumfmt()	whether valid numeric %*fmt*
	st_isstrfmt()	whether valid string %*fmt*
abbrev( )	abbrev()	abbreviate strings
strtoname( )	strtoname()	translate strings to Stata names

Description

The above functions interface with Stata.

Remarks

The following manual entries have to do with getting data from or putting data into Stata:

[M-5]	**st_data( )**	Load copy of current Stata dataset
[M-5]	**st_view( )**	Make matrix that is a view onto current Stata dataset
[M-5]	**st_store( )**	Modify values stored in current Stata dataset
[M-5]	**st_nvar( )**	Numbers of variables and observations

In some cases, you may find yourself needing to translate variable names into variable indices and vice versa:

[M-5] **st_varname( )**   Obtain variable names from variable indices

[M-5] **st_varindex( )**   Obtain variable indices from variable names

[M-5] **st_tsrevar( )**   Create time-series op.varname variables

The other functions mostly have to do with getting and putting Stata's scalars, matrices, and returned results:

[M-5] **st_local( )**   Obtain strings from and put strings into Stata

[M-5] **st_global( )**   Obtain strings from and put strings into global macros

[M-5] **st_numscalar( )**   Obtain values from and put values into Stata scalars

[M-5] **st_matrix( )**   Obtain and put Stata matrices

The stata() function, documented in

[M-5] **stata( )**   Execute Stata command

allows you to cause Stata to execute a command that you construct in a string.

Reference

Gould, W. W. 2008. Mata Matters: Macros. *Stata Journal* 8: 401–412.

Also see

[M-4] **intro** — Index and guide to functions

Title

Contents

Factorial & combinations		
factorial( )	`factorial()`	factorial
	`lnfactorial()`	natural logarithm of factorial
	`gamma()`	gamma function
	`lngamma()`	natural logarithm of gamma function
	`digamma()`	derivative of `lngamma()`
	`trigamma()`	second derivative of `lngamma()`
comb( )	`comb()`	combinatorial function n choose k
cvpermute( )	`cvpermutesetup()`	permutation setup
	`cvpermute()`	return permutations, one at a time

Densities & distributions		
normal( )	`normalden()`	normal density
	`normal()`	cumulative normal dist.
	`invnormal()`	inverse cumulative normal
	`lnnormalden()`	logarithm of the normal density
	`lnnormal()`	logarithm of the cumulative normal dist.
	`binormal()`	cumulative binormal dist.
	`betaden()`	beta density
	`ibeta()`	cumulative beta dist.; a.k.a. incomplete beta function
	`ibetatail()`	reverse cumulative beta dist.
	`invibeta()`	inverse cumulative beta
	`invibetatail()`	inverse reverse cumulative beta
	`binomialp()`	binomial probability
	`binomial()`	cumulative of binomial
	`binomialtail()`	reverse cumulative of binomial
	`invbinomial()`	inverse binomial (lower tail)
	`invbinomialtail()`	inverse binomial (upper tail)
	`chi2()`	cumulative chi-squared dist.
	`chi2tail()`	reverse cumulative chi-squared dist.
	`invchi2()`	inverse cumulative chi-squared
	`invchi2tail()`	inverse reverse cumulative chi-squared
	`Fden()`	F density
	`F()`	cumulative F dist.
	`Ftail()`	reverse cumulative F dist.
	`invF()`	inverse cumulative F
	`invFtail()`	inverse reverse cumulative F

normal( ), *continued*

`gammaden()`	gamma density
`gammap()`	cumulative gamma dist.;
	a.k.a. incomplete gamma function
`gammaptail()`	reverse cumulative gamma dist.;
`invgammap()`	inverse cumulative gamma
`invgammaptail()`	inverse reverse cumulative gamma
`dgammapda()`	$\partial P(a,x)/\partial a$, where $P(a,x) = $ `gammap`(a,x)
`dgammapdx()`	$\partial P(a,x)/\partial x$, where $P(a,x) = $ `gammap`(a,x)
`dgammapdada()`	$\partial^2 P(a,x)/\partial a^2$, where $P(a,x) = $ `gammap`(a,x)
`dgammapdadx()`	$\partial^2 P(a,x)/\partial a\partial x$, where $P(a,x) = $ `gammap`(a,x)
`dgammapdxdx()`	$\partial^2 P(a,x)/\partial x^2$, where $P(a,x) = $ `gammap`(a,x)
`hypergeometricp()`	hypergeometric probability
`hypergeometric()`	cumulative of hypergeometric
`nbetaden()`	noncentral beta density
`nibeta()`	cumulative noncentral beta dist.
`invnibeta()`	inverse cumulative noncentral beta
`nbinomialp()`	negative binomial probability
`nbinomial()`	negative binomial cumulative
`nbinomialtail()`	reverse cumulative of negative binomial
`invnbinomial()`	inverse negative binomial (lower tail)
`invnbinomialtail()`	inverse negative binomial (upper tail)
`nchi2()`	noncentral cumulative chi-squared dist.
`invnchi2()`	inverse noncentral cumulative chi-squared
`npnchi2()`	noncentrality parameter of `nchi2()`
`nFden()`	noncentral F density
`nFtail()`	reverse cumulative noncentral F dist.
`invnFtail()`	inverse reverse cumulative noncentral F
`poissonp()`	Poisson probability
`poisson()`	cumulative of Poisson
`poissontail()`	reverse cumulative of Poisson
`invpoisson()`	inverse Poisson (lower tail)
`invpoissontail()`	inverse Poisson (upper tail)
`tden()`	Student t density
`ttail()`	reverse cumulative t dist.
`invttail()`	inverse reverse cumulative t

Maximization & minimization		
optimize( )	optimize()	function maximization and minimization
	optimize_init()	begin optimization
	optimize_init_*()	set details
	optimize()	perform optimization
	optimize_result_*()	access results
	optimize_query()	report settings
moptimize( )	moptimize()	function optimization
	moptimize_ado_cleanup()	perform cleanup after ado
	moptimize_evaluate()	evaluate function at initial values
	moptimize_init()	begin setup of optimization problem
	moptimize_init_*()	set details
	moptimize_result_*()	access maximize() results
	moptimize_query()	report settings
	moptimize_util_*()	utility functions for writing evaluators and processing results

Logits, odds, & related		
logit( )	logit()	log of the odds ratio
	invlogit()	inverse log of the odds ratio
	cloglog()	complementary log-log
	invcloglog()	inverse complementary log-log

Multivariate normal		
ghk( )	ghk()	GHK multivariate normal (MVN) simulator
	ghk_init()	GHK MVN initialization
	ghk_init_*()	set details
	ghk()	perform simulation
	ghk_query_npts()	return number of simulation points
ghkfast( )	ghkfast()	GHK MVN simulator
	ghkfast_init()	GHK MVN initialization
	ghkfast_init_*()	set details
	ghkfast()	perform simulation
	ghkfast_i()	results for the ith observation
	ghk_query_*()	display settings

Description

The above functions are statistical, probabilistic, or designed to work with data matrices.

Remarks

Concerning data matrices, see

        [M-4] **stata**   Stata interface functions

and especially

        [M-5] **st_data( )**   Load copy of current Stata dataset
        [M-5] **st_view( )**   Make matrix that is a view onto current Stata dataset

For other mathematical functions, see

        [M-4] **matrix**   Matrix mathematical functions
        [M-4] **scalar**   Scalar mathematical functions
        [M-4] **mathematical**   Important mathematical functions

Also see

[M-4] **intro** — Index and guide to functions

Title

[M-4] string — String manipulation functions

Contents

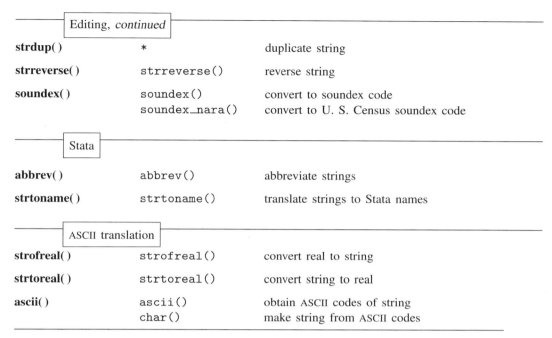

Editing, continued		
strdup( )	*	duplicate string
strreverse( )	strreverse()	reverse string
soundex( )	soundex()	convert to soundex code
	soundex_nara()	convert to U. S. Census soundex code

Stata		
abbrev( )	abbrev()	abbreviate strings
strtoname( )	strtoname()	translate strings to Stata names

ASCII translation		
strofreal( )	strofreal()	convert real to string
strtoreal( )	strtoreal()	convert string to real
ascii( )	ascii()	obtain ASCII codes of string
	char()	make string from ASCII codes

Description

The above functions are for manipulating strings. Strings in Mata are strings of ASCII characters, usually the printable characters, but Mata enforces no such restriction. In particular, strings may contain binary 0.

Remarks

In addition to the above functions, two operators are especially useful for dealing with strings.

The first is +. Addition is how you concatenate strings:

```
: "abc" + "def"
abcdef
: command = "list"
: args = "mpg weight"
: result = command + " " + args
: result
list mpg weight
```

The second is *. Multiplication is how you duplicate strings:

```
: 5*"a"
aaaaa
: "b"*3
bbb
: indent = 20
: title = indent*" " + "My Title"
: title
                    My Title
```

Also see

[M-4] **intro** — Index and guide to functions

Title

[M-4] utility — Matrix utility functions

Contents

| Missing values | | |

missing( )	missing( )	count of missing values
	rowmissing( )	count of missing values, by row
	colmissing( )	count of missing values, by column
	nonmissing( )	count of nonmissing values
	rownonmissing( )	count of nonmissing values, by row
	colnonmissing( )	count of nonmissing values, by column
	hasmissing( )	whether matrix has missing values
missingof( )	missingof( )	appropriate missing value

| Range, sums, & cross products | | |

minmax( )	rowmin( )	minimum, by row
	colmin( )	minimum, by column
	min( )	minimum, overall
	rowmax( )	maximum, by row
	colmax( )	maximum, by column
	max( )	maximum, overall
	rowminmax( )	minimum and maximum, by row
	colminmax( )	minimum and maximum, by column
	minmax( )	minimum and maximum, overall
	rowmaxabs( )	rowmax(abs( ))
	colmaxabs( )	colmax(abs( ))
minindex( )	minindex( )	indices of minimums
	maxindex( )	indices of maximums
sum( )	rowsum( )	sum of each row
	colsum( )	sum of each column
	sum( )	overall sum
	quadrowsum( )	quad-precision sum of each row
	quadcolsum( )	quad-precision sum of each column
	quadsum( )	quad-precision overall sum
runningsum( )	runningsum( )	running sum of vector
	quadrunningsum( )	quad-precision runningsum( )
cross( )	cross( )	$X'X$, $X'Z$, etc.
crossdev( )	crossdev( )	$(X\!:\!-x)'(X\!:\!-x)$, $(X\!:\!-x)'(Z\!:\!-z)$, etc.
quadcross( )	quadcross( )	quad-precision cross( )
	quadcrossdev( )	quad-precision crossdev( )

Programming		
reldif( )	reldif()	relative difference
	mreldif()	max. relative difference between matrices
	mreldifsym()	max. relative difference from symmetry
	mreldifre()	max. relative difference from real
all( )	all()	sum($!L$)==0
	any()	sum(L)!=0
	allof()	all(P:==s)
	anyof()	any(P:==s)
panelsetup( )	panelsetup()	initialize panel-data processing
	panelstats()	summary statistics on panels
	panelsubmatrix()	obtain matrix for panel i
	panelsubview()	obtain view matrix for panel i
_negate( )	_negate()	fast negation of matrix

Constants & tolerances		
mindouble( )	mindouble()	minimum nonmissing value
	maxdouble()	maximum nonmissing value
	smallestdouble()	smallest $e > 0$
epsilon( )	epsilon()	unit roundoff error
floatround( )	floatround()	round to float precision
solve_tol( )	solve_tol()	tolerance used by solvers and inverters

Description

Matrix utility functions tell you something about the matrix, such as the number of rows or whether it is diagonal.

Remarks

There is a thin line between utility and manipulation; also see

[M-4] **manipulation**   Matrix manipulation functions

Also see

[M-4] **intro** — Index and guide to functions

[M-5] Mata functions

Title

[M-5] **intro** — Mata functions

Contents

See [M-4] **intro**

Description

The following entries, alphabetically arranged, document all the Mata functions.

Remarks

See [M-4] **intro** for an index, grouped logically, of the functions presented in this section.

Also see

[M-4] **intro** — Index and guide to functions

[M-0] **intro** — Introduction to the Mata manual

Title

[M-5] abbrev( ) — Abbreviate strings

Syntax

> *string matrix* abbrev(*string matrix s*, *real matrix n*)

Description

abbrev(*s*, *n*) returns the *n*-character abbreviation of *s* such that the abbreviation uniquely identifies a variable in Stata, if there are variables in Stata. Otherwise, abbrev(*s*, *n*) returns the *n*-character abbreviation of *s*.

1. *n* is the abbreviation length and is assumed to contain integer values in the range 5, 6, ..., 32.

2. If *s* contains a period, ., and $n < 8$, then the value *n* defaults to 8. Otherwise, if $n < 5$, then *n* defaults to 5.

3. If *n* is missing, the entire string (up to the first binary 0) is returned.

If there is a binary 0 in *s*, the abbreviation is derived from the beginning of the string up to but not including the binary 0.

When arguments are not scalar, abbrev( ) returns element-by-element results.

Conformability

abbrev(*s*, *n*):

s:	$r_1 \times c_1$
n:	$r_2 \times c_2$; *s* and *n* r-conformable
result:	$\max(r_1, r_2) \times \max(c_1, c_2)$

Diagnostics

abbrev( ) returns "" if *s* is "". abbrev( ) aborts with error if *s* is not a string.

Also see

[M-4] **string** — String manipulation functions

Title

[M-5] **abs( )** — Absolute value (length)

Syntax

> *real matrix*   abs(*numeric matrix Z*)

Description

For Z real, abs(Z) returns the elementwise absolute values of Z.

For Z complex, abs(Z) returns the elementwise length of each element. If $Z = a + bi$, returned is sqrt($a^2 + b^2$), although the calculation is not made in that way. The method actually used prevents overflow.

Conformability

abs(Z):

Z:	$r \times c$
result:	$r \times c$

Diagnostics

abs(.) returns . (missing).

Also see

[M-4] **scalar** — Scalar mathematical functions

Title

[M-5] **adosubdir( )** — Determine ado-subdirectory for file

Syntax

> *string scalar* adosubdir(*string scalar filename*)

Description

adosubdir(*filename*) returns the subdirectory in which Stata would search for *filename*. Typically, the subdirectory will be simply the first letter of *filename*. However, certain files may result in a different subdirectory, depending on their extension.

Remarks

adosubdir("xyz.ado") returns "x" because Stata ado-files are located in subdirectories with name given by their first letter.

adosubdir("xyz.style") returns "style" because Stata style files are located in subdirectories named style.

Conformability

adosubdir(*filename*):
> *filename*: 1×1
> *result*: 1×1

Diagnostics

adosubdir( ) returns the first letter of the filename if the filetype is unknown to Stata, thus treating unknown filetypes as if they were ado-files.

adosubdir( ) aborts with error if the filename is too long for the operating system; nothing else causes abort with error.

Also see

[M-4] **io** — I/O functions

Title

Syntax

real scalar   all(*real matrix L*)

real scalar   any(*real matrix L*)

real scalar   allof(*transmorphic matrix P*, *transmorphic scalar s*)

real scalar   anyof(*transmorphic matrix P*, *transmorphic scalar s*)

Description

all(*L*) is equivalent to sum(!*L*)==0 but is significantly faster.

any(*L*) is equivalent to sum(*L*)!=0 but is slightly faster.

allof(*P*, *s*) returns 1 if every element of *P* equals *s* and returns 0 otherwise. allof(*P*, *s*) is faster and consumes less memory than the equivalent construction all(*P*:==*s*).

anyof(*P*, *s*) returns 1 if any element of *P* equals *s* and returns 0 otherwise. anyof(*P*, *s*) is faster and consumes less memory than the equivalent any(*P*:==*s*).

Remarks

These functions are fast, so their use is encouraged over alternative constructions.

all() and any() are typically used with logical expressions to detect special cases, such as

```
if (any(x :< 0)) {
        . . .
}
```

or

```
if (all(x :>= 0)) {
        . . .
}
```

allof() and anyof() are used to look for special values:

```
if (allof(x, 0)) {
        . . .
}
```

or

```
if (anyof(x, 0)) {
        ...
}
```

Do not use `allof()` and `anyof()` to check for missing values—e.g., `anyof(x, .)`—because to really check, you would have to check not only . but also .a, .b, ..., .z. Instead use `missing()`; see [M-5] **missing( )**.

Conformability

`all(L)`, `any(L)`:

L:	$r \times c$
result:	1×1

`allof(P, s)`, `anyof(P, s)`:

P:	$r \times c$
s:	1×1
result:	1×1

Diagnostics

`all(L)` and `any(L)` treat missing values in *L* as true.

`all(L)` and `any(L)` return 0 (false) if *L* is $r \times 0$, $0 \times c$, or 0×0.

`allof(P, s)` and `anyof(P, s)` return 0 (false) if *P* is $r \times 0$, $0 \times c$, or 0×0.

Also see

[M-4] **utility** — Matrix utility functions

Title

[M-5] args( ) — Number of arguments

Syntax

> *real scalar* args()

Description

args() returns the number of arguments actually passed to the function; see [M-2] **optargs**.

Conformability

args():

> *result*: 1×1

Diagnostics

None.

Also see

[M-4] **programming** — Programming functions

Title

Syntax

$$A = \texttt{asarray\_create} \left(\left[\ keytype \right. \right. \qquad declare\ A$$

$$\left[\ , \ keydim \right.$$

$$\left[\ , \ minsize \right. \qquad \text{,}$$

$$\left[\ , \ minratio \right.$$

$$\left. \left. \left. \left. \left[\ , \ maxratio \right] \right] \right] \right] \right)$$

$\texttt{asarray}(A, \ key, \ a)$	$A[key] = a$
$a = \texttt{asarray}(A, \ key)$	$a = A[key]\ or\ a = notfound$
$\texttt{asarray\_remove}(A, \ key)$	$delete\ A[key]\ if\ it\ exists$
$bool = \texttt{asarray\_contains}(A, \ key)$	$A[key]\ exists?$
$N = \texttt{asarray\_elements}(A)$	$\#\ of\ elements\ in\ A$
$keys = \texttt{asarray\_keys}(A)$	$all\ keys\ in\ A$
$loc = \texttt{asarray\_first}(A)$	$location\ of\ first\ element\ or\ NULL$
$loc = \texttt{asarray\_next}(A, \ loc)$	$location\ of\ next\ element\ or\ NULL$
$key = \texttt{asarray\_key}(A, \ loc)$	$key\ at\ loc$
$a = \texttt{asarray\_contents}(A, \ loc)$	$contents\ a\ at\ loc$
$\texttt{asarray\_notfound}(A, \ notfound)$	$set\ notfound\ value$
$notfound = \texttt{asarray\_notfound}(A)$	$query\ notfound\ value$

where

   A: Associative array $A[key]$. Created by $\texttt{asarray\_create}()$ and passed to the other functions. If A is declared, it is declared *transmorphic*.

   keytype: Element type of keys; `"string"`, `"real"`, `"complex"`, or `"pointer"`. Optional; default `"string"`.

   keydim: Dimension of key; $1 \leq keydim \leq 50$. Optional; default 1.

minsize: Initial size of hash table used to speed locating keys in *A*; *real scalar*; $5 \leq minsize \leq 1,431,655,764$. Optional; default 100.

minratio: Fraction filled at which hash table is automatically downsized; *real scalar*; $0 \leq minratio \leq 1$. Optional; default 0.5.

maxratio: Fraction filled at which hash table is automatically upsized; *real scalar*; $1 < maxratio \leq .$ (meaning infinity). Optional; default 1.5.

key: Key under which an element is stored in *A*; *string scalar* by default; type and dimension are declared using `asarray_create()`.

a: Element of *A*; *transmorphic*; may be anything of any size; different *A[key]* elements may have different types of contents.

bool: Boolean logic value; *real scalar*; equal to 1 or 0 meaning true or false.

N: Number of elements stored in *A*; *real scalar*; $0 \leq N \leq 2,147,483,647$.

keys: List of all keys that exist in *A*. Each row is a key. Thus *keys* is a *string colvector* if keys are *string scalars*, a *string matrix* if keys are *string vectors*, a *real colvector* if keys are *real scalars*, etc. Note that `rows`(*keys*) = *N*.

loc: A location in *A*; *transmorphic*. The first location is returned by `asarray_first()`, subsequent locations by `asarray_next()`. *loc*==NULL when there are no more elements.

notfound: Value returned by `asarray(`*A*, *key*`)` when *key* does not exist in *A*. *notfound* = `J(0,0,.)` by default.

Description

`asarray()` provides one- and multi-dimensional associative arrays, also known as containers, maps, dictionaries, indices, and hash tables. In associative arrays, rather than being dense integers, the indices can be anything, even strings. So you might have *A["Frank Smith"]* equal to something and *A["Mary Jones"]* equal to something else. In Mata, you write that as `asarray(`*A*, `"Frank Smith"`, *something*`)` and `asarray(`*A*, `"Mary Jones"`, *somethingelse*`)` to define the elements and `asarray(`*A*, `"Frank Smith")` and `asarray(`*A*, `"Mary Jones")` to obtain their values.

A = `asarray_create()` declares (creates) an associative array. The function allows arguments, but they are optional. Without arguments, `asarray_create()` declares an associative array with string scalar keys, corresponding to the *A["Frank Smith"]* and *A["Mary Jones"]* example above.

A = `asarray_create(`*keytype*, *keydim*`)` declares an associative array with *keytype* keys each of dimension $1 \times keydim$. `asarray_create("string", 1)` is equivalent to `asarray_create()` without arguments. `asarray_create("string", 2)` declares the keys to be string, as before, but now they are 1×2 rather than 1×1, so array elements would be of the form *A["Frank Smith", "Mary Jones"]*. *A["Mary Jones", "Frank Smith"]* would be a different element. `asarray_create("real", 2)` declares the keys to be real 1×2, which would somewhat correspond to our ordinary idea of a matrix, namely *A[i,j]*. The difference would be that to store, say, *A[100,980]*, it would not be necessary to store the interior elements, and in addition to storing *A[100,980]*, we could store *A[3.14159,2.71828]*.

`asarray_create()` has three more optional arguments: *minsize*, *minratio*, and *maxratio*. We recommend that you do not specify them. They are discussed in *Setting the efficiency parameters* under *Remarks* below.

asarray(A, *key*, *a*) sets or resets element $A[key] = a$. Note that if you declare *key* to be 1×2, you must use the parentheses vector notation to specify key literals, such as asarray(A, (100,980), 2.15). Alternatively, if k = (100,980), then you can omit the parentheses in asarray(A, k, 2.15).

asarray(A, *key*) returns element *A[key]* or it returns *notfound* if the element does not exist. By default, *notfound* is J(0,0,.), but you can change that using asarray_notfound(). If you redefined *notfound* to be 0 and defined keys to be real 1×2, you would be on your way to recording sparse matrices efficiently.

asarray_remove(A, *key*) removes *A[key]*, or it does nothing if *A[key]* is already undefined.

asarray_contains(A, *key*) returns 1 if *A[key]* is defined, and it returns 0 otherwise.

asarray_elements(A) returns the number of elements stored in A.

asarray_keys(A) returns a vector or matrix containing all the keys, one to a row. The keys are not in alphabetical or numerical order. If you want them that way, code sort(asarray_keys(A), 1) if your keys are scalar, or in general, code sort(asarray_keys(A), *idx*); see [M-5] **sort( )**.

asarray_first(A) and asarray_next(A, *loc*) provide a way of obtaining the names one at a time. Code

```
        for (loc=asarray_first(A); loc!=NULL; loc=asarray_next(A, loc)) {
                ...
        }
```

asarray_key(A, *loc*) and asarray_contents(A, *loc*) return the key and contents at *loc*, so the loop becomes

```
        for (loc=asarray_first(A); loc!=NULL; loc=asarray_next(A, loc)) {
                ...
                ... asarray_key(A, loc) ...
                ...
                ... asarray_contents(A, loc) ...
                ...
        }
```

asarray_notfound(A, *notfound*) defines what asarray(A, *key*) returns when the element does not exist. By default, *notfound* is J(0,0,.), which is to say, a 0×0 real matrix. You can reset *notfound* at any time. asarray_notfound(A) returns the current value of *notfound*.

Remarks

Before writing a program using asarray(), you should try it interactively. Remarks are presented under the following headings:

> *Example 1: Scalar keys and scalar contents*
> *Example 2: Scalar keys and matrix contents*
> *Example 3: Vector keys and scalar contents; sparse matrix*
> *Setting the efficiency parameters*

Example 1: Scalar keys and scalar contents

```
: A = asarray_create()
: asarray(A, "bill", 1.25)
: asarray(A, "mary", 2.75)
: asarray(A, "dan",  1.50)
: asarray(A, "bill")
  1.25
: asarray(A, "mary")
  2.75
: asarray(A, "mary", 3.25)
: asarray(A, "mary")
  3.25
: sum = 0
: for (loc=asarray_first(A); loc!=NULL; loc=asarray_next(A, loc)) {
>         sum = sum + asarray_contents(A, loc)
> }
: sum
  6
: sum/asarray_elements(A)
  2
```

Example 2: Scalar keys and matrix contents

```
: A = asarray_create()
: asarray(A, "Count", (1,2\3,4))
: asarray(A, "Hilbert", Hilbert(3))
: asarray(A, "Count")
        1    2

  1 │  1    2
  2 │  3    4

: asarray(A, "Hilbert")
[symmetric]
               1              2            3

  1 │          1
  2 │         .5    .3333333333
  3 │ .3333333333           .25          .2
```

Example 3: Vector keys and scalar contents; sparse matrix

```
: A = asarray_create("real", 2)
: asarray_notfound(A, 0)
: asarray(A, (   1,    1), 1)
: asarray(A, (1000,  999), .5)
: asarray(A, (1000, 1000), 1)
: asarray(A, (1000, 1001), .5)
: asarray(A, (1,1))
  1
```

```
: asarray(A, (2,2))
  0

: // one way to get the trace:
: trace = 0

: for (i=1; i<=1000; i++) trace = trace + asarray(A, (i,i))

: trace
  2

: // another way to get the trace
: trace = 0

: for (loc=asarray_first(A); loc!=NULL; loc=asarray_next(A, loc)) {
>         index = asarray_key(A, loc)
>         if (index[1]==index[2]) {
>                 trace = trace + asarray_contents(A, loc)
>         }
> }

: trace
  2
```

Setting the efficiency parameters

The syntax `asarray_create()` is

$$A = \texttt{asarray\_create}(\textit{keytype},\ \textit{keydim},\ \textit{minsize},\ \textit{minratio},\ \textit{maxratio})$$

All arguments are optional. The first two specify the characteristics of the key and their use has already been illustrated. The last three are efficiency parameters. In most circumstances, we recommend you do not specify them. The default values have been chosen to produce reasonable execution times with reasonable memory consumption.

`asarray()` works via hash tables. Say we wish to record n entries. The idea is to allocate a hash table of N rows, where N can be less than, equal to, or greater than n. When one needs to find the element corresponding to a key, one calculates a function of the key, called a hash function, that returns an integer h from 1 to N. One first looks in row h. If row h is already in use and the keys are different, we have a collision. In that case, we have to allocate a duplicates list for the hth row and put the duplicate keys and contents into that list. Collisions are bad because, when they occur, `asarray()` has to allocate a duplicates list, requiring both time and memory, though it does not require much. When fetching results, if row h has a duplicates list, `asarray()` has to search the list, which it does sequentially, and that takes extra time, too. Hash tables work best when collisions happen rarely.

Obviously, collisions are certain to occur if $N < n$. Note, however, that although performance suffers, the method does not break. A hash table of N can hold any number of entries, even if $N < n$.

Performance depends on details of implementation. We have examined the behavior of `asarray()` and discovered that collisions rarely occur when $n/N \leq 0.75$. When $n/N = 1.5$, performance suffers, but not by as much as you might expect. Around $n/N = 2$, performance degrades considerably.

When you add or remove an element, `asarray()` examines n/N and considers rebuilding the table with a larger or smaller N; it rebuilds the table when n/N is large to preserve efficiency. It rebuilds the table when n/N is small to conserve memory. Rebuilding the table is a computer-intensive operation, and so should not be performed too often.

In making these decisions, `asarray()` uses three parameters:

maxratio: When $n/N \geq maxratio$, the table is upsized to $N = 1.5n$.

minratio: When $n/N \leq minratio/1.5$, the table is downsized to $N = 1.5n$. (For an exception, see *minsize*.)

minsize: If the new $N < 1.5minsize$, the table is downsized to $N = 1.5minsize$ if it is not already that size.

The default values of the three parameters are 1.5, 0.5, and 100. You can reset them, though you are unlikely to improve on the default values of *minratio* and *maxratio*.

You can improve on *minsize* when you know the number of elements that will be in the table and that number is greater than 100. For instance, if you know the table will contain at least 1,000 elements, starting *minsize* at 1,000, which implies $N = 1,500$, will prevent two rescalings, namely, from 150 to 451, and from 451 to 1,354. This saves a little time.

You can also turn off the resizing features. Setting *minratio* to 0 turns off downsizing. Setting *maxratio* to . (missing) turns off upsizing. You might want to turn off both downsizing and upsizing if you set *minsize* sufficiently large for your problem.

We would never recommend turning off upsizing alone, and we seldom would recommend turning off downsizing alone. In a program where it is known that the array will exist for only a short time, however, turning off downsizing can be efficient. In a program where the array might exist for a considerable time, turning off downsizing is dangerous because then the array could only grow (and probably will).

Conformability

`asarray_create(`*keytype*, *keydim*, *minsize*, *minratio*, *maxratio*`):`

keytype:	1×1	(optional)
keydim:	1×1	(optional)
minsize:	1×1	(optional)
minratio:	1×1	(optional)
maxratio:	1×1	(optional)
result:	*transmorphic*	

`asarray(`*A*, *key*, *a*`):`

A:	*transmorphic*
key:	$1 \times keydim$
a:	$r_{key} \times c_{key}$
result:	*void*

`asarray(`*A*, *key*`):`

A:	*transmorphic*
key:	$1 \times keydim$
result:	$r_{key} \times c_{key}$

```
asarray_remove(A, key):
```
A:	*transmorphic*
key:	$1 \times keydim$
result:	*void*

```
asarray_contains(A, key), asarray_elements(A, key):
```
A:	*transmorphic*
key:	$1 \times keydim$
result:	1×1

```
asarray_keys(A, key):
```
A:	*transmorphic*
key:	$1 \times keydim$
result:	$n \times keydim$

```
asarray_first(A):
```
A:	*transmorphic*
result:	*transmorphic*

```
asarray_first(A, loc):
```
A:	*transmorphic*
loc:	*transmorphic*
result:	*transmorphic*

```
asarray_key(A, loc):
```
A:	*transmorphic*
loc:	*transmorphic*
result:	$1 \times keydim$

```
asarray_contents(A, loc):
```
A:	*transmorphic*
loc:	*transmorphic*
result:	$r_{key} \times c_{key}$

```
asarray_notfound(A, notfound):
```
A:	*transmorphic*
notfound:	$r \times c$
result:	*void*

```
asarray_notfound(A):
```
A:	*transmorphic*
result:	$r \times c$

Diagnostics

None.

Also see

[M-5] **hash1( )** — Jenkins' one-at-a-time hash function

[M-4] **manipulation** — Matrix manipulation

[M-4] **programming** — Programming functions

Title

[M-5] **ascii( )** — Manipulate ASCII codes

Syntax

> *real rowvector* ascii(*string scalar s*)
>
> *string scalar* char(*real rowvector c*)

Description

ascii(*s*) returns a row vector containing the ASCII codes corresponding to *s*. For instance, ascii("abc") returns (97, 98, 99).

char(*c*) returns a string consisting of the specified ASCII codes. For instance, char((97, 98, 99)) returns "abc".

Conformability

```
ascii(s):
        s:      1 × 1
    result:     1 × strlen(s)

char(c):
        c:      1 × n,   n ≥ 0
    result:     1 × 1
```

Diagnostics

ascii(*s*) returns J(1,0,.) if strlen(*s*)==0.

In char(*c*), if any element of *c* is outside the range 0 to 255, the returned string is terminated at that point. For instance, char((97,98,99,1000,97,98,99))="abc".

char(J(1,0,.)) returns "".

Also see

[M-4] **string** — String manipulation functions

Title

[M-5] **assert( )** — Abort execution if false

Syntax

>*void* assert(*real scalar r*)

>*void* asserteq(*transmorphic matrix A*, *transmorphic matrix B*)

Description

assert(*r*) produces the error message "assertion is false" and aborts with error if $r == 0$.

asserteq(*A*, *B*) is logically equivalent to assert(*A==B*). If the assertion is false, however, information is presented on the number of mismatches.

Remarks

In the midst of complicated code, you know that a certain calculation must produce a result greater than 0, but you worry that perhaps you have an error in your code:

```
        . . .
        assert(n>0)
        . . .
```

In another spot, you have produced matrix A and know every element of A should be positive or zero:

```
        . . .
        assert(A:>=0)
        . . .
```

Once you are convinced that your function works, these verifications should be removed. In a third part of your code, however, the problem is different if the number of rows *r* exceed the number of columns *c*. In all the cases you need to use it, however, *r* will be less than *c*, so you are not much interested in programming the alternative solution:

```
        . . .
        assert(rows(PROBLEM) < cols(PROBLEM))
        . . .
```

Leave that one in.

Conformability

assert(*r*):
>>*r*: 1×1
>*result*:   *void*

313

asserteq(A, B):

A:	$r_1 \times c_1$
B:	$r_2 \times c_2$
result:	*void*

Diagnostics

assert(r) aborts with error if $r == 0$.

asserteq(A, B) aborts with error if $A \neq B$.

Also see

[M-4] **programming** — Programming functions

Title

[M-5] blockdiag( ) — Block-diagonal matrix

Syntax

$\textit{numeric matrix}$ $\texttt{blockdiag}(\textit{numeric matrix } Z_1, \textit{ numeric matrix } Z_2)$

Description

$\texttt{blockdiag}(Z_1, Z_2)$ returns a block-diagonal matrix with Z_1 in the upper-left corner and Z_2 in the lower right, i.e.,

$$\begin{bmatrix} Z_1 & \mathbf{0} \\ \mathbf{0} & Z_2 \end{bmatrix}$$

Z_1 and Z_2 may be either real or complex and need not be of the same type.

Remarks

To create a block diagonal matrix of Z_1, Z_2, Z_3, code

```
: blockdiag(Z1, blockdiag(Z2,Z3))
```

Conformability

$\texttt{blockdiag}(Z_1, Z_2)$:

$$\begin{array}{ll} Z_1: & r_1 \times c_1 \\ Z_2: & r_2 \times c_2 \\ \textit{result}: & r_1 + r_2 \times c_1 + c_2 \end{array}$$

Diagnostics

None. Either or both Z_1 and Z_2 may be void.

Also see

[M-4] **standard** — Functions to create standard matrices

Title

Syntax

$colvector\ C$ = bufio()

real scalar	bufbyteorder(C)
void	bufbyteorder(C, *real scalar byteorder*)
real scalar	bufmissingvalue(C)
void	bufmissingvalue(C, *real scalar version*)
void	bufput(C, B, *offset*, *bfmt*, X)
scalar	bufget(C, B, *offset*, *bfmt*)
rowvector	bufget(C, B, *offset*, *bfmt*, c)
matrix	bufget(C, B, *offset*, *bfmt*, r, c)
void	fbufput(C, *fh*, *bfmt*, X)
scalar	fbufget(C, *fh*, *bfmt*)
rowvector	fbufget(C, *fh*, *bfmt*, c)
matrix	fbufget(C, *fh*, *bfmt*, r, c)
real scalar	bufbfmtlen(*string scalar bfmt*)
real scalar	bufbfmtisnum(*string scalar bfmt*)

where

C:	*colvector* returned by bufio()
B:	*string scalar* (buffer)
offset:	*real scalar* (buffer position, starts at 0)
fh:	file handle returned by fopen()
bfmt:	*string scalar* (binary format; see below)
r:	*string scalar*
c:	*string scalar*
X:	value to be written; see *Remarks*

316

bfmt may contain

bfmt	meaning
%$\{$ 8|4 $\}$z	8-byte double or 4-byte double
%$\{$ 4|2|1 $\}$b$\left[\,$s|u$\,\right]$	4-, 2-, or 1-byte integer; Stata, signed, or unsigned
%#s	text string
%#S	binary string

Description

These functions manipulate buffers (string scalars) containing binary data and, optionally, perform I/O.

bufio( ) returns a control vector, C, that you pass to the other buffer functions. C specifies the byte order of the buffer and specifies how missing values are to be encoded. Despite its name, bufio( ) opens no files and performs no I/O. bufio( ) merely returns a vector of default values for use with the remaining buffer functions.

bufbyteorder( ) and bufmissingvalue( ) allow changing the defaults in C.

bufput( ) and bufget( ) copy elements into and out of buffers. No I/O is performed. Buffers can then be written by using fwrite( ) and read by using fread( ); see [M-5] **fopen( )**.

fbufput( ) and fbufget( ) do the same, and they perform the corresponding I/O by using fwrite( ) or fread( ).

bufbfmtlen(*bfmt*) and bufbfmtisnum(*bfmt*) are utility routines for processing *bfmt*s; they are rarely used. bufbfmtlen(*bfmt*) returns the implied length, in bytes, of the specified *bfmt*, and bufbfmtisnum(*bfmt*) returns 1 if the *bfmt* is numeric, 0 if string.

Remarks

If you wish simply to read and write matrices, etc., see fgetmatrix( ) and fputmatrix( ) and the other functions in [M-5] **fopen( )**.

The functions documented here are of interest if

1. you wish to create your own binary-data format because you are writing routines in low-level languages such as FORTRAN or C and need to transfer data between your new routines and Stata, or

2. you wish to write a program to read and write the binary format of another software package.

These are advanced and tedious programming projects.

Remarks are presented under the following headings:

> *Basics*
> *Argument C*
> *Arguments B and offset*
> *Argument fh*
> *Argument bfmt*
> *bfmts for numeric data*
> *bfmts for string data*
> *Argument X*
> *Arguments r and c*
> *Advanced issues*

Basics

Let's assume that you wish to write a matrix to disk so you can move it back and forth from FORTRAN. You settle on a file format in which the number of rows and number of columns are first written as 4-byte integers, and then the values of the matrix are written as 8-byte doubles, by row:

One solution to writing matrices in such a format is

```
fh = fopen("filename", "w")
C  = bufio()
fbufput(C, fh, "%4b", rows(X))
fbufput(C, fh, "%4b", cols(X))
fbufput(C, fh, "%8z", X)
fclose(fh)
```

The code to read the matrix back is

```
fh   = fopen("filename", "r")
C    = bufio()
rows = fbufget(C, fh, "%4b")
cols = fbufget(C, fh, "%4b")
X    = fbufget(C, fh, "%8z", rows, cols)
fclose(fh)
```

Another solution, which would be slightly more efficient, is

```
fh = fopen("filename", "w")
C   = bufio()
buf = 8*char(0)
bufput(C, buf, 0, "%4b", rows(X))
bufput(C, buf, 4, "%4b", cols(X))
fwrite(C, buf)
fbufput(C, fh, "%8z", X)
fclose(fh)
```

and

```
fh   = fopen("filename", "r")
C    = bufio()
buf  = fread(fh, 8)
rows = bufget(C, buf, 0, "%4b")
cols = bufget(C, buf, 4, "%4b")
X    = fbufget(C, fh, "%8z", rows, cols)
fclose(fh)
```

What makes the above approach more efficient is that, rather than writing 4 bytes (the number of rows), and 4 bytes again (the number of columns), we created one 8-byte buffer and put the two 4-byte fields in it, and then we wrote all 8 bytes at once. We did the reverse when we read back the data: we read all 8 bytes and then broke out the fields. The benefit is minuscule here but, in general, writing longer buffers results in faster I/O.

In all the above examples, we wrote and read the entire matrix with one function call,

```
fbufput(C, fh, "%8z", X)
```

and

```
X    = fbufget(C, fh, "%8z", rows, cols)
```

Perhaps you would have preferred our having coded

```
for (i=1; i<=rows(X); i++) {
        for (j=1; j<=cols(X); j++) {
                fbufput(C, fh, "%8z", X[i,j])
        }
}
```

and perhaps you would have preferred our having coded something similar to read back the matrix. Had we done so, the results would have been the same.

If you are familiar with FORTRAN, you know that it records matrices in column-dominant order, rather than the row-dominant order used by Mata. It would be a little easier to code the FORTRAN side of things if we changed our file-format design to write columns first:

# rows	# cols	X[1,1]	X[2,1]	...
4 bytes	4 bytes	8 bytes	8 bytes	

One way we could do that would be to write the loops ourselves:

```
fh = fopen("filename", "w")
C  = bufio()
fbufput(C, fh, "%4b", rows(X))
fbufput(C, fh, "%4b", cols(X))
for (j=1; j<=cols(X); i++) {
        for (i=1; i<=rows(X); j++) {
                fbufput(C, fh, "%8z", X[i,j])
        }
}
```

and

```
fh   = fopen("filename", "r")
C    = bufio()
rows = fbufget(C, fh, "%4b")
cols = fbufget(C, fh, "%4b")
X    = J(rows, cols, .)
for (j=1; j<=cols(X); i++) {
        for (i=1; i<=rows(X); j++) {
                X[i,j] = fbufget(C, fh, "%8z")
        }
}
```

We could do that, but there are more efficient and easier ways to proceed. For instance, we could simply transpose the matrix before writing and after reading, and if we do that transposition in place, our code will use no extra memory:

```
fh = fopen("filename", "w")
C  = bufio()
fbufput(C, fh, "%4b", rows(X))
fbufput(C, fh, "%4b", cols(X))
_transpose(X)
fbufput(C, fh, "%8z", X)
_transpose(X)
fclose(fh)
```

The code to read the matrices back is

```
fh   = fopen("filename", "r")
C    = bufio()
rows = fbufget(C, fh, "%4b")
cols = fbufget(C, fh, "%4b")
X    = fbufget(C, fh, "%8z", cols, rows)
_transpose(X)
fclose(fh)
```

Argument C

Argument C in

> bufput$(C, B, offset, bfmt, X)$,
>
> bufget$(C, B, offset, bfmt, \ldots)$,
>
> fbufput$(C, fh, bfmt, X)$, and
>
> fbufget$(C, fh, bfmt, \ldots)$

specifies the control vector. You obtain C by coding

```
C = bufio()
```

bufio() returns C, which is nothing more than a vector filled in with default values. The other buffer routines know how to interpret the vector. The vector contains two pieces of information:

1. The byte order to be used

2. The missing-value coding scheme to be used

Some computer hardware writes numbers left to right (e.g., Sun), and other computer hardware writes numbers right to left (e.g., Intel); see [M-5] **byteorder( )**. If you are going to write binary files, and if you want your files to be readable on all computers, you must write code to deal with this issue.

Many programmers ignore the issue because the programs they write are intended for one computer or on computers like the one they use. If that is the case, you can ignore the issue, too. The default byte order filled in by bufio() is the byte order of the computer you are using.

If you intend to read and write files across different computers, however, you will need to concern yourself with byte order, and how you do that is described in *Advanced issues* below.

The other issue you may need to consider is missing values. If you are writing a binary format that is intended to be used outside Stata, it is best if the values you write simply do not include missing values. Not all packages have them, and the packages that do don't agree on how they are encoded. In such cases, if the data you are writing include missing values, change the values to another value such as -1, 99, 999, or -9999.

If, however, you are writing binary files in Stata to be read back in Stata, you can allow Stata's missing values ., .a, .b, ..., .z. No special action is required. The missing-value scheme in *C* specifies how those missing values are encoded, and there is only one way right now, so there is in fact no issue at all. *C* includes the extra information in case Stata ever changes the way it encodes missing values so that you will have a way to read and write old-format files. How this process works is described in *Advanced issues*.

Arguments B and offset

Functions

> bufput(*C*, *B*, *offset*, *bfmt*, *X*) and
>
> bufget(*C*, *B*, *offset*, *bfmt*, ...)

do not perform I/O; they copy values into and out of the buffer. *B* specifies the buffer, and *offset* specifies the position within it.

B is a string scalar.

offset is an integer scalar specifying the position within *B*. Offset 0 specifies the first byte of *B*.

For bufput(), *B* must already exist and be long enough to receive the result, and it is usual to code something like

```
B = (4 + 4 + rows(X)*cols(X)*8) * char(0)
bufput(C, B, 0, "%4b", rows(X))
bufput(C, B, 4, "%4b", cols(X))
bufput(C, B, 8, "%8z", X)
```

Argument fh

Argument *fh* in

> fbufput(C, *fh*, *bfmt*, X) and
>
> fbufget(C, *fh*, *bfmt*, ...)

plays the role of arguments B and *offset* in bufput() and bufget(). Rather than copy into or out of a buffer, data are written to, or read from, file *fh*. *fh* is obtained from fopen(); see [M-5] **fopen( )**.

Argument bfmt

Argument *bfmt* in

> bufput(C, B, *offset*, *bfmt*, X),
>
> bufget(C, B, *offset*, *bfmt*, ...),
>
> fbufput(C, *fh*, *bfmt*, X), and
>
> fbufget(C, *fh*, *bfmt*, ...)

specifies how the elements are to be written or read.

bfmts for numeric data

The numeric *bfmts* are

bfmt	interpretation
%8z	8-byte floating point
%4z	4-byte floating point
%4bu	4-byte unsigned integer
%4bs	4-byte signed integer
%4b	4-byte Stata integer
%2bu	2-byte unsigned integer
%2bs	2-byte signed integer
%2b	2-byte Stata integer
%1bu	1-byte unsigned integer
%1bs	1-byte signed integer
%1b	1-byte Stata integer

A Stata integer is the same as a signed integer, except that the largest 27 values are given the interpretation ., .a, .b, ..., .z.

bfmts for string data

The string *bfmts* are

bfmt	interpretation
%#s	text string
%#S	binary string

where # represents the length of the string field. Examples include %8s and %639876S.

When writing, it does not matter whether you use %#s or %#S, the same actions are taken:

1. If the string being written is shorter than #, the field is padded with char(0).

2. If the string being written is longer than #, only the first # bytes of the string are written.

When reading, the distinction between %#s and %#S is important:

1. When reading with %#s, if char(0) appears within the first # bytes, the returned result is truncated at that point.

2. When reading with %#S, a full # bytes are returned in all cases.

Argument X

Argument X in

$$\texttt{bufput}(C, B, \textit{offset}, \textit{bfmt}, X) \text{ and}$$

$$\texttt{fbufput}(C, \textit{fh}, \textit{bfmt}, X)$$

specifies the value to be written. X may be real or string and may be a scalar, vector, or matrix. If X is a vector, the elements are written one after the other. If X is a matrix, the elements of the first row are written one after the other, followed by the elements of the second row, and so on.

In

$$X = \texttt{bufget}(C, B, \textit{offset}, \textit{bfmt}, \ldots) \text{ and}$$

$$X = \texttt{fbufget}(C, \textit{fh}, \textit{bfmt}, \ldots)$$

X is returned.

Arguments r and c

Arguments r and c are optional in the following:

$$X = \texttt{bufget}(C, B, \textit{offset}, \textit{bfmt}),$$

$$X = \texttt{bufget}(C, B, \textit{offset}, \textit{bfmt}, c),$$

$$X = \texttt{bufget}(C, B, \textit{offset}, \textit{bfmt}, r, c),$$

$$X = \texttt{fbufget}(C, \textit{fh}, \textit{bfmt}),$$

$X = \texttt{fbufget}(C, \textit{fh}, \textit{bfmt}, c)$, and

$X = \texttt{fbufget}(C, \textit{fh}, \textit{bfmt}, r, c)$.

If r is not specified, results are as if $r = 1$.

If c is not specified, results are as if $c = 1$.

Thus

$X = \texttt{bufget}(C, B, \textit{offset}, \textit{bfmt})$ and

$X = \texttt{fbufget}(C, \textit{fh}, \textit{bfmt})$

read one element and return it, whereas

$X = \texttt{bufget}(C, B, \textit{offset}, \textit{bfmt}, c)$ and

$X = \texttt{fbufget}(C, \textit{fh}, \textit{bfmt}, c)$

read c elements and return them in a column vector, and

$X = \texttt{bufget}(C, B, \textit{offset}, \textit{bfmt}, r, c)$ and

$X = \texttt{fbufget}(C, \textit{fh}, \textit{bfmt}, r, c)$

read $r * c$ elements and return them in an $r \times c$ matrix.

Advanced issues

A properly designed binary-file format includes a signature line first thing in the file:

```
fh = fopen(filename, "w")
fwrite(fh, "MyFormat For Mats v. 1.0")
          /* ----+----1----+----2---- */
```

and

```
fh = fopen(filename, "r")
if (fread(fh, 24) != "MyFormat For Mats v. 1.0") {
        errprintf("%s not My-Format file\n", filename)
        exit(610)
}
```

If you are concerned with byte order and mapping of missing values, you should write the byte order and missing-value mapping in the file, write in natural byte order, and be prepared to read back in either byte order.

The code for writing is

```
fh = fopen(filename, "w")
fwrite(fh, "MyFormat For Mats v. 1.0")

C = bufio()
fbufput(C, fh, "%1bu", bufbyteorder(C))
fbufput(C, fh, "%2bu", bufmissingvalue(C))
```

and the corresponding code for reading the file is

```
fh = fopen(filename, "r")
if (fread(fh, 24) != "MyFormat For Mats v. 1.0") {
        errprintf("%s not My-Format file\n", filename)
        exit(610)
}

C = bufio()
bufbyteorder(C, fbufget(C, "%1bu"))
bufmissingvalue(C, fbufget(C, "%2bu"))
```

All we write in the file before recording the byte order are strings and bytes. This way, when we read the file later, we can set the byte order before reading any 2-, 4-, or 8-byte fields.

bufbyteorder(*C*)—bufbyteorder( ) with one argument—returns the byte-order encoding recorded in *C*. It returns 1 (meaning HILO) or 2 (meaning LOHI).

bufbyteorder(*C, value*)—bufbyteorder( ) with two arguments—resets the byte order recorded in *C*. Once reset, all buffer functions will automatically reverse bytes if necessary.

bufmissingvalue( ) works the same way. With one argument, it returns a code for the encoding scheme recorded in *C* (said code being the Stata release number multiplied by 100). With two arguments, it resets the code. Once the code is reset, all buffer routines used will automatically take the appropriate action.

Conformability

bufio( ):

result:	*colvector*	

bufbyteorder(*C*):

C:	*colvector*	made by bufio( )
result:	1 × 1	containing 1 (HILO) or 2 (LOHI)

bufbyteorder(*C, byteorder*):

C:	*colvector*	made by bufio( )
byteorder:	1 × 1	containing 1 (HILO) or 2 (LOHI)
result:	*void*	

bufmissingvalue(*C*):

C:	*colvector*	made by bufio( )
result:	1 × 1	

bufmissingvalue(*C, version*):

C:	*colvector*	made by bufio( )
version:	1 × 1	
result:	*void*	

bufput(C, B, *offset*, *bfmt*, X):

C:	*colvector*	made by bufio( )
B:	1×1	
offset:	1×1	
bfmt:	1×1	
X:	$r \times c$	
result:	*void*	

bufget(C, B, *offset*, *bfmt*):

C:	*colvector*	made by bufio( )
B:	1×1	
offset:	1×1	
bfmt:	1×1	
result:	1×1	

bufget(C, B, *offset*, *bfmt*, r):

C:	*colvector*	made by bufio( )
B:	1×1	
offset:	1×1	
bfmt:	1×1	
r:	1×1	
result:	$1 \times c$	

bufget(C, B, *offset*, *bfmt*, r, c):

C:	*colvector*	made by bufio( )
B:	1×1	
offset:	1×1	
bfmt:	1×1	
r:	1×1	
c:	1×1	
result:	$r \times c$	

fbufput(C, *fh*, *bfmt*, X):

C:	*colvector*	made by bufio( )
fh:	1×1	
bfmt:	1×1	
X:	$r \times c$	
result:	*void*	

fbufget(C, *fh*, *bfmt*):

C:	*colvector*	made by bufio( )
fh:	1×1	
bfmt:	1×1	
result:	1×1	

fbufget(C, *fh*, *bfmt*, r):

C:	*colvector*	made by bufio( )
fh:	1×1	
bfmt:	1×1	
r:	1×1	
result:	$1 \times c$	

fbufget(C, fh, $bfmt$, r, c):

C:	*colvector*	made by bufio( )
fh:	1×1	
$bfmt$:	1×1	
r:	1×1	
c:	1×1	
result:	$r \times c$	

bufbfmtlen($bfmt$):

$bfmt$:	1×1
result:	1×1

bufbfmtisnum($bfmt$):

$bfmt$:	1×1
result:	1×1

Diagnostics

bufio( ) cannot fail.

bufbyteorder(C) cannot fail. bufbyteorder(C, *byteorder*) aborts with error if *byteorder* is not 1 or 2.

bufmissingvalue(C) cannot fail. bufmissingvalue(C, *version*) aborts with error if *version* < 100 or *version* > stataversion( ).

bufput(C, B, *offset*, *bfmt*, X) aborts with error if B is too short to receive the result, *offset* is out of range, *bfmt* is invalid, or *bfmt* is a string format and X is numeric or vice versa. Putting a void matrix results in 0 bytes being inserted into the buffer and is not an error.

bufget(C, B, *offset*, *bfmt*, . . .) aborts with error if B is too short, *offset* is out of range, or *bfmt* is invalid. Reading zero rows or columns results in a void returned result and is not an error.

fbufput(C, fh, *bfmt*, X) aborts with error if fh is invalid, *bfmt* is invalid, or *bfmt* is a string format and X is numeric or vice versa. Putting a void matrix results in 0 bytes being written and is not an error. I/O errors are possible; use fstatus( ) to detect them.

fbufget(C, fh, *bfmt*, . . .) aborts with error if fh is invalid or *bfmt* is invalid. Reading zero rows or columns results in a void returned result and is not an error. End-of-file and I/O errors are possible; use fstatus( ) to detect them.

bufbfmtlen($bfmt$) and bufbfmtisnum($bfmt$) abort with error if *bfmt* is invalid.

Also see

[M-5] **fopen( )** — File I/O

[M-4] **io** — I/O functions

Title

[M-5] **byteorder( )** — Byte order used by computer

Syntax

> *real scalar* byteorder( )

Description

byteorder( ) returns 1 if the computer is HILO (records most significant byte first) and returns 2 if LOHI (records least significant byte first).

Remarks

Pretend that the values 00 and 01 are recorded at memory positions 58 and 59 and that you know what is recorded there is a 2-byte integer. How should the 2-byte number be interpreted: as 0001 (meaning 1) or 0100 (meaning 256 in decimal)? For different computers, the answer varies. For HILO computers, the number is to be interpreted as 0001. For LOHI computers, the number is interpreted as 0100.

Regardless, it does not matter because the computer is consistent with itself. An issue arises, however, when we write binary files that may be read on computers using a different byte order or when we read files from computers that used a different byte order.

Stata and Mata automatically handle these problems for you, so you may wish to stop reading. byteorder( ), however, is the tool on which the solution is based. If you intend to write code based on your own binary-file format or to write code to process the binary files of others, then you may need to use it.

There are two popular solutions to the byte-order problem: (1) write the file in a known byte order or (2) write the file by using whichever byte order is convenient and record the byte order used in the file. StataCorp tends to use the second, but others who have worried about this problem have used both solutions.

In solution (1), it is usually agreed that the file will be written in HILO order. If you are using a HILO computer, you write and read files the ordinary way, ignoring the problem altogether. If you are on a LOHI computer, however, you must reverse the bytes before placing them in the file. If you are writing code designed to execute on both kinds of computers, you must write code for both circumstances, and you must consider the problem when both reading and writing the files.

In solution (2), files are written LOHI or HILO, depending solely on the computer being used. Early in the file, however, the byte order is recorded. When reading the file, you compare the order in which the file is recorded with the order of the computer and, if they are different, you reverse bytes.

Mata-buffered I/O utilities will automatically reverse bytes for you. See [M-5] **bufio( )**.

Conformability

```
byteorder():
```
> *result*: 1×1

Diagnostics

None.

Also see

[M-5] **bufio( )** — Buffered (binary) I/O

[M-5] **stataversion( )** — Version of Stata being used

[M-4] **programming** — Programming functions

Title

[M-5] C( ) — Make complex

Syntax

complex matrix C(*numeric matrix A*)

complex matrix C(*real matrix R*, *real matrix I*)

Description

C(*A*) returns *A* converted to complex. C(*A*) returns *A* if *A* is already complex. If *A* is real, C(*A*) returns *A+0i*—*A* cast up to complex. Coding C(*A*) is thus how you ensure that the matrix is treated as complex.

C(*R*, *I*) returns the complex matrix *R+Ii* and is faster than the alternative R + I:*1i.

Remarks

Many of Mata's functions are overloaded, meaning they return a real when given real arguments and a complex when given complex arguments. Given real arguments, if the result cannot be expressed as a real, missing value is returned. Thus sqrt(-1) evaluates to missing, whereas sqrt(-1+0i) is 1*i*.

C( ) is the fast way to make arguments that might be real into complex. You can code

result = sqrt(C(*x*))

If *x* already is complex, C( ) does nothing; if *x* is real, C(*x*) returns the complex equivalent.

The two-argument version of C( ) is less frequently used. C(*R*, *I*) is literally equivalent to R :+ I*1i, meaning that *R* and *I* need only be c-conformable.

For instance, C(1, (1,2,3)) evaluates to (1+1i, 1+2i, 1+3i).

Conformability

C(*A*):

A:	$r \times c$
result:	$r \times c$

C(*R*, *I*):

R:	$r_1 \times c_1$
I:	$r_2 \times c_2$, *R* and *I* c-conformable
result:	$\max(r_1,r_2) \times \max(c_1,c_2)$

Diagnostics

$C(Z)$, if Z is complex, literally returns Z and not a copy of Z. This makes execution of C() applied to complex arguments instant.

In $C(R, I)$, the i,j element of the result will be missing anywhere $R[i,j]$ or $I[i,j]$ is missing. For instance, C((1,3,.), (.,2,4)) results in (., 3+2i, .). If $R[i,j]$ and $I[i,j]$ are both missing, then the $R[i,j]$ value will be used; e.g., C(.a, .b) results in .a.

Also see

[M-5] **Re( )** — Extract real or imaginary part

[M-4] **scalar** — Scalar mathematical functions

[M-4] **utility** — Matrix utility functions

Title

> **[M-5] c( )** — Access c( ) value

Syntax

> *scalar*   c(*string scalar name*)

returned is either a real or string scalar, depending on the value of *name*.

Description

c(*name*) returns Stata's c-class value; see [P] **creturn**.

Do not confuse c( ) with C( ), which makes complex out of real arguments; see [M-5] **C( )**.

Remarks

See [P] **creturn** or, in Stata, type

> . creturn list

to see what is stored in c( ). Among the especially useful c( ) values are

> *string* c("current_date")
> *string* c("current_time")
> *string* c("os")
> *string* c("dirsep")

Conformability

c(*name*):
>   *name*: 1×1
>   *result*: 1×1

Diagnostics

c(*name*) returns a string or real depending on the particular c( ) value requested. If *name* is an invalid name or contains a name for which no c( ) value is defined, returned is "".

Also see

[M-4] **programming** — Programming functions

Title

[M-5] **callersversion( )** — Obtain version number of caller

Syntax

> *real scalar* `callersversion()`

Description

`callersversion()` returns the version set by the caller (see [M-2] **version**), or if the caller did not set the version, it returns the version of Stata under which the caller was compiled.

Remarks

`callersversion()` is how [M-2] **version** is made to work. Say that you have written function

> *real matrix* useful(*real matrix A*, *real scalar k*)

and assume that useful() aborts with error if *A* is void. You wrote useful() in the days of Stata 10. For Stata 11, you want to change useful() so that it returns J(0,0,.) if *A* is void, but you want to maintain the current behavior for old Stata 10 callers and programs. You do that as follows:

```
real matrix useful(real matrix A, real scalar k)
{
        ...
        if (callersversion()>=10) {
                if (rows(A)==0 | cols(A)==0) return(J(0,0,.))
        }
        ...
}
```

Conformability

`callersversion()`:
> *result*: 1×1

Diagnostics

None.

Also see

[M-2] **version** — Version control

[M-4] **programming** — Programming functions

Title

[M-5] **cat( )** — Load file into string matrix

Syntax

string colvector cat(*string scalar filename* [, *real scalar line1* [, *real scalar line2*]])

Description

cat(*filename*) returns a column vector containing the lines from ASCII file *filename*.

cat(*filename*, *line1*) returns a column vector containing the lines from ASCII file *filename* starting with line number *line1*.

cat(*filename*, *line1*, *line2*) returns a column vector containing the lines from ASCII file *filename* starting with line number *line1* and ending with line number *line2*.

Remarks

cat(*filename*) removes new-line characters at the end of lines.

Conformability

cat(*filename*, *line1*, *line2*):

     filename: 1×1

        line1: 1×1 (optional)

        line2: 1×1 (optional)

       result: $r \times 1$, $r \geq 0$

Diagnostics

cat(*filename*) aborts with error if *filename* does not exist.

cat( ) returns a 0×1 result if *filename* contains 0 bytes.

Also see

[M-4] **io** — I/O functions

Title

Syntax

string scalar pwd()

void   chdir(*string scalar dirpath*)

real scalar   _chdir(*string scalar dirpath*)

void   mkdir(*string scalar dirpath*)

void   mkdir(*string scalar dirpath*, *real scalar public*)

real scalar   _mkdir(*string scalar dirpath*)

real scalar   _mkdir(*string scalar dirpath*, *real scalar public*)

void   rmdir(*string scalar dirpath*)

real scalar   _rmdir(*string scalar dirpath*)

Description

pwd() returns the full name (path) of the current working directory.

chdir(*dirpath*) changes the current working directory to *dirpath*. chdir() aborts with error if the directory does not exist or the operating system cannot change to it.

_chdir(*dirpath*) does the same thing but returns 170 (a return code) when chdir() would abort. _chdir() returns 0 if it is successful.

mkdir(*dirpath*) and mkdir(*dirpath*, *public*) create directory *dirpath*. mkdir() aborts with error if the directory already exists or cannot be created. If *public* $\neq 0$ is specified, the directory is given permissions so that everyone can read it; otherwise, it is given the usual permissions.

_mkdir(*dirpath*) and _mkdir(*dirpath*, *public*) do the same thing but return 693 (a return code) when mkdir() would abort. _mkdir() returns 0 if it is successful.

rmdir(*dirpath*) removes directory *dirpath*. rmdir() aborts with error if the directory does not exist, is not empty, or the operating system refuses to remove it.

_rmdir(*dirpath*) does the same thing but returns 693 (a return code) when rmdir() would abort. _rmdir() returns 0 if it is successful.

Conformability

pwd():
        result: 1×1

chdir(*dirpath*):
        dirpath: 1×1
        result: *void*

_chdir(*dirpath*):
        dirpath: 1×1
        result: 1×1

mkdir(*dirpath*, *public*):
        dirpath: 1×1
        public: 1×1 (optional)
        result: *void*

_mkdir(*dirpath*, *public*):
        dirpath: 1×1
        public: 1×1 (optional)
        result: 1×1

rmdir(*dirpath*):
        dirpath: 1×1
        result: *void*

_rmdir(*dirpath*):
        dirpath: 1×1
        result: 1×1

Diagnostics

pwd() never aborts with error, but it can return "" if the operating system does not know or does not have a name for the current directory (which happens when another process removes the directory in which you are working).

chdir(*dirpath*) aborts with error if the directory does not exist or the operating system cannot change to it.

_chdir(*dirpath*) never aborts with error; it returns 0 on success and 170 on failure.

mkdir(*dirpath*) and mkdir(*dirpath*, *public*) abort with error if the directory already exists or the operating system cannot change to it.

_mkdir(*dirpath*) and _mkdir(*dirpath*, *public*) never abort with error; they return 0 on success and 693 on failure.

rmdir(*dirpath*) aborts with error if the directory does not exist, is not empty, or the operating system cannot remove it.

_rmdir(*dirpath*) never aborts with error; it returns 0 on success and 639 on failure.

Also see

[M-4] **io** — I/O functions

Title

[M-5] cholesky( ) — Cholesky square-root decomposition

Syntax

> *numeric matrix*   cholesky(*numeric matrix A*)
>
> *void*   _cholesky(*numeric matrix A*)

Description

cholesky(*A*) returns the Cholesky decomposition *G* of symmetric (Hermitian), positive-definite matrix *A*. cholesky() returns a lower-triangular matrix of missing values if *A* is not positive definite.

_cholesky(*A*) does the same thing, except that it overwrites *A* with the Cholesky result.

Remarks

The Cholesky decomposition *G* of a symmetric, positive-definite matrix *A* is

$$A = GG'$$

where *G* is lower triangular. When *A* is complex, *A* must be Hermitian, and G', of course, is the conjugate transpose of *G*.

Decomposition is performed via [M-1] **LAPACK**.

Conformability

cholesky(*A*):

> $A: n \times n$
> *result*: $n \times n$

_cholesky(*A*):

> *input*:
>> $A: n \times n$
> *output*:
>> $A: n \times n$

Diagnostics

cholesky() returns a lower-triangular matrix of missing values if *A* contains missing values or if *A* is not positive definite.

_cholesky(*A*) overwrites *A* with a lower-triangular matrix of missing values if *A* contains missing values or if *A* is not positive definite.

Both functions use the elements from the lower triangle of A without checking whether A is symmetric or, in the complex case, Hermitian.

André-Louis Cholesky (1875–1918) was born near Bordeaux in France. He studied at the Ecole Polytechnique and then joined the French army. Cholesky served in Tunisia and Algeria and then worked in the Geodesic Section of the Army Geographic Service, where he invented his now-famous method. In the war of 1914–1918, he served in the Vosges and in Romania but after return to the Western front was fatally wounded. Cholesky's method was written up posthumously by one of his fellow officers but attracted little attention until the 1940s.

Reference

Chabert, J.-L., É. Barbin, J. Borowczyk, M. Guillemot, and A. Michel-Pajus. 1999. *A History of Algorithms: From the Pebble to the Microchip*. Trans. C. Weeks. Berlin: Springer.

Also see

[M-5] **lud( )** — LU decomposition

[M-4] **matrix** — Matrix functions

Title

Syntax

numeric matrix	cholinv(*numeric matrix A*)
numeric matrix	cholinv(*numeric matrix A, real scalar tol*)
void	_cholinv(*numeric matrix A*)
void	_cholinv(*numeric matrix A, real scalar tol*)

Description

cholinv(*A*) and cholinv(*A, tol*) return the inverse of real or complex, symmetric (Hermitian), positive-definite, square matrix *A*.

_cholinv(*A*) and _cholinv(*A, tol*) do the same thing except that, rather than returning the inverse matrix, they overwrite the original matrix *A* with the inverse.

In all cases, optional argument *tol* specifies the tolerance for determining singularity; see *Remarks* below.

Remarks

These routines calculate the inverse of a symmetric, positive-definite square matrix *A*. See [M-5] **luinv( )** for the inverse of a general square matrix.

A is required to be square and positive definite. See [M-5] **qrinv( )** and [M-5] **pinv( )** for generalized inverses of nonsquare or rank-deficient matrices. See [M-5] **invsym( )** for generalized inverses of real, symmetric matrices.

cholinv(*A*) is logically equivalent to cholsolve(*A*, I(rows(*A*))); see [M-5] **cholsolve( )** for details and for use of the optional *tol* argument.

Conformability

cholinv(*A, tol*):

A:	$n \times n$	
tol:	1×1	(optional)
result:	$n \times n$	

_cholinv(*A, tol*):

input:

A:	$n \times n$	
tol:	1×1	(optional)

output:

A:	$n \times n$

Diagnostics

The inverse returned by these functions is real if A is real and is complex if A is complex. If you use these functions with a non–positive definite matrix, or a matrix that is too close to singularity, returned will be a matrix of missing values. The determination of singularity is made relative to *tol*. See *Tolerance* under *Remarks* in [M-5] **cholsolve( )** for details.

cholinv(A) and _cholinv(A) return a result containing all missing values if A is not positive definite or if A contains missing values.

_cholinv(A) aborts with error if A is a view.

See [M-5] **cholsolve( )** and [M-1] **tolerance** for information on the optional *tol* argument.

Both functions use the elements from the lower triangle of A without checking whether A is symmetric or, in the complex case, Hermitian.

Also see

[M-5] **invsym( )** — Symmetric real matrix inversion

[M-5] **luinv( )** — Square matrix inversion

[M-5] **qrinv( )** — Generalized inverse of matrix via QR decomposition

[M-5] **pinv( )** — Moore–Penrose pseudoinverse

[M-5] **cholsolve( )** — Solve AX=B for X using Cholesky decomposition

[M-4] **matrix** — Matrix functions

[M-4] **solvers** — Functions to solve AX=B and to obtain A inverse

Title

[M-5] cholsolve( ) — Solve AX=B for X using Cholesky decomposition

Syntax

numeric matrix   cholsolve(*numeric matrix A*, *numeric matrix B*)

numeric matrix   cholsolve(*numeric matrix A*, *numeric matrix B*, *real scalar tol*)

void   _cholsolve(*numeric matrix A*, *numeric matrix B*)

void   _cholsolve(*numeric matrix A*, *numeric matrix B*, *real scalar tol*)

Description

cholsolve(A, B) solves $AX = B$ and returns X for symmetric (Hermitian), positive-definite A.
cholsolve( ) returns a matrix of missing values if A is not positive definite or if A is singular.

cholsolve(A, B, *tol*) does the same thing; it allows you to specify the tolerance for declaring that
A is singular; see *Tolerance* under *Remarks* below.

_cholsolve(A, B) and _cholsolve(A, B, *tol*) do the same thing except that, rather than returning
the solution X, they overwrite B with the solution, and in the process of making the calculation, they
destroy the contents of A.

Remarks

The above functions solve $AX = B$ via Cholesky decomposition and are accurate. When A is not
symmetric and positive definite, [M-5] **lusolve( )**, [M-5] **qrsolve( )**, and [M-5] **svsolve( )** are alternatives
based on the LU decomposition, the QR decomposition, and the singular value decomposition (SVD).
The alternatives differ in how they handle singular A. Then the LU-based routines return missing
values, whereas the QR-based and SVD-based routines return generalized (least-squares) solutions.

Remarks are presented under the following headings:

    Derivation
    Relationship to inversion
    Tolerance

Derivation

We wish to solve for X

$$AX = B \tag{1}$$

when A is symmetric and positive definite. Perform the Cholesky decomposition of A so that we have
$A = GG'$. Then (1) can be written

$$GG'X = B \tag{2}$$

Define
$$Z = G'X \tag{3}$$
Then (2) can be rewritten
$$GZ = B \tag{4}$$

It is easy to solve (4) for Z because G is a lower-triangular matrix. Once Z is known, it is easy to solve (3) for X because G' is upper triangular.

Relationship to inversion

See *Relationship to inversion* in [M-5] **lusolve( )** for a discussion of the relationship between solving the linear system and matrix inversion.

Tolerance

The default tolerance used is
$$\eta = \frac{\texttt{(1e-13)*trace(abs(}G\texttt{))}}{n}$$

where G is the lower-triangular Cholesky factor of A: $n \times n$. A is declared to be singular if `cholesky()` (see [M-5] **cholesky( )**) finds that A is not positive definite, or if A is found to be positive definite, if any diagonal element of G is less than or equal to η. Mathematically, positive definiteness implies that the matrix is not singular. In the numerical method used, two checks are made: `cholesky()` makes one and then the η rule is applied to ensure numerical stability in the use of the result `cholesky()` returns.

If you specify *tol* > 0, the value you specify is used to multiply η. You may instead specify *tol* ≤ 0 and then the negative of the value you specify is used in place of η; see [M-1] **tolerance**.

See [M-5] **lusolve( )** for a detailed discussion of the issues surrounding solving nearly singular systems. The main point to keep in mind is that if A is ill conditioned, then small changes in A or B can lead to radically large differences in the solution for X.

Conformability

cholsolve(A, B, *tol*):

> *input*:

A:	$n \times n$	
B:	$n \times k$	
tol:	1×1	(optional)
result:	$n \times k$	

_cholsolve(A, B, *tol*):

> *input*:

A:	$n \times n$	
B:	$n \times k$	
tol:	1×1	(optional)

> *output*:

A:	0×0	
B:	$n \times k$	

Diagnostics

cholsolve(A, B, ...), and _cholsolve(A, B, ...) return a result of all missing values if A is not positive definite or if A contains missing values.

_cholsolve(A, B, ...) also aborts with error if A or B is a view.

All functions use the elements from the lower triangle of A without checking whether A is symmetric or, in the complex case, Hermitian.

Also see

[M-5] **cholesky( )** — Cholesky square-root decomposition

[M-5] **cholinv( )** — Symmetric, positive-definite matrix inversion

[M-5] **solvelower( )** — Solve AX=B for X, A triangular

[M-5] **lusolve( )** — Solve AX=B for X using LU decomposition

[M-5] **qrsolve( )** — Solve AX=B for X using QR decomposition

[M-5] **svsolve( )** — Solve AX=B for X using singular value decomposition

[M-5] **solve_tol( )** — Tolerance used by solvers and inverters

[M-4] **matrix** — Matrix functions

[M-4] **solvers** — Functions to solve AX=B and to obtain A inverse

Title

Syntax

$real\ matrix$ comb($real\ matrix\ n$, $real\ matrix\ k$)

Description

comb(n, k) returns the elementwise combinatorial function n-choose-k, the number of ways to choose k items from n items, regardless of order.

Conformability

comb(n, k):

n:	$r_1 \times c_1$	
k:	$r_2 \times c_2$,	n and k r-conformable
$result$:	$\max(r_1,r_2) \times \max(c_1,c_2)$	

Diagnostics

comb(n, k) returns missing when either argument is missing or when the result would be larger than 10^{300}.

Also see

[M-4] **statistical** — Statistical functions

Title

[M-5] **cond( )** — Condition number

Syntax

> *real scalar* cond(*numeric matrix A*)

> *real scalar* cond(*numeric matrix A*, *real scalar p*)

Description

cond(*A*) returns cond(*A*, 2).

cond(*A*, *p*) returns the value of the condition number of *A* for the specified norm *p*, where *p* may be 0, 1, 2, or . (missing).

Remarks

The condition number of a matrix *A* is

$$cond = \texttt{norm}(A, p) \times \texttt{norm}(A^{-1}, p)$$

These functions return missing when *A* is singular.

Values near 1 indicate that the matrix is well conditioned, and large values indicate ill conditioning.

Conformability

cond(*A*):

A:	$r \times c$
result:	1×1

cond(*A*, *p*):

A:	$r \times c$
p:	1×1
result:	1×1

Diagnostics

cond(*A*, *p*) aborts with error if *p* is not 0, 1, 2, or . (missing).

cond(*A*) and cond(*A*, *p*) return missing when *A* is singular or if *A* contains missing values.

cond(*A*) and cond(*A*, *p*) return 1 when *A* is void.

cond(*A*) and cond(*A*, 2) return missing if the SVD algorithm fails to converge, which is highly unlikely; see [M-5] **svd( )**.

Also see

[M-5] **norm( )** — Matrix and vector norms

[M-4] **matrix** — Matrix functions

Title

[M-5] conj( ) — Complex conjugate

Syntax

> *numeric matrix*   conj(*numeric matrix Z*)
>
> *void*   _conj(*numeric matrix A*)

Description

conj(Z) returns the elementwise complex conjugate of Z, i.e., conj($a+bi$) = $a - bi$. conj() may be used with real or complex matrices. If Z is real, Z is returned unmodified.

_conj(A) replaces A with conj(A). Coding _conj(A) is equivalent to coding A = conj(A), except that less memory is used.

Remarks

Given $m \times n$ matrix Z, conj(Z) returns an $m \times n$ matrix; it does not return the transpose. To obtain the conjugate transpose matrix, also known as the adjoint matrix, adjugate matrix, Hermitian adjoin, or Hermitian transpose, code

> Z'

See [M-2] **op_transpose**.

A matrix equal to its conjugate transpose is called Hermitian or self-adjoint, although in this manual, we often use the term symmetric.

Conformability

```
conj(Z):
            Z:      r × c
        result:     r × c
_conj(A):
    input:
            A:      r × c
    output:
            A:      r × c
```

Diagnostics

conj(*Z*) returns a real matrix if *Z* is real and a complex matrix if *Z* is complex.

conj(*Z*), if *Z* is real, returns *Z* itself and not a copy. This makes conj() execute instantly when applied to real matrices.

_conj(*A*) does nothing if *A* is real (and hence, does not abort if *A* is a view).

Also see

[M-5] **_transpose**( ) — Transposition in place

[M-4] **scalar** — Scalar mathematical functions

Title

[M-5] corr( ) — Make correlation matrix from variance matrix

Syntax

> *real matrix*   corr(*real matrix V*)
>
> *void*   _corr(*real matrix V*)

Description

corr(*V*) returns the correlation matrix corresponding to variance matrix *V*.

_corr(*V*) changes the contents of *V* from being a variance matrix to being a correlation matrix.

Remarks

See function variance() in [M-5] **mean( )** for obtaining a variance matrix from data.

Conformability

corr(*V*):
   input:
               V: $k \times k$
       result: $k \times k$

_corr(*V*):
   input:
               V: $k \times k$
   output:
               V: $k \times k$

Diagnostics

corr() and _corr() abort with error if *V* is not square. *V* should also be symmetric, but this is not checked.

Also see

[M-5] **mean( )** — Means, variances, and correlations

[M-4] **statistical** — Statistical functions

Title

Syntax

> *real matrix* cross(X, Z)
>
> *real matrix* cross(X, w, Z)
>
> *real matrix* cross(X, xc, Z, zc)
>
> *real matrix* cross(X, xc, w, Z, zc)

where

X:	*real matrix X*
xc:	*real scalar xc*
w:	*real vector w*
Z:	*real matrix Z*
zc:	*real scalar zc*

Description

cross() makes calculations of the form

$$X'X$$
$$X'Z$$
$$X' \operatorname{diag}(w) X$$
$$X' \operatorname{diag}(w) Z$$

cross() is designed for making calculations that often arise in statistical formulas. In one sense, cross() does nothing that you cannot easily write out in standard matrix notation. For instance, cross(X, Z) calculates $X'Z$. cross(), however, has the following differences and advantages over the standard matrix-notation approach:

1. cross() omits the rows in X and Z that contain missing values, which amounts to dropping observations with missing values.

2. cross() uses less memory and is especially efficient when used with views.

3. cross() watches for special cases and makes calculations in those special cases more efficiently. For instance, if you code cross(X, X), cross() observes that the two matrices are the same and makes the calculation for a symmetric matrix result.

cross(X, Z) returns $X'Z$. Usually rows(X)==rows(Z), but X is also allowed to be a scalar, which is then treated as if J(rows(Z), 1, 1) were specified. Thus cross$(1, Z)$ is equivalent to colsum(Z).

cross(X, w, Z) returns $X'\operatorname{diag}(w)Z$. Usually, rows$(w)$==rows$(Z)$ or cols(w)==rows(Z), but w is also allowed to be a scalar, which is treated as if J(rows(Z), 1, w) were specified. Thus cross$(X,1,Z)$ is the same as cross(X,Z). Z may also be a scalar, just as in the two-argument case.

351

cross(X, xc, Z, zc) is similar to cross(X, Z) in that $X'Z$ is returned. In the four-argument case, however, X is augmented on the right with a column of 1s if xc!=0 and Z is similarly augmented if zc!=0. cross(X, 0, Z, 0) is equivalent to cross(X, Z). Z may be specified as a scalar.

cross(X, xc, w, Z, zc) is similar to cross(X, w, Z) in that X'diag(w)Z is returned. As with the four-argument cross(), X is augmented on the right with a column of 1s if xc!=0 and Z is similarly augmented if zc!=0. Both Z and w may be specified as scalars. cross(X, 0, 1, Z, 0) is equivalent to cross(X, Z).

Remarks

In the following examples, we are going to calculate linear regression coefficients using $b = (X'X)^{-1}X'y$, which means using $\sum x/n$, and variances using $n/(n-1) \times (\sum x^2/n - mean^2)$. See [M-5] **crossdev( )** for examples of the same calculations made in a more numerically stable way.

The examples use the automobile data. Since we are using the absolute form of the calculation equations, it would be better if all variables had values near 1 (in which case the absolute form of the calculation equations are perfectly adequate). Thus we suggest

```
. sysuse auto
. replace weight = weight/1000
```

Some of the examples use a weight w. For that, you might try

```
. gen w = int(4*runiform())+1
```

▷ Example 1: Linear regression, the traditional way

```
: y = X = .
: st_view(y, ., "mpg")
: st_view(X, ., "weight foreign")
:
: X = X, J(rows(X),1,1)
: b = invsym(X'X)*X'y
```

Comments: Does not handle missing values and uses much memory if X is large.

◁

▷ Example 2: Linear regression using cross( )

```
: y = X = .
: st_view(y, ., "mpg")
: st_view(X, ., "weight foreign")
:
: XX = cross(X,1 , X,1)
: Xy = cross(X,1 , y,0)
: b  = invsym(XX)*Xy
```

Comments: There is still an issue with missing values; mpg might not be missing everywhere weight and foreign are missing.

◁

▷ Example 3: Linear regression using cross( ) and one view

```
: // We will form
: //
: //   (y X)'(y X)  =  (y'y, y'X \ X'y, X'X)
:
: M = .
: st_view(M, ., "mpg weight foreign", 0)
:
: CP = cross(M,1 , M,1)
: XX = CP[|2,2 \ .,.|]
: Xy = CP[|2,1 \ .,1|]
: b  = invsym(XX)*Xy
```

Comments: Using one view handles all missing-value issues (we specified fourth argument 0 to st_view( ); see [M-5] **st_view( )**).

◁

▷ Example 4: Linear regression using cross( ) and subviews

```
: M = X = y = .
: st_view(M, ., "mpg weight foreign", 0)
: st_subview(y, M, ., 1)
: st_subview(X, M, ., (2\.))
:
: XX = cross(X,1 , X,1)
: Xy = cross(X,1 , y,0)
: b  = invsym(XX)*Xy
```

Comments: Using subviews also handles all missing-value issues; see [M-5] **st_subview( )**. The subview approach is a little less efficient than the previous solution but is perhaps easier to understand. The efficiency issue concerns only the extra memory required by the subviews y and X, which is not much.

Also, this subview solution could be used to handle the missing-value problems of calculating linear regression coefficients the traditional way, shown in example 1:

```
: M = X = y = .
: st_view(M, ., "mpg weight foreign", 0)
: st_subview(y, M, ., 1)
: st_subview(X, M, ., (2\.))
:
: X = X, J(rows(X), 1, 1)
: b = invsym(X'X)*X'y
```

◁

▷ Example 5: Weighted linear regression, the traditional way

```
: M = w = y = X = .
: st_view(M, ., "w mpg weight foreign", 0)
: st_subview(w, M, ., 1)
: st_subview(y, M, ., 2)
: st_subview(X, M, ., (3\.))
:
: X = X, J(rows(X), 1, 1)
: b = invsym(X'diag(w)*X)*X'diag(w)'y
```

Comments: The memory requirements are now truly impressive because diag(w) is an $N \times N$ matrix! That is, the memory requirements are truly impressive when N is large. Part of the power of Mata is that you can write things like invsym(X'diag(w)*X)*X'diag(w)'y and obtain solutions. We do not mean to be dismissive of the traditional approach; we merely wish to emphasize its memory requirements and note that there are alternatives.

◁

▷ Example 6: Weighted linear regression using cross( )

```
: M = w = y = X = .
: st_view(M, ., "w mpg weight foreign", 0)
: st_subview(w, M, ., 1)
: st_subview(y, M, ., 2)
: st_subview(X, M, ., (3\.))
:
: XX = cross(X,1 ,w, X,1)
: Xy = cross(X,1 ,w, y,0)
: b  = invsym(XX)*Xy
```

Comments: The memory requirements here are no greater than they were in example 4, which this example closely mirrors. We could also have mirrored the logic of example 3:

```
: M = w = M2 = .
: st_view(M, ., "w mpg weight foreign", 0)
: st_subview(w, M, ., 1)
: st_subview(M2, M, ., (2\.))
:
: CP = cross(M,1 ,w, M,1)
: XX = CP[|3,3 \ .,.|]
: Xy = CP[|3,1 \ .,2|]
: b  = invsym(XX)*Xy
```

Note how similar these solutions are to their unweighted counterparts. The only important difference is the appearance of w as the middle argument of cross(). Since specifying the middle argument as a scalar 1 is also allowed and produces unweighted estimates, the above code could be modified to produce unweighted or weighted estimates, depending on how w is defined.

◁

▷ Example 7: Mean of one variable

```
: x = .
: st_view(x, ., "mpg", 0)
:
: CP = cross(1,0 , x,1)
: mean = CP[1]/CP[2]
```

Comments: An easier and every bit as good a solution would be

```
: x = .
: st_view(x, ., "mpg", 0)
:
: mean = mean(x,1)
```

mean() is implemented in terms of cross(). Actually, mean() is implemented using the quad-precision version of cross(); see [M-5] **quadcross( )**. We could implement our solution in terms of quadcross():

```
: x = .
: st_view(x, ., "mpg", 0)
:
: CP = quadcross(1,0 , x,1)
: mean = CP[1]/CP[2]
```

quadcross( ) returns a double-precision result just as does cross( ). The difference is that quad-cross( ) uses quad precision internally in calculating sums.

◁

▷ Example 8: Means of multiple variables

```
: X = .
: st_view(X, ., "mpg weight displ", 0)
:
: CP = cross(1,0 , X,1)
: n  = cols(CP)
: means = CP[|1\n-1|] :/ CP[n]
```

Comments: The above logic will work for one variable, too. With mean( ), the solution would be

```
: X = .
: st_view(X, ., "mpg weight displ", 0)
:
: means = mean(X, 1)
```

◁

▷ Example 9: Weighted means of multiple variables

```
: M = w = X = .
: st_view(M, ., "w mpg weight displ", 0)
: st_subview(w, M, ., 1)
: st_subview(X, M, ., (2\.))
:
: CP = cross(1,0, w,  X,1)
: n  = cols(CP)
: means = CP[|1\n-1|] :/ CP[n]
```

Comments: Note how similar this solution is to the unweighted solution: w now appears as the middle argument of cross( ). The line CP = cross(1,0, w, X,1) could also be coded CP = cross(w,0, X,1); it would make no difference.

The mean( ) solution to the problem is

```
: M = w = X = .
: st_view(M, ., "w mpg weight displ", 0)
: st_subview(w, M, ., 1)
: st_subview(X, M, ., (2\.))
:
: means = mean(X, w)
```

◁

▷ Example 10: Variance matrix, traditional approach 1

```
: X = .
: st_view(X, ., "mpg weight displ", 0)
:
: n     = rows(X)
: means = mean(X, 1)
: cov   = (X'X/n - means'means)*(n/(n-1))
```

Comments: This above is not 100% traditional since we used `mean()` to obtain the means, but that does make the solution more understandable. The solution requires calculating X', requiring that the data matrix be duplicated. Also, we have used a numerically poor calculation formula.

◁

▷ Example 11: Variance matrix, traditional approach 2

```
: X = .
: st_view(X, ., "mpg weight displ", 0)
:
: n     = rows(X)
: means = mean(X, 1)
: cov   = (X:-means)'(X:-means) :/ (n-1)
```

Comments: We use a better calculation formula and, in the process, increase our memory usage substantially.

◁

▷ Example 12: Variance matrix using cross()

```
: X = .
: st_view(X, ., "mpg weight displ", 0)
:
: n     = rows(X)
: means = mean(X, 1)
: XX    = cross(X, X)
: cov   = ((XX:/n)-means'means)*(n/(n-1))
```

Comments: The above solution conserves memory but uses the numerically poor calculation formula. A related function, `crossdev()`, will calculate deviation crossproducts:

```
: X = .
: st_view(X, ., "mpg weight displ", 0)
:
: n     = rows(X)
: means = mean(X, 1)
: xx    = crossdev(X,means, X,means)
: cov   = xx:/(n-1)
```

See [M-5] **crossdev( )**. The easiest solution, however, is

```
: X = .
: st_view(X, ., "mpg weight displ", 0)
:
: cov = variance(X, 1)
```

See [M-5] **mean( )** for a description of the variance() function. variance() is implemented in terms of crossdev().

◁

▷ Example 13: Weighted variance matrix, traditional approaches

```
: M = w = X = .
: st_view(M, ., "w mpg weight displ", 0)
: st_subview(w, M, ., 1)
: st_subview(X, M, ., (2\.))
:
: n     = colsum(w)
: means = mean(X, w)
: cov   = (X'diag(w)*X:/n - means'means)*(n/(n-1))
```

Comments: Above we use the numerically poor formula. Using the better deviation formula, we would have

```
: M = w = X = .
: st_view(M, ., "w mpg weight displ", 0)
: st_subview(w, M, ., 1)
: st_subview(X, M, ., (2\.))
:
: n     = colsum(w)
: means = mean(X, w)
: cov   = (X:-means)'diag(w)*(X:-means) :/ (n-1)
```

The memory requirements include making a copy of the data with the means removed and making an $N \times N$ diagonal matrix.

◁

▷ Example 14: Weighted variance matrix using cross()

```
: M = w = X = .
: st_view(M, ., "w mpg weight displ", 0)
: st_subview(w, M, ., 1)
: st_subview(X, M, ., (2\.))
:
: n     = colsum(w)
: means = mean(X, w)
: cov   = (cross(X,w,X):/n - means'means)*(n/(n-1))
```

Comments: As in example 12, the above solution conserves memory but uses a numerically poor calculation formula. Better is to use crossdev():

```
: M = w = X = .
: st_view(M, ., "w mpg weight displ", 0)
: st_subview(w, M, ., 1)
: st_subview(X, M, ., (2\.))
:
: n     = colsum(w)
: means = mean(X, w)
: cov   = crossdev(X,means, w, X,means) :/ (n-1)
```

and easiest is to use `variance( )`:

```
: M = w = X = .
: st_view(M, ., "w mpg weight displ", 0)
: st_subview(w, M, ., 1)
: st_subview(X, M, ., (2\.))
:
: cov = variance(X, w)
```

See [M-5] **crossdev( )** and [M-5] **mean( )**.

◁

Comment concerning cross( ) and missing values

`cross( )` automatically omits rows containing missing values in making its calculation. Depending on this feature, however, is considered bad style because so many other Mata functions do not provide that feature and it is easy to make a mistake.

The right way to handle missing values is to exclude them when constructing views and subviews, as we have done above. When we constructed a view, we invariably specified fourth argument 0 to `st_view( )`. In formal programming situations, you will probably specify the name of the `touse` variable you have previously constructed in your ado-file that calls your Mata function.

Conformability

$\text{cross}(X, xc, w, Z, zc)$:

X:	$n \times v_1$ or 1×1,	1×1 treated as if $n \times 1$
xc:	1×1	(optional)
w:	$n \times 1$ or $1 \times n$ or 1×1	(optional)
Z:	$n \times v_2$	
zc:	1×1	(optional)
$result$:	$(v_1+(xc!=0)) \times (v_2+(zc!=0))$	

Diagnostics

$\text{cross}(X, xc, w, Z, zc)$ omits rows in X and Z that contain missing values.

Also see

[M-5] **crossdev( )** — Deviation cross products

[M-5] **quadcross( )** — Quad-precision cross products

[M-5] **mean( )** — Means, variances, and correlations

[M-4] **statistical** — Statistical functions

[M-4] **utility** — Matrix utility functions

Title

[M-5] **crossdev( )** — Deviation cross products

Syntax

real matrix crossdev(X, x, Z, z)

real matrix crossdev(X, x, w, Z, z)

real matrix crossdev(X, xc, x, Z, zc, z)

real matrix crossdev(X, xc, x, w, Z, zc, z)

where

X:	*real matrix X*
xc:	*real scalar xc*
x:	*real rowvector x*
w:	*real vector w*
Z:	*real matrix Z*
zc:	*real scalar zc*
z:	*real rowvector z*

Description

crossdev( ) makes calculations of the form

$$(X: -x)' (X: -x)$$
$$(X: -x)' (Z: -z)$$
$$(X: -x)' \operatorname{diag}(w) (X: -x)$$
$$(X: -x)' \operatorname{diag}(w) (Z: -z)$$

crossdev( ) is a variation on [M-5] **cross( )**. crossdev( ) mirrors cross( ) in every respect except that it has two additional arguments: x and z. x and z record the amount by which X and Z are to be deviated. x and z usually contain the (appropriately weighted) column means of X and Z.

Remarks

x usually contains the same number of rows as X but, if $xc \neq 0$, x may contain an extra element on the right recording the amount from which the constant 1 should be deviated.

The same applies to z: it usually contains the same number of rows as Z but, if $zc \neq 0$, z may contain an extra element on the right.

▷ Example 1: Linear regression using one view

```
: M = .
: st_view(M, ., "mpg weight foreign", 0)
:
: means = mean(M, 1)
: CP = crossdev(M,means , M,means)
: xx = CP[|2,2 \ .,.|]
: xy = CP[|2,1 \ .,1|]
: b  = invsym(xx)*xy
: b  = b \ means[1]-means[|2\.|]*b
```

Compare this solution with example 3 in [M-5] **cross( )**.

◁

▷ Example 2: Linear regression using subviews

```
: M = X = y = .
: st_view(M, ., "mpg weight foreign", 0)
: st_subview(y, M, ., 1)
: st_subview(X, M, ., (2\.))
:
: xmean = mean(X, 1)
: ymean = mean(y, 1)
: xx    = crossdev(X,xmean , X,xmean)
: xy    = crossdev(X,xmean , y,ymean)
: b     = invsym(xx)*xy
: b     = b \ ymean-xmean*b
```

Compare this solution with example 4 in [M-5] **cross( )**.

◁

▷ Example 3: Weighted linear regression

```
: M = X = y = w = .
: st_view(M, ., "w mpg weight foreign", 0)
: st_subview(w, M, ., 1)
: st_subview(y, M, ., 2)
: st_subview(X, M, ., (3\.))
:
: xmean = mean(X, w)
: ymean = mean(y, w)
: xx    = crossdev(X,xmean, w, X,xmean)
: xy    = crossdev(X,xmean, w, y,ymean)
: b     = invsym(xx)*xy
: b     = b \ ymean-xmean*b
```

Compare this solution with example 6 in [M-5] **cross( )**.

◁

▷ Example 4: Variance matrix

```
: X = .
: st_view(X, ., "mpg weight displ", 0)
:
: n     = rows(X)
: means = mean(X, 1)
: xx    = crossdev(X,means , X,means)
: cov   = xx:/(n-1)
```

This is exactly what variance() does; see [M-5] **mean( )**. Compare this solution with example 12 in [M-5] **cross( )**.

◁

▷ Example 5: Weighted variance matrix

```
: M = w = X = .
: st_view(M, ., "w mpg weight displ", 0)
: st_subview(w, M, ., 1)
: st_subview(X, M, ., (2\.))
:
: n     = colsum(w)
: means = mean(X, w)
: cov   = crossdev(X,means, w, X,means) :/ (n-1)
```

This is exactly what variance() does with weighted data; see [M-5] **mean( )**. Compare this solution with example 14 in [M-5] **cross( )**.

◁

Conformability

crossdev(X, xc, x, w, Z, zc, z):

$$
\begin{array}{rl}
X: & n \times v_1 \quad \text{or} \quad 1 \times 1, \quad 1 \times 1 \text{ treated as if } n \times 1 \\
xc: & 1 \times 1 \hspace{4cm} \text{(optional)} \\
x: & 1 \times v_1 \quad \text{or} \quad 1 \times v_1 + (xc \neq 0) \\
w: & n \times 1 \quad \text{or} \quad 1 \times n \quad \text{or} \quad 1 \times 1 \text{ (optional)} \\
Z: & n \times v_2 \\
zc: & 1 \times 1 \hspace{4cm} \text{(optional)} \\
z: & 1 \times v_2 \quad \text{or} \quad 1 \times v_2 + (zc \neq 0) \\
result: & (v_1 + (xc \neq 0)) \times (v_2 + (zc \neq 0))
\end{array}
$$

Diagnostics

crossdev(X, xc, x, w, Z, zc, z) omits rows in X and Z that contain missing values.

Also see

[M-5] **cross( )** — Cross products

[M-5] **quadcross( )** — Quad-precision cross products

[M-4] **utility** — Matrix utility functions

[M-4] **statistical** — Statistical functions

Title

[M-5] **cvpermute( )** — Obtain all permutations

Syntax

$$info = \texttt{cvpermutesetup}(real\ colvector\ V\ \left[\ ,\ real\ scalar\ unique\right])$$

$$real\ colvector\ \texttt{cvpermute}(info)$$

where *info* should be declared *transmorphic*.

Description

cvpermute( ) returns all permutations of the values of column vector V, one at a time. If $V = (1\backslash2\backslash3)$, there are six permutations and they are $(1\backslash2\backslash3)$, $(1\backslash3\backslash2)$, $(2\backslash1\backslash3)$, $(2\backslash3\backslash1)$, $(3\backslash1\backslash2)$, and $(3\backslash2\backslash1)$. If $V = (1\backslash2\backslash1)$, there are three permutations and they are $(1\backslash1\backslash2)$, $(1\backslash2\backslash1)$, and $(2\backslash1\backslash1)$.

Vector V is specified by calling cvpermutesetup( ),

$$info = \texttt{cvpermutesetup}(V)$$

info holds information that is needed by cvpermute( ) and it is *info*, not V, that is passed to cvpermute( ). To obtain the permutations, repeated calls are made to cvpermute( ) until it returns J(0,1,.):

```
info = cvpermutesetup(V)
while ((p=cvpermute(info)) != J(0,1,.)) {
      ... p ...
}
```

Column vector p will contain a permutation of V.

cvpermutesetup( ) may be specified with one or two arguments:

$$info = \texttt{cvpermutesetup}(V)$$

$$info = \texttt{cvpermutesetup}(V,\ unique)$$

unique is usually not specified. If *unique* is specified, it should be 0 or 1. Not specifying *unique* is equivalent to specifying *unique* = 0. Specifying *unique* = 1 states that the elements of V are unique or, at least, are to be treated that way.

When the arguments of V are unique—for instance, $V = (1\backslash2\backslash3)$—specifying *unique* = 1 will make cvpermute( ) run faster. The same permutations will be returned, although usually in a different order.

When the arguments of V are not unique—for instance, $V = (1\backslash2\backslash1)$—specifying *unique* = 1 will make cvpermute( ) treat them as if they were unique. With *unique* = 0, there are three permutations of $(1\backslash2\backslash1)$. With *unique* = 1, there are six permutations, just as there are with $(1\backslash2\backslash3)$.

Remarks

▷ Example 1

You have the following data:

```
v1      v2
───────────
22      29
17      33
21      26
20      32
16      35
───────────
```

You wish to do an exact permutation test for the correlation between v1 and v2.

That is, you wish to (1) calculate the correlation between v1 and v2—call that value r—and then (2) calculate the correlation between v1 and v2 for all permutations of v1, and count how many times the result is more extreme than r.

For the first step,

```
: X = (22, 29 \
>       17, 33 \
>       21, 26 \
>       20, 32 \
>       16, 35)
:
: correlation(X)
[symmetric]
                1               2
     ┌─────────────────────────────────┐
   1 │           1                      │
   2 │  -.8468554653            1       │
     └─────────────────────────────────┘
```

The correlation is −.846855. For the second step,

```
: V1 = X[,1]
: V2 = X[,2]
: num = den = 0
: info = cvpermutesetup(V1)
: while ((V1=cvpermute(info)) != J(0,1,.)) {
>       rho = correlation((V1,V2))[2,1]
>       if (rho<=-.846 | rho>=.846) num++
>       den++
> }
: (num, den, num/den)
                1               2               3
     ┌─────────────────────────────────────────────┐
   1 │          13             120     .1083333333  │
     └─────────────────────────────────────────────┘
```

Of the 120 permutations, 13 (10.8%) were outside .846855 or −.846855.

◁

▷ Example 2

You now wish to do the same thing but using the Spearman rank-correlation coefficient. Mata has no function that will calculate that, but Stata has a command that does—see [R] **spearman**—so we will use the Stata command as our subroutine.

This time, we will assume that the data have been loaded into a Stata dataset:

```
. list

     | var1   var2 |
     |-------------|
  1. |  22     29  |
  2. |  17     33  |
  3. |  21     26  |
  4. |  20     32  |
  5. |  16     35  |
```

For the first step,

```
. spearman var1 var2
 Number of obs =          5
Spearman's rho =    -0.9000

Test of Ho: var1 and var2 are independent
      Prob > |t| =     0.0374
```

For the second step,

```
. mata
------------------------------------------------- mata (type end to exit) ------
: V1 = st_data(., "var1")

: info = cvpermutesetup(V1)

: num = den = 0

: while ((V1=cvpermute(info)) != J(0,1,.)) {
>         st_store(., "var1", V1)
>         stata("quietly spearman var1 var2")
>         rho = st_numscalar("r(rho)")
>         if (rho<=-.9 | rho>=.9) num++
>         den++
> }

: (num, den, num/den)
                 1            2            3
    +----------------------------------------+
  1 |         2          120    .0166666667  |
    +----------------------------------------+
```

Only two of the permutations resulted in a rank correlation of at least .9 in magnitude.

In the code above, we obtained the rank correlation from r(rho) which, we learned from [R] **spearman**, is where spearman stores it.

Also note how we replaced the contents of var1 by using st_store(). Our code leaves the dataset changed and so could be improved.

◁

Conformability

cvpermutesetup(V, *unique*):

V:	$n \times 1$	
unique:	1×1	(optional)
result:	$1 \times L$	

cvpermute(*info*):

info:	$1 \times L$
result:	$n \times 1$ or 0×1

where

$$L \;=\; \begin{cases} 3 & \text{if } n = 0 \\ 4 & \text{if } n = 1 \\ (n+3)(n+2)/2 - 6 & \text{otherwise} \end{cases}$$

The value of L is not important except that the *info* vector returned by cvpermutesetup( ) and then passed to cvpermute( ) consumes memory. For instance,

n	L	Total memory ($8 * L$)
5	22	176 bytes
10	72	576
50	1,372	10,560
100	5,247	41,976
1,000	502,497	4,019,976

Diagnostics

cvpermute( ) returns J(0,1,.) when there are no more permutations.

Also see

[M-4] **statistical** — Statistical functions

Title

[M-5] date( ) — Date and time manipulation

Syntax

$tc = $ clock(*strdatetime*, *pattern* [, *year*])

$tc = $ mdyhms(*month*, *day*, *year*, *hour*, *minute*, *second*)

$tc = $ dhms(*td*, *hour*, *minute*, *second*)

$tc = $ hms(*hour*, *minute*, *second*)

hour = hh(*tc*)

minute = mm(*tc*)

second = ss(*tc*)

$td = $ dofc(*tc*)

$tC = $ Cofc(*tc*)

$tC = $ Clock(*strdatetime*, *pattern* [, *year*])

$tC = $ Cmdyhms(*month*, *day*, *year*, *hour*, *minute*, *second*)

$tC = $ Cdhms(*td*, *hour*, *minute*, *second*)

$tC = $ Chms(*hour*, *minute*, *second*)

hour = hhC(*tC*)

minute = mmC(*tC*)

second = ssC(*tC*)

$td = $ dofC(*tC*)

$tc = $ cofC(*tC*)

$td = $ date(*strdate*, *dpattern* [, *year*])

$td = $ mdy(*month*, *day*, *year*)

$tw = $ yw(*year*, *week*)

$tm = $ ym(*year*, *month*)

$tq = $ yq(*year*, *quarter*)

$th = $ yh(*year*, *half*)

$tc = $ cofd(*td*)

$tC = $ Cofd(*td*)

$month$ = month(td)

day = day(td)

$year$ = year(td)

dow = dow(td)

$week$ = week(td)

$quarter$ = quarter(td)

$half$ = halfyear(td)

doy = doy(td)

ty = yearly($strydate$, $ypattern$ [, $year$])

ty = yofd(td)

td = dofy(ty)

th = halfyearly($strhdate$, $hpattern$ [, $year$])

th = hofd(td)

td = dofh(th)

tq = quarterly($strqdate$, $qpattern$ [, $year$])

tq = qofd(td)

td = dofq(tq)

tm = monthly($strmdate$, $mpattern$ [, $year$])

tm = mofd(td)

td = dofm(tm)

tw = weekly($strwdate$, $wpattern$ [, $year$])

tw = wofd(td)

td = dofw(tw)

$hours$ = hours(ms)

$minutes$ = minutes(ms)

$seconds$ = seconds(ms)

ms = msofhours($hours$)

ms = msofminutes($minutes$)

ms = msofseconds($seconds$)

where

tc:	number of milliseconds from 01jan1960 00:00:00.000, unadjusted for leap seconds
tC:	number of milliseconds from 01jan1960 00:00:00.000, adjusted for leap seconds
strdatetime:	string-format date, time, or date/time, e.g., "15jan1992", "1/15/1992", "15-1-1992", "January 15, 1992", "9:15", "13:42", "1:42 p.m.", "1:42:15.002 pm", "15jan1992 9:15", "1/15/1992 13:42", "15-1-1992 1:42 p.m.", "January 15, 1992 1:42:15.002 pm"
pattern:	order of month, day, year, hour, minute, and seconds in *strdatetime*, plus optional default century, e.g., "DMYhms" (meaning day, month, year, hour, minute, second), "DMYhm", "MDYhm", "hmMDY", "hms", "hm", "MDY", "MD19Y", "MDY20Yhm"
td:	number of days from 01jan1960
strdate:	string-format date, e.g., "15jan1992", "1/15/1992", "15-1-1992", "January 15, 1992"
dpattern:	order of month, day, and year in *strdate*, plus optional default century, e.g., "DMY" (meaning day, month, year), "MDY", "MD19Y"
hour:	hour, 0–23
minutes:	minutes, 0–59
seconds:	seconds, 0.000–59.999 (maximum 60.999 in case of leap second)
month:	month number, 1–12
day:	day-of-month number, 1–31
year:	year, e.g., 1942, 1995, 2008
week:	week within year, 1–52
quarter:	quarter within year, 1–4
half:	half within year, 1–2
dow:	day of week, 0–6, 0 = Sunday
doy:	day within year, 1–366
ty:	calendar year
strydate:	string-format calendar year, e.g., "1980", "80"
ypattern:	pattern of *strydate*, e.g., "Y", "19Y"
th:	number of halves from 1960h1
strhdate:	string-format *hdate*, e.g., "1982-1", "1982h2", "2 1982"
hpattern:	pattern of *strhdate*, e.g., "YH", "19YH", "HY"

tq:	number of quarters from 1960q1
strqdate:	string-format *qdate*, e.g., "1982-3", "1982q2", "3 1982"
qpattern:	pattern of *strqdate*, e.g., "YQ", "19YQ", "QY"
tm:	number of months from 1960m1
strmdate:	string-format *mdate*, e.g., "1982-3", "1982m2", "3/1982"
mpattern:	pattern of *strmdate*, e.g., "YM", "19YM", "MR"
tw:	number of weeks from 1960w1
strwdate:	string-format *wdate*, e.g., "1982-3", "1982w2", "1982-15"
wpattern:	pattern of *strwdate*, e.g., "YW", "19YW", "WY"
hours:	interval of time in hours (positive or negative, real)
minutes:	interval of time in minutes (positive or negative, real)
seconds:	interval of time in seconds (positive or negative, real)
ms:	interval of time in milliseconds (positive or negative, integer)

Functions return an element-by-element result. Functions are usually used with scalars.

All variables are *real matrix* except the *str** and **pattern* variables, which are *string matrix*.

Description

These functions mirror Stata's date functions; see [D] **dates and times**.

Conformability

clock(*strdatetime*, *pattern*, *year*), Clock(*strdatetime*, *pattern*, *year*):

strdatetime:	$r_1 \times c_1$	
pattern:	$r_2 \times c_2$	(c-conformable with *strdatetime*)
year:	$r_3 \times c_3$	(optional, c-conformable)
result:	$\max(r_1, r_2, r_3) \times \max(c_1, c_2, c_3)$	

mdyhms(*month*, *day*, *year*, *hour*, *minute*, *second*),
Cmdyhms(*month*, *day*, *year*, *hour*, *minute*, *second*):

month:	$r_1 \times c_1$	
day:	$r_2 \times c_2$	
year:	$r_3 \times c_3$	
hour:	$r_4 \times c_4$	
minute:	$r_5 \times c_5$	
second:	$r_6 \times c_6$	(all variables c-conformable)
result:	$\max(r_1, r_2, r_3, r_4, r_5, r_6) \times \max(c_1, c_2, c_3 c_4, c_5, c_6)$	

hms(*hour*, *minute*, *second*), Chms(*hour*, *minute*, *second*):

hour:	$r_1 \times c_1$	
minute:	$r_2 \times c_2$	
second:	$r_3 \times c_3$	
result:	$\max(r_1, r_2, r_3) \times \max(c_1, c_2, c_3)$	

dhms(td, $hour$, $minute$, $second$), Cdhms(td, $hour$, $minute$, $second$):

td:	$r_1 \times c_1$
hour:	$r_2 \times c_2$
minute:	$r_3 \times c_3$
second:	$r_4 \times c_4$ (all variables c-conformable)
result:	$\max(r_1,r_2,r_3,r_4) \times \max(c_1,c_2,c_3c_4)$

hh(x), mm(x), ss(x), hhC(x), mmC(x), ssC(x),

x:	$r \times c$
result:	$r \times c$

date($strdate$, $dpattern$, $year$):

strdate:	$r_1 \times c_1$
dpattern:	$r_2 \times c_2$ (c-conformable with *strdate*)
year:	$r_3 \times c_3$ (optional, c-conformable)
result:	$\max(r_1,r_2,r_3) \times \max(c_1,c_2,c_3)$

mdy($month$, day, $year$):

month:	$r_1 \times c_1$
day:	$r_2 \times c_2$
year:	$r_3 \times c_3$ (all variables c-conformable)
result:	$\max(r_1,r_2,r_3) \times \max(c_1,c_2,c_3)$

yw($year$, $detail$), ym($year$, $detail$), yq($year$, $detail$), yh($year$, $detail$):

year:	$r_1 \times c_1$
detail:	$r_2 \times c_2$ (c-conformable with *year*)
result:	$\max(r_1,r_2) \times \max(c_1,c_2)$

month(td), day(td), year(td), dow(td), week(td), quarter(td), halfyear(td), doy(td):

td:	$r \times c$
result:	$r \times c$

yearly(str, pat, $year$), halfyearly(str, pat, $year$), quarterly(str, pat, $year$), monthly(str, pat, $year$), weekly(str, pat, $year$):

str:	$r_1 \times c_1$
pat:	$r_2 \times c_2$ (c-conformable with *str*)
year:	$r_3 \times c_3$ (optional, c-conformable)
result:	$\max(r_1,r_2,r_3) \times \max(c_1,c_2,c_3)$

Cofc(x), cofC(x), dofc(x), dofC(x), cofd(x), Cofd(x), yofd(x), dofy(x), hofd(x), dofh(x), qofd(x), dofq(x), mofd(x), dofm(x), wofd(x), dofw(x):

x:	$r \times c$
result:	$r \times c$

hours(x), minutes(x), seconds(x), msofhours(x), msofminutes(x): msofseconds(x):

x:	$r \times c$
result:	$r \times c$

Diagnostics

None.

Also see

[M-4] **scalar** — Scalar mathematical functions

Title

Syntax

D = deriv_init()

(varies)	deriv_init_evaluator(D [, &*function*()])
(varies)	deriv_init_evaluatortype(D [, *evaluatortype*])
(varies)	deriv_init_params(D [, *real rowvector parameters*])
(varies)	deriv_init_argument(D, *real scalar k* [, *X*])
(varies)	deriv_init_narguments(D [, *real scalar K*])
(varies)	deriv_init_weights(D [, *real colvector weights*])
(varies)	deriv_init_h(D [, *real rowvector h*])
(varies)	deriv_init_scale(D [, *real matrix scale*])
(varies)	deriv_init_bounds(D [, *real rowvector minmax*])
(varies)	deriv_init_search(D [, *search*])
(varies)	deriv_init_verbose(D [, {"on" | "off"}])

(varies)	deriv(D, {0 | 1 | 2})
real scalar	_deriv(D, {0 | 1 | 2})

real scalar	deriv_result_value(D)
real vector	deriv_result_values(D)
void	_deriv_result_values(D, v)
real rowvector	deriv_result_gradient(D)
void	_deriv_result_gradient(D, g)
real matrix	deriv_result_scores(D)
void	_deriv_result_scores(D, S)
real matrix	deriv_result_Jacobian(D)
void	_deriv_result_Jacobian(D, J)
real matrix	deriv_result_Hessian(D)
void	_deriv_result_Hessian(D, H)

real rowvector   deriv_result_h(*D*)

real matrix   deriv_result_scale(*D*)

real matrix   deriv_result_delta(*D*)

real scalar   deriv_result_errorcode(*D*)

string scalar   deriv_result_errortext(*D*)

real scalar   deriv_result_returncode(*D*)

void   deriv_query(*D*)

where *D*, if it is declared, should be declared

    transmorphic *D*

and where *evaluatortype* optionally specified in deriv_init_evaluatortype() is

evaluatortype	description
"d"	*function*( ) returns *scalar* value
"v"	*function*( ) returns *colvector* value
"t"	*function*( ) returns *rowvector* value

The default is "d" if not set.

and where *search* optionally specified in deriv_init_search() is

search	description
"interpolate"	use linear and quadratic interpolation to search for an optimal delta
"bracket"	use a bracketed quadratic formula to search for an optimal delta
"off"	do not search for an optimal delta

The default is "interpolate" if not set.

Description

These functions compute derivatives of the real function $f(p)$ at the real parameter values p.

deriv_init() begins the definition of a problem and returns D, a problem-description handle set that contains default values.

The deriv_init_*(D, ...) functions then allow you to modify those defaults. You use these functions to describe your particular problem: to set the identity of function $f()$, to set parameter values, and the like.

deriv(D, *todo*) then computes derivatives depending upon the value of *todo*.

deriv(D, 0) returns the function value without computing derivatives.

deriv(D, 1) returns the first derivatives, also known as the gradient vector for scalar-valued functions (type d and v) or the Jacobian matrix for vector-valued functions (type t).

deriv(D, 2) returns the matrix of second derivatives, also known as the Hessian matrix; the gradient vector is also computed. This syntax is not allowed for type t evaluators.

The deriv_result_*(D) functions can then be used to access other values associated with the solution.

Usually you would stop there. In other cases, you could compute derivatives at other parameter values:

> deriv_init_params(D, *p_alt*)
>
> deriv(D, *todo*)

Aside: The deriv_init_*(D, ...) functions have two modes of operation. Each has an optional argument that you specify to set the value and that you omit to query the value. For instance, the full syntax of deriv_init_params() is

> *void* deriv_init_params(D, *real rowvector parameters*)
>
> *real rowvector* deriv_init_params(D)

The first syntax sets the parameter values and returns nothing. The second syntax returns the previously set (or default, if not set) parameter values.

All the deriv_init_*(D, ...) functions work the same way.

Remarks

Remarks are presented under the following headings:

> *First example*
> *Notation and formulas*
> *Type d evaluators*
> *Example of a type d evaluator*
> *Type v evaluators*
> *User-defined arguments*
> *Example of a type v evaluator*
> *Type t evaluators*
> *Example of a type t evaluator*
> *Functions*
> > *deriv_init( )*
> > *deriv_init_evaluator( ) and deriv_init_evaluatortype( )*
> > *deriv_init_argument( ) and deriv_init_narguments( )*
> > *deriv_init_weights( )*
> > *deriv_init_params( )*
> > *Advanced init functions*
> > > *deriv_init_h( ), ... _scale( ), ... _bounds( ), and ... _search( )*
> > > *deriv_init_verbose( )*
> > *deriv( )*
> > *_deriv( )*
> > *deriv_result_value( )*
> > *deriv_result_values( ) and _deriv_result_values( )*
> > *deriv_result_gradient( ) and _deriv_result_gradient( )*
> > *deriv_result_scores( ) and _deriv_result_scores( )*
> > *deriv_result_Jacobian( ) and _deriv_result_Jacobian( )*
> > *deriv_result_Hessian( ) and _deriv_result_Hessian( )*
> > *deriv_result_h( ), ... _scale( ), and ... _delta( )*
> > *deriv_result_errorcode( ), ... _errortext( ), and ... _returncode( )*
> > *deriv_query( )*

First example

The derivative functions may be used interactively.

Below we use the functions to compute $f'(x)$ at $x = 0$, where the function is

$$f(x) = \exp(-x^2 + x - 3)$$

```
: void myeval(x, y)
> {
>          y = exp(-x^2 + x - 3)
> }
: D = deriv_init()
: deriv_init_evaluator(D, &myeval())
: deriv_init_params(D, 0)
: dydx = deriv(D, 1)
: dydx
  .0497870683
: exp(-3)
  .0497870684
```

The derivative, given the above function, is $f'(x) = (-2 \times x + 1) \times \exp(-x^2 + x - 3)$, so $f'(0) = \exp(-3)$.

Notation and formulas

We wrote the above in the way that mathematicians think, i.e., differentiate $y = f(x)$. Statisticians, on the other hand, think differentiate $s = f(b)$. To avoid favoritism, we will write $v = f(p)$ and write the general problem with the following notation:

Differentiate $v = f(p)$ with respect to p, where

> v: a scalar
> p: $1 \times np$

The gradient vector is $g = f'(p) = df/dp$, where

> g: $1 \times np$

and the Hessian matrix is $H = f''(p) = d^2f/dpdp'$, where

> H: $np \times np$

`deriv()` can also work with vector-valued functions. Here is the notation for vector-valued functions:

Differentiate $v = f(p)$ with respect to p, where

> v: $1 \times nv$, a vector
> p: $1 \times np$

The Jacobian matrix is $J = f'(p) = df/dp$, where

> J: $nv \times np$

and where

$$J[i,j] = dv[i]/dp[j]$$

Second-order derivatives are not computed by `deriv()` when used with vector-valued functions.

deriv() uses the following formula for computing the numerical derivative of $f()$ at p

$$f'(p) = \frac{f(p+d) - f(p-d)}{2d}$$

where we refer to d as the delta used for computing numerical derivatives. To search for an optimal delta, we decompose d into two parts.

$$d = h \times scale$$

By default, h is a fixed value that depends on the parameter value.

$$h = (\texttt{abs}(p)\texttt{+1e-3})\texttt{*1e-3}$$

deriv() searches for a value of *scale* that will result in an optimal numerical derivative, i.e., one where d is as small as possible subject to the constraint that $f(x+d) - f(x-d)$ will be calculated to at least half the accuracy of a double-precision number. This is accomplished by searching for *scale* such that $f(x+d)$ and $f(x-d)$ fall between *v0* and *v1*, where

$$v0 = (\texttt{abs}(f(x))\texttt{+1e-8})\texttt{*1e-8}$$
$$v1 = (\texttt{abs}(f(x))\texttt{+1e-7})\texttt{*1e-7}$$

Use deriv_init_h() to change the default h values. Use deriv_init_scale() to change the default initial *scale* values. Use deriv_init_bounds() to change the default bounds (1e-8, 1e-7) used for determining the optimal *scale*.

Type d evaluators

You must write an evaluator function to calculate $f()$ before you can use the derivative functions. The example we showed above was of what is called a type d evaluator. Let's stay with that.

The evaluator function we wrote was

```
void myeval(x,  y)
{
        y = exp(-x^2 + x - 3)
}
```

All type d evaluators open the same way,

> *void evaluator(x, y)*

although what you name the arguments is up to you. We named the arguments the way that mathematicians think, although we could just as well have named them the way that statisticians think:

> *void evaluator(b, s)*

To avoid favoritism, we will write them as

> *void evaluator(p, v)*

That is, we will think in terms of computing the derivative of $v = f(p)$ with respect to the elements of p.

Here is the full definition of a type d evaluator:

---

void evaluator(real rowvector p, v)

where *v* is the value to be returned:

    v: *real scalar*

evaluator() is to fill in *v* given the values in *p*.

evaluator() may return *v* = . if *f*() cannot be evaluated at *p*.

Example of a type d evaluator

We wish to compute the gradient of the following function at $p_1 = 1$ and $p_2 = 2$:

$$v = \exp(-p_1^2 - p_2^2 - p_1 p_2 + p_1 - p_2 - 3)$$

Our numerical solution to the problem is

```
: void eval_d(p, v)
> {
>            v = exp(-p[1]^2 - p[2]^2 - p[1]*p[2] + p[1] - p[2] - 3)
> }
: D = deriv_init()
: deriv_init_evaluator(D, &eval_d())
: deriv_init_params(D, (1,2))
: grad = deriv(D, 1)
: grad
```

	1	2
1	-.0000501051	-.0001002102

```
: (-2*1 - 2 + 1)*exp(-1^2 - 2^2 - 1*2 + 1 - 2 - 3)
  -.0000501051
: (-2*2 - 1 - 1)*exp(-1^2 - 2^2 - 1*2 + 1 - 2 - 3)
  -.0001002102
```

For this problem, the elements of the gradient function are given by the following formulas, and we see that deriv() computed values that are in agreement with the analytical results (to the number of significant digits being displayed).

$$\frac{dv}{dp_1} = (-2p_1 - p_2 + 1)\exp(-p_1^2 - p_2^2 - p_1 p_2 + p_1 - p_2 - 3)$$

$$\frac{dv}{dp_2} = (-2p_2 - p_1 - 1)\exp(-p_1^2 - p_2^2 - p_1 p_2 + p_1 - p_2 - 3)$$

Type v evaluators

In some statistical applications, you will find type v evaluators more convenient to code than type d evaluators.

In statistical applications, one tends to think of a dataset of values arranged in matrix X, the rows of which are observations. The function $h(p, X[i, .])$ can be calculated for each row separately, and it is the sum of those resulting values that forms the function $f(p)$ from which we would like to compute derivatives.

Type v evaluators are for such cases.

In a type d evaluator, you return scalar $v = f(p)$.

In a type v evaluator, you return a column vector, v, such that $\texttt{colsum}(v) = f(p)$.

The code outline for type v evaluators is the same as those for d evaluators. All that differs is that v, which is a *real scalar* in the d case, is now a *real colvector* in the v case.

User-defined arguments

The type v evaluators arise in statistical applications and, in such applications, there are data; i.e., just knowing p is not sufficient to calculate v, g, and H. Actually, that same problem can also arise when coding type d evaluators.

You can pass extra arguments to evaluators. The first line of all evaluators, regardless of type, is

> *void evaluator(p, v)*

If you code

```
deriv_init_argument(D, 1, X)
```

the first line becomes

> *void evaluator(p, X, v)*

If you code

```
deriv_init_argument(D, 1, X)
deriv_init_argument(D, 2, Y)
```

the first line becomes

> *void evaluator(p, X, Y, v)*

and so on, up to nine extra arguments. That is, you can specify extra arguments to be passed to your function.

(Continued on next page)

Example of a type v evaluator

You have the following data:

```
: x
              1

     1      .35
     2      .29
     3       .3
     4       .3
     5      .65
     6      .56
     7      .37
     8      .16
     9      .26
    10      .19
```

You believe that the data are the result of a beta distribution process with fixed parameters alpha and beta, and you wish to compute the gradient vector and Hessian matrix associated with the log likelihood at some values of those parameters alpha and beta (a and b in what follows). The formula for the density of the beta distribution is

$$\text{density}(x) = \frac{\Gamma(a+b)}{\Gamma(a)\Gamma(b)} \, x^{a-1} \, (1-x)^{b-1}$$

In our type v solution to this problem, we compute the gradient and Hessian at $a = 0.5$ and $b = 2$.

```
: void lnbetaden_v(p, x, lnf)
> {
>         a = p[1]
>         b = p[2]
>         lnf = lngamma(a+b)  :- lngamma(a)  :- lngamma(b)  :+
>               (a-1)*log(x)  :+ (b-1)*log(1:-x)
> }
: D = deriv_init()
: deriv_init_evaluator(D, &lnbetaden_v())
: deriv_init_evaluatortype(D, "v")
: deriv_init_params(D, (0.5, 2))
: deriv_init_argument(D, 1, x)                          ←   important
: deriv(D, 2)
[symmetric]
                 1                2

     1    -116.4988089
     2      8.724410052    -1.715062542


: deriv_result_gradient(D)
                 1                2

     1      15.12578465    -1.701917722
```

Note the following:

1. Rather than calling the returned value v, we called it lnf. You can name the arguments as you please.

2. We arranged for an extra argument to be passed by coding deriv_init_argument(D, 1, x). The extra argument is the vector x, which we listed previously for you. In our function, we received the argument as x, but we could have used a different name just as we used lnf rather than v.

3. We set the evaluator type to "v".

Type t evaluators

Type t evaluators are for when you need to compute the Jacobian matrix from a vector-valued function.

Type t evaluators are different from type v evaluators in that the resulting vector of values should not be summed. One example is when the function $f()$ performs a nonlinear transformation from the domain of p to the domain of v.

Example of a type t evaluator

Let's compute the Jacobian matrix for the following transformation:

$$v_1 = p_1 + p_2$$

$$v_2 = p_1 - p_2$$

Here is our numerical solution, evaluating the Jacobian at $p = (0,0)$:

```
: void eval_t1(p, v)
> {
>         v = J(1,2,.)
>         v[1] = p[1] + p[2]
>         v[2] = p[1] - p[2]
> }
: D = deriv_init()
: deriv_init_evaluator(D, &eval_t1())
: deriv_init_evaluatortype(D, "t")
: deriv_init_params(D, (0,0))
: deriv(D, 1)
[symmetric]
          1     2

    1 │   1
    2 │   1    -1
```

Now let's compute the Jacobian matrix for a less trivial transformation:

$$v_1 = p_1^2$$

$$v_2 = p_1 p_2$$

Here is our numerical solution, evaluating the Jacobian at $p = (1, 2)$:

```
: void eval_t2(p, v)
> {
>         v = J(1,2,.)
>         v[1] = p[1]^2
>         v[2] = p[1] * p[2]
> }
: D = deriv_init()
: deriv_init_evaluator(D, &eval_t2())
: deriv_init_evaluatortype(D, "t")
: deriv_init_params(D, (1,2))
: deriv(D, 1)
                   1              2

    1 |  1.999999998           0
    2 |            2           1
```

Functions

deriv_init( )

> *transmorphic* deriv_init()

deriv_init() is used to begin a derivative problem. Store the returned result in a variable name of your choosing; we have used D in this documentation. You pass D as the first argument to the other deriv*() functions.

deriv_init() sets all deriv_init_*() values to their defaults. You may use the query form of deriv_init_*() to determine an individual default, or you can use deriv_query() to see them all.

The query form of deriv_init_*() can be used before or after calling deriv().

deriv_init_evaluator( ) and deriv_init_evaluatortype( )

> *void* deriv_init_evaluator(D, *pointer(function) scalar fptr*)

> *void* deriv_init_evaluatortype(D, *evaluatortype*)

> *pointer(function) scalar* deriv_init_evaluator(D)

> *string scalar*   deriv_init_evaluatortype(D)

deriv_init_evaluator(D, *fptr*) specifies the function to be called to evaluate $f(p)$. Use of this function is required. If your function is named myfcn(), you code deriv_init_evaluator(D, &myfcn()).

deriv_init_evaluatortype(D, *evaluatortype*) specifies the capabilities of the function that has been set using deriv_init_evaluator(). Alternatives for *evaluatortype* are "d", "v", and "t". The default is "d" if you do not invoke this function.

deriv_init_evaluator(D) returns a pointer to the function that has been set.

deriv_init_evaluatortype(D) returns the evaluator type currently set.

deriv_init_argument( ) and deriv_init_narguments( )

      void   deriv_init_argument(*D*, *real scalar k*, *X*)

      void   deriv_init_narguments(*D*, *real scalar K*)

    pointer scalar deriv_init_argument(*D*, *real scalar k*)

    real scalar   deriv_init_narguments(*D*)

deriv_init_argument(*D*, *k*, *X*) sets the *k*th extra argument of the evaluator function to be *X*. *X* can be anything, including a view matrix or even a pointer to a function. No copy of *X* is made; it is a pointer to *X* that is stored, so any changes you make to *X* between setting it and *X* being used will be reflected in what is passed to the evaluator function.

deriv_init_narguments(*D*, *K*) sets the number of extra arguments to be passed to the evaluator function. This function is useless and included only for completeness. The number of extra arguments is automatically set when you use deriv_init_argument().

deriv_init_argument(*D*, *k*) returns a pointer to the object that was previously set.

deriv_init_narguments(*D*) returns the number of extra arguments that were passed to the evaluator function.

deriv_init_weights( )

      void   deriv_init_weights(*D*, *real colvector weights*)

    pointer scalar deriv_init_weights(*D*)

deriv_init_weights(*D*, *weights*) sets the weights used with type v evaluators to produce the function value. By default, deriv() with a type v evaluator uses colsum(*v*) to compute the function value. With weights, deriv() uses cross(*weights*, *v*). *weights* must be row conformable with the column vector returned by the evaluator.

deriv_init_weights(*D*) returns a pointer to the weight vector that was previously set.

deriv_init_params( )

      void   deriv_init_params(*D*, *real rowvector params*)

    real rowvector deriv_init_params(*D*)

deriv_init_params(*D*, *params*) sets the parameter values at which the derivatives will be computed. Use of this function is required.

deriv_init_params(*D*) returns the parameter values at which the derivatives were computed.

Advanced init functions

The rest of the deriv_init_*() functions provide finer control of the numerical derivative taker.

deriv_init_h( ), . . . _scale( ), . . . _bounds( ), . . . _search( )

void	deriv_init_h(*D*, *real rowvector h*)
void	deriv_init_scale(*D*, *real rowvector s*)
void	deriv_init_bounds(*D*, *real rowvector minmax*)
void	deriv_init_search(*D*, *search*)
real rowvector	deriv_init_h(*D*)
real rowvector	deriv_init_scale(*D*)
real rowvector	deriv_init_bounds(*D*)
string scalar	deriv_init_search(*D*)

deriv_init_h(*D, h*) sets the *h* values used to compute numerical derivatives.

deriv_init_scale(*D, s*) sets the starting scale values used to compute numerical derivatives.

deriv_init_bounds(*D, minmax*) sets the minimum and maximum values used to search for optimal scale values. The default is *minmax* = (1e-8, 1e-7).

deriv_init_search(*D*, "interpolate") causes deriv( ) to use linear and quadratic interpolation to search for an optimal delta for computing the numerical derivatives. This is the default search method.

deriv_init_search(*D*, "bracket") causes deriv( ) to use a bracketed quadratic formula to search for an optimal delta for computing the numerical derivatives.

deriv_init_search(*D*, "off") prevents deriv( ) from searching for an optimal delta.

deriv_init_h(*D*) returns the user-specified *h* values.

deriv_init_scale(*D*) returns the user-specified starting scale values.

deriv_init_bounds(*D*) returns the user-specified search bounds.

deriv_init_search(*D*) returns the currently set search method.

deriv_init_verbose( )

void	deriv_init_verbose(*D*, *verbose*)
string scalar	deriv_init_verbose(*D*)

deriv_init_verbose(*D, verbose*) sets whether error messages that arise during the execution of deriv( ) or _deriv( ) are to be displayed. Setting *verbose* to "on" means that they are displayed; "off" means that they are not displayed. The default is "on". Setting *verbose* to "off" is of interest only to users of _deriv( ).

deriv_init_verbose(*D*) returns the current value of *verbose*.

deriv( )

> *(varies)* deriv(*D*, *todo*)

deriv(*D*, *todo*) invokes the derivative process. If something goes wrong, deriv() aborts with error.

> deriv(*D*, 0) returns the function value without computing derivatives.

> deriv(*D*, 1) returns the gradient vector; the Hessian matrix is not computed.

> deriv(*D*, 2) returns the Hessian matrix; the gradient vector is also computed.

Before you can invoke deriv(), you must have defined your evaluator function, *evaluator*(), and you must have set the parameter values at which deriv() is to compute derivatives:

> *D* = deriv_init()

> deriv_init_evaluator(*D*, &*evaluator*())

> deriv_init_params(*D*, (...))

The above assumes that your evaluator function is type d. If your evaluator function type is v (i.e., it returns a column vector of values instead of a scalar value), you will also have coded

> deriv_init_evaluatortype(*D*, "v")

and you may have coded other deriv_init_*() functions as well.

Once deriv() completes, you may use the deriv_result_*() functions. You may also continue to use the deriv_init_*() functions to access initial settings, and you may use them to change settings and recompute derivatives (i.e., invoke deriv() again) if you wish.

_deriv( )

> *real scalar* _deriv(*D*, *todo*)

_deriv(*D*) performs the same actions as deriv(*D*) except that, rather than returning the requested derivatives, _deriv() returns a real scalar and, rather than aborting if numerical issues arise, _deriv() returns a nonzero value. _deriv() returns 0 if all went well. The returned value is called an error code.

deriv() returns the requested result. It can work that way because the numerical derivative calculation must have gone well. Had it not, deriv() would have aborted execution.

_deriv() returns an error code. If it is 0, the numerical derivative calculation went well, and you can obtain the gradient vector by using deriv_result_gradient(). If things did not go well, you can use the error code to diagnose what went wrong and take the appropriate action.

Thus _deriv(*D*) is an alternative to deriv(*D*). Both functions do the same thing. The difference is what happens when there are numerical difficulties.

deriv() and _deriv() work around most numerical difficulties. For instance, the evaluator function you write is allowed to return *v* equal to missing if it cannot calculate the $f()$ at $p+d$. If that happens while computing the derivative, deriv() and _deriv() will search for a better *d* for taking the derivative. deriv(), however, cannot tolerate that happening at *p* (the parameter values you set using deriv_init_params()) because the function value must exist at the point when you want deriv() to compute the numerical derivative. deriv() issues an error message and aborts, meaning that execution is stopped. There can be advantages in that. The calling program need not include complicated code for such instances, figuring that stopping is good enough because a human will know to address the problem.

_deriv(), however, does not stop execution. Rather than aborting, _deriv() returns a nonzero value to the caller, identifying what went wrong. The only exception is that _deriv() will return a zero value to the caller even when the evaluator function returns v equal to missing at p, allowing programmers to handle this special case without having to turn deriv_init_verbose() off.

Programmers implementing advanced systems will want to use _deriv() instead of deriv(). Everybody else should use deriv().

Programmers using _deriv() will also be interested in the functions deriv_init_verbose(), deriv_result_errorcode(), deriv_result_errortext(), and deriv_result_returncode().

The error codes returned by _deriv() are listed below, under the heading *deriv_result_errorcode( )*, *. . . _errortext( ), and . . . _returncode( )*.

deriv_result_value( )

> *real scalar* deriv_result_value(*D*)

deriv_result_value(*D*) returns the value of $f()$ evaluated at p.

deriv_result_values( ) and _deriv_result_values( )

> *real matrix* deriv_result_values(*D*)
>
> *void*   _deriv_result_values(*D*, *v*)

deriv_result_values(*D*) returns the vector values returned by the evaluator. For type v evaluators, this is the column vector that sums to the value of $f()$ evaluated at p. For type t evaluators, this is the rowvector returned by the evaluator.

_deriv_result_values(*D*, *v*) uses swap() (see [M-5] **swap( )**) to interchange *v* with the vector values stored in *D*. This destroys the vector values stored in *D*.

These functions should be called only with type v evaluators.

deriv_result_gradient( ) and _deriv_result_gradient( )

> *real rowvector* deriv_result_gradient(*D*)
>
> *void*   _deriv_result_gradient(*D*, *g*)

deriv_result_gradient(*D*) returns the gradient vector evaluated at p.

_deriv_result_gradient(*D*, *g*) uses swap() to interchange *g* with the gradient vector stored in *D*. This destroys the gradient vector stored in *D*.

deriv_result_scores( ) and _deriv_result_scores( )

> *real matrix* deriv_result_scores(*D*)
>
> *void* _deriv_result_scores(*D*, *S*)

deriv_result_scores(*D*) returns the matrix of the scores evaluated at *p*. The matrix of scores can be summed over the columns to produce the gradient vector.

_deriv_result_scores(*D*, *S*) uses swap() to interchange *S* with the scores matrix stored in *D*. This destroys the scores matrix stored in *D*.

These functions should be called only with type v evaluators.

deriv_result_Jacobian( ) and _deriv_result_Jacobian( )

> *real matrix* deriv_result_Jacobian(*D*)
>
> *void* _deriv_result_Jacobian(*D*, *J*)

deriv_result_Jacobian(*D*) returns the Jacobian matrix evaluated at *p*.

_deriv_result_Jacobian(*D*, *J*) uses swap() to interchange *J* with the Jacobian matrix stored in *D*. This destroys the Jacobian matrix stored in *D*.

These functions should be called only with type t evaluators.

deriv_result_Hessian( ) and _deriv_result_Hessian( )

> *real matrix* deriv_result_Hessian(*D*)
>
> *void* _deriv_result_Hessian(*D*, *H*)

deriv_result_Hessian(*D*) returns the Hessian matrix evaluated at *p*.

_deriv_result_Hessian(*D*, *H*) uses swap() to interchange *H* with the Hessian matrix stored in *D*. This destroys the Hessian matrix stored in *D*.

These functions should not be called with type t evaluators.

deriv_result_h( ), ..._scale( ), and ..._delta( )

> *real rowvector* deriv_result_h(*D*)
>
> *real rowvector* deriv_result_scale(*D*)
>
> *real rowvector* deriv_result_delta(*D*)

deriv_result_h(*D*) returns the vector of *h* values that was used to compute the numerical derivatives.

deriv_result_scale(*D*) returns the vector of scale values that was used to compute the numerical derivatives.

deriv_result_delta(*D*) returns the vector of delta values used to compute the numerical derivatives.

deriv_result_errorcode( ), ... _errortext( ), and ... _returncode( )

> *real scalar*   deriv_result_errorcode(*D*)
>
> *string scalar* deriv_result_errortext(*D*)
>
> *real scalar*   deriv_result_returncode(*D*)

These functions are for use after _deriv().

deriv_result_errorcode(*D*) returns the same error code as _deriv(). The value will be zero if there were no errors. The error codes are listed in the table directly below.

deriv_result_errortext(*D*) returns a string containing the error message corresponding to the error code. If the error code is zero, the string will be "".

deriv_result_returncode(*D*) returns the Stata return code corresponding to the error code. The mapping is listed in the table directly below.

In advanced code, these functions might be used as

```
(void) _deriv(D, todo)
... if (ec = deriv_result_code(D)) {
        errprintf("{p}\n")
        errprintf("%s\n", deriv_result_errortext(D))
        errprintf("{p_end}\n")
        exit(deriv_result_returncode(D))
        /*NOTREACHED*/
}
```

The error codes and their corresponding Stata return codes are

Error code	Return code	Error text
1	198	invalid todo argument
2	111	evaluator function required
3	459	parameter values required
4	459	parameter values not feasible
5	459	could not calculate numerical derivatives—discontinuous region with missing values encountered
6	459	could not calculate numerical derivatives—flat or discontinuous region encountered
16	111	*function*() not found
17	459	Hessian calculations not allowed with type t evaluators

NOTE: Error 4 can occur only when evaluating $f()$ at the parameter values.
This error occurs only with deriv().

deriv_query( )

> *void* deriv_query(*D*)

deriv_query(*D*) displays a report on the current deriv_init_*() values and some of the deriv_result_*() values. deriv_query(*D*) may be used before or after deriv(), and it is useful when using `deriv()` interactively or when debugging a program that calls `deriv()` or `_deriv()`.

Conformability

All functions have 1×1 inputs and have 1×1 or *void* outputs, except the following:

`deriv_init_params(D, params)`:

D:	*transmorphic*
params:	$1 \times np$
result:	*void*

`deriv_init_params(D)`:

D:	*transmorphic*
result:	$1 \times np$

`deriv_init_argument(D, k, X)`:

D:	*transmorphic*
k:	1×1
X:	*anything*
result:	*void*

`deriv_init_weights(D, params)`:

D:	*transmorphic*
params:	$N \times 1$
result:	*void*

`deriv_init_h(D, h)`:

D:	*transmorphic*
h:	$1 \times np$
result:	*void*

`deriv_init_h(D)`:

D:	*transmorphic*
result:	$1 \times np$

`deriv_init_scale(D, scale)`:

D:	*transmorphic*
scale:	$1 \times np$ (type d and v evaluator)
	$nv \times np$ (type t evaluator)
void:	*void*

`deriv_init_bounds(D, minmax)`:

D:	*transmorphic*
minmax:	1×2
result:	*void*

`deriv_init_bounds(D)`:

D:	*transmorphic*
result:	$1 \times w$

deriv(*D*, 0):

         D: *transmorphic*
      result: 1×1
               $1 \times nv$ (type t evaluator)

deriv(*D*, 1):

         D: *transmorphic*
      result: $1 \times np$
               $nv \times np$ (type t evaluator)

deriv(*D*, 2):

         D: *transmorphic*
      result: $np \times np$

deriv_result_values(*D*):

         D: *transmorphic*
      result: $N \times 1$
               $1 \times nv$ (type t evaluator)
               $N \times 1$ (type v evaluator)

_deriv_result_values(*D*, *v*):

         D: *transmorphic*
         v: $N \times 1$
               $1 \times nv$ (type t evaluator)
               $N \times 1$ (type v evaluator)
      result: *void*

deriv_result_gradient(*D*):

         D: *transmorphic*
      result: $1 \times np$

_deriv_result_gradient(*D*, *g*):

         D: *transmorphic*
         g: $1 \times np$
      result: *void*

deriv_result_scores(*D*):

         D: *transmorphic*
      result: $N \times np$

_deriv_result_scores(*D*, *S*):

         D: *transmorphic*
         S: $N \times np$
      result: *void*

deriv_result_Jacobian(*D*):

         D: *transmorphic*
      result: $nv \times np$

_deriv_result_Jacobian(*D*, *J*):

         D: *transmorphic*
         J: $nv \times np$
      result: *void*

```
deriv_result_Hessian(D):
```

$$
\begin{array}{ll}
D: & \textit{transmorphic} \\
\textit{result}: & np \times np
\end{array}
$$

```
_deriv_result_Hessian(D, H):
```

$$
\begin{array}{ll}
D: & \textit{transmorphic} \\
H: & np \times np \\
\textit{result}: & \textit{void}
\end{array}
$$

```
deriv_result_h(D):
```

$$
\begin{array}{ll}
D: & \textit{transmorphic} \\
\textit{result}: & 1 \times np
\end{array}
$$

```
deriv_result_scale(D):
```

$$
\begin{array}{ll}
D: & \textit{transmorphic} \\
\textit{result}: & 1 \times np \text{ (type d and v evaluator)} \\
& nv \times np \text{ (type t evaluator)}
\end{array}
$$

```
deriv_result_delta(D):
```

$$
\begin{array}{ll}
D: & \textit{transmorphic} \\
\textit{result}: & 1 \times np \text{ (type d and v evaluator)} \\
& nv \times np \text{ (type t evaluator)}
\end{array}
$$

Diagnostics

All functions abort with error when used incorrectly.

deriv() aborts with error if it runs into numerical difficulties. _deriv() does not; it instead returns a nonzero error code.

Methods and formulas

See sections 1.3.4 and 1.3.5 of Gould, Pitblado, and Sribney (2006) for an overview of the methods and formulas deriv() uses to compute numerical derivatives.

Reference

Gould, W. W., J. S. Pitblado, and W. M. Sribney. 2006. *Maximum Likelihood Estimation with Stata*. 3rd ed. College Station, TX: Stata Press.

Also see

[M-4] **mathematical** — Important mathematical functions

Title

[M-5] **designmatrix( )** — Design matrices

Syntax

real matrix designmatrix(*real colvector v*)

Description

designmatrix(*v*) returns a rows(*v*) $\times$ colmax(*v*) matrix with ones in the indicated columns and zero everywhere else.

Remarks

designmatrix((1\2\3)) is equal to I(3), the 3×3 identity matrix.

Conformability

designmatrix(*v*):
       v: $r \times 1$
   result: $r \times$ colmax(*v*) $(0 \times 0$ if $r = 0)$

Diagnostics

designmatrix(*v*) aborts with error if any element of *v* is <1.

Also see

[M-4] **standard** — Functions to create standard matrices

Title

[M-5] det( ) — Determinant of matrix

Syntax

> *numeric scalar* det(*numeric matrix A*)

> *numeric scalar* dettriangular(*numeric matrix A*)

Description

det(*A*) returns the determinant of *A*.

dettriangular(*A*) returns the determinant of *A*, treating *A* as if it were triangular (even if it is not).

Remarks

Calculation of the determinant is made by obtaining the LU decomposition of *A* and then calculating the determinant of *U*:

$$
\begin{aligned}
\det(A) &= \det(PLU) \\
&= \det(P) \times \det(L) \times \det(U) \\
&= \pm 1 \times 1 \times \det(U) \\
&= \pm \det(U)
\end{aligned}
$$

Since *U* is (upper) triangular, det(*U*) is simply the product of its diagonal elements. See [M-5] **lud( )**.

Conformability

det(*A*), dettriangular(*A*):
```
        A:     n × n
   result:     1 × 1
```

Diagnostics

det(*A*) and dettriangular(*A*) return 1 if *A* is 0×0.

det(*A*) aborts with error if *A* is not square and returns missing if *A* contains missing values.

dettriangular(*A*) aborts with error if *A* is not square and returns missing if any element on the diagonal of *A* is missing.

Both det(*A*) and dettriangular(*A*) will return missing value if the determinant exceeds 8.99e+307.

Also see

[M-5] **lud( )** — LU decomposition

[M-4] **matrix** — Matrix functions

Title

> **[M-5] _diag( )** — Replace diagonal of a matrix

Syntax

$$void \ \_diag(numeric \ matrix \ Z, \ numeric \ vector \ v)$$

Description

$\_diag(Z, v)$ replaces the diagonal of the matrix Z with v. Z need not be square.

1. If v is a vector, the vector replaces the principal diagonal.

2. If v is 1×1, each element of the principal diagonal is replaced with v.

3. If v is a void vector (1×0 or 0×1), Z is left unchanged.

Conformability

$\_diag(Z, v)$:

> *input*:
>
> | | |
> |---|---|
> | Z: | $n \times m, n \le m$ |
> | v: | $1 \times 1, 1 \times n,$ or $n \times 1$ |
>
> or
>
> | | |
> |---|---|
> | Z: | $n \times m, n > m$ |
> | v: | $1 \times 1, 1 \times m,$ or $m \times 1$ |
>
> *output*:
>
> | | |
> |---|---|
> | Z: | $n \times m$ |

Diagnostics

$\_diag(Z, v)$ aborts with error if Z or v is a view.

Also see

[M-5] **diag( )** — Create diagonal matrix

[M-4] **manipulation** — Matrix manipulation

Title

[M-5] **diag( )** — Create diagonal matrix

Syntax

numeric matrix diag(*numeric matrix Z*)

numeric matrix diag(*numeric vector z*)

Description

diag() creates diagonal matrices.

diag(Z), Z a matrix, extracts the principal diagonal of Z to create a new matrix. Z must be square.

diag(z), z a vector, creates a new matrix with the elements of z on its diagonal.

Remarks

Do not confuse diag() with its functional inverse, diagonal(); see [M-5] **diagonal**( ). diag() creates a matrix from a vector (or matrix); diagonal() extracts the diagonal of a matrix into a vector.

Use of diag() should be avoided because it wastes memory. The colon operators will allow you to use vectors directly:

Desired calculation	Equivalent
diag(v)*X,	
$\quad v$ is a column	v:*X
$\quad v$ is a row	v':*X
$\quad v$ is a matrix	diagonal(v):*X
X*diag(v)	
$\quad v$ is a column	X:*v'
$\quad v$ is a row	X:*v
$\quad v$ is a matrix	X:*diagonal(v)$'$

In the above table, it is assumed that v is real. If v might be complex, the transpose operators that appear must be changed to transposeonly() calls, because we do not want the conjugate. For instance, v':*X would become transposeonly(v):*X.

Conformability

diag(Z):

$$Z: \quad m \times n$$
$$result: \quad \min(m,n) \times \min(m,n)$$

diag(z):

$$z: \quad 1 \times n \quad \text{or} \quad n \times 1$$
$$result: \quad n \times n$$

Diagnostics

None.

Also see

[M-5] _diag( ) — Replace diagonal of a matrix

[M-5] diagonal( ) — Extract diagonal into column vector

[M-5] isdiagonal( ) — Whether matrix is diagonal

[M-4] manipulation — Matrix manipulation

Title

Syntax

$$real\ scalar \quad \texttt{diag0cnt}(real\ matrix\ X)$$

Description

$\texttt{diag0cnt}(X)$ returns the number of principal diagonal elements of X that are 0.

Remarks

$\texttt{diag0cnt}()$ is often used after [M-5] **invsym( )** to count the number of columns dropped because of collinearity.

Conformability

$\texttt{diag0cnt}(X)$:

$$
\begin{array}{rl}
X: & r \times c \\
result: & 1 \times 1
\end{array}
$$

Diagnostics

$\texttt{diag0cnt}(X)$ returns 0 if X is void.

Also see

[M-5] **invsym( )** — Symmetric real matrix inversion

[M-4] **utility** — Matrix utility functions

Title

[M-5] **diagonal( )** — Extract diagonal into column vector

Syntax

numeric colvector   diagonal(*numeric matrix A*)

Description

diagonal(*A*) extracts the diagonal of *A* and returns it in a column vector.

Remarks

diagonal() may be used with nonsquare matrices.

Do not confuse diagonal() with its functional inverse, diag(); see [M-5] **diag( )**. diagonal() extracts the diagonal of a matrix into a vector; diag() creates a diagonal matrix from a vector.

Conformability

diagonal(*A*):

          A: $r \times c$
    result: $\min(r, c) \times 1$

Diagnostics

None.

Also see

[M-5] **diag( )** — Create diagonal matrix

[M-5] **isdiagonal( )** — Whether matrix is diagonal

[M-5] **blockdiag( )** — Block-diagonal matrix

[M-4] **manipulation** — Matrix manipulation

Title

[M-5] dir( ) — File list

Syntax

string colvector   dir(*dirname*, *filetype*, *pattern*)

string colvector   dir(*dirname*, *filetype*, *pattern*, *prefix*)

where

dirname:	*string scalar* containing directory name
filetype:	*string scalar* containing "files", "dirs", or "other"
pattern:	*string scalar* containing match pattern
prefix:	*real scalar* containing 0 or 1

Description

dir(*dirname*, *filetype*, *pattern*) returns a column vector containing the names of the files in *dir* that match *pattern*.

dir(*dirname*, *filetype*, *pattern*, *prefix*) does the same thing but allows you to specify whether you want a simple list of files (*prefix* = 0) or a list of filenames prefixed with *dirname* (*prefix* ≠ 0). dir(*dirname*, *filetype*, *pattern*) is equivalent to dir(*dirname*, *filetype*, *pattern*, 0).

pattern is interpreted by [M-5] **strmatch( )**.

Remarks

Examples:

    dir(".", "dirs", "*")
        returns a list of all directories in the current directory.

    dir(".", "files", "*")
        returns a list of all regular files in the current directory.

    dir(".", "files", "*.sthlp")
        returns a list of all *.sthlp files found in the current directory.

Conformability

dir(*dirname*, *filetype*, *pattern*, *prefix*):

dirname:	1×1	
filetype:	1×1	
pattern:	1×1	
prefix:	1×1	(optional)
result:	$k \times 1$,	*k* number of files matching pattern

Diagnostics

dir(*dirname*, *filetype*, *pattern*, *prefix*) returns J(0,1,"") if

1. no files matching *pattern* are found,

2. directory *dirname* does not exist, or

3. *filetype* is misspecified (is not equal to "files", "dirs", or "others").

dirname may be specified with or without the directory separator on the end.

dirname = "" is interpreted the same as *dirname* = "."; the current directory is searched.

Also see

[M-4] **io** — I/O functions

Title

Syntax

> *real scalar*   direxists(*string scalar dirname*)

Description

direxists(*dirname*) returns 1 if *dirname* contains a valid path to a directory and returns 0 otherwise.

Conformability

```
direxists(dirname):
      dirname:      1 × 1
       result:      1 × 1
```

Diagnostics

None.

Also see

[M-4] **io** — I/O functions

Title

[M-5] **direxternal( )** — Obtain list of existing external globals

Syntax

> *string colvector* direxternal(*string scalar pattern*)

Description

direxternal(*pattern*) returns a column vector containing the names matching *pattern* of the existing external globals. direxternal() returns J(0,1,"") if there are no such globals.

pattern is interpreted by [M-5] **strmatch( )**.

Remarks

See [M-5] **findexternal( )** for the definition of a global.

A list of all globals can be obtained by direxternal("*").

Conformability

direxternal(*pattern*):
      pattern: 1×1
        result: $n \times 1$

Diagnostics

direxternal(*pattern*) returns J(0,1,"") when there are no globals matching *pattern*.

Also see

[M-5] **findexternal( )** — Find, create, and remove external globals

[M-4] **programming** — Programming functions

Title

Syntax

> *void* display(*string colvector s*)
>
> *void* display(*string colvector s*, *real scalar asis*)

Description

display(*s*) displays the string or strings contained in *s*.

display(*s*, *asis*) does the same thing but allows you to control how SMCL codes are treated. display(*s*, 0) is equivalent to display(*s*); any SMCL codes are honored.

display(*s*, *asis*), *asis* ≠ 0, displays the contents of *s* exactly as they are. For instance, when *asis* ≠ 0, "{it}" is just the string of characters {, i, t, and } and those characters are displayed; {it} is not given the SMCL interpretation of enter italic mode.

Remarks

When *s* is a scalar, the differences between coding

```
: display(s)
```

and coding

```
: s
```

are

1. display(*s*) will not indent *s*; *s* by itself causes *s* to be indented by two spaces.

2. display(*s*) will honor any SMCL codes contained in *s*; s by itself is equivalent to display(*s*, 1). For example,

    ```
    : s = "this is an {it:example}"
    : display(s)
    this is an example
    : s
      this is an {it:example}
    ```

3. When *s* is a vector, display(*s*) simply displays the lines, whereas *s* by itself adorns the lines with row and column numbers:

```
: s = ("this is line 1" \ "this is line 2")
: display(s)
this is line 1
this is line 2

: s
```

```
         1 │  this is line 1
         2 │  this is line 2
```

Another alternative to display( ) is printf( ); see [M-5] **printf( )**. When *s* is a scalar, display( ) and printf( ) do the same thing:

```
: display("this is an {it:example}")
this is an example

: printf("%s\n", "this is an {it:example}")
this is an example
```

printf( ), however, will not allow *s* to be nonscalar; it has other capabilities.

Conformability

display(*s*, *asis*)

s:	$k \times 1$	
asis:	1×1	(optional)
result:	*void*	

Diagnostics

None.

Also see

[M-5] **displayas( )** — Set display level

[M-5] **displayflush( )** — Flush terminal-output buffer

[M-5] **printf( )** — Format output

[M-4] **io** — I/O functions

Title

[M-5] **displayas( )** — Set display level

Syntax

> *void* displayas(*string scalar level*)

where *level* may be

level	minimum abbreviation
"result"	"res"
"text"	"txt"
"error"	"err"
"input"	"inp"

Description

displayas(*level*) sets whether and how subsequent output is to be displayed.

Remarks

If this function is never invoked, then the output level is result. Say that Mata was invoked in such a way that all output except error messages is being suppressed (e.g., quietly was coded in front of the mata command or in front of the ado-file that called your Mata function). If output is being suppressed, then Mata output is being suppressed, including any output created by your program. Say that you reach a point in your program where you wish to output an error message. You coded

```
printf("{err:you made a mistake}\n")
```

Even though you coded the SMCL directive {err:}, the error message will still be suppressed. SMCL determines how something is rendered, not whether it is rendered. What you need to code is

```
displayas("err")
printf("{err:you made a mistake}\n")
```

Actually, you could code

```
displayas("err")
printf("you made a mistake\n")
```

because, in addition to setting the output level (telling Stata that all subsequent output is of the specified level), it also sets the current SMCL rendering to what is appropriate for that kind of output. Hence, if you coded

```
displayas("err")
printf("{res:you made a mistake}\n")
```

the text you made a mistake would appear in the style of results despite any quietlys attempting to suppress output. Coding the above is considered bad style.

Conformability

displayas(*level*):
       level: 1×1
     result: *void*

Diagnostics

displayas(*level*) aborts with error if *level* contains an inappropriate string.

Also see

[M-5] **printf( )** — Format output

[M-5] **display( )** — Display text interpreting SMCL

[M-4] **io** — I/O functions

Title

[M-5] **displayflush( )** — Flush terminal-output buffer

Syntax

> *void* displayflush()

Description

To achieve better performance, Stata buffers terminal output, so, within a program, output may not appear when a display() or printf() command is executed. The output might be held in a buffer and displayed later.

displayflush() forces Stata to display all pending output at the terminal. displayflush() is rarely used.

Remarks

See [M-5] **printf( )** for an example of the use of displayflush().

Use of displayflush() slows execution. Use displayflush() only when it is important that output be displayed at the terminal now, such as when providing messages indicating what your program is doing.

Diagnostics

None.

Also see

[M-5] **printf( )** — Format output

[M-5] **display( )** — Display text interpreting SMCL

[M-4] **io** — I/O functions

Title

[M-5] **Dmatrix( )** — Duplication matrix

Syntax

> *real matrix* Dmatrix(*real scalar n*)

Description

Dmatrix(*n*) returns the $n^2 \times n(n+1)/2$ duplication matrix D for which D*vech(*X*) = vec(*X*), where *X* is an arbitrary $n \times n$ symmetric matrix.

Remarks

Duplication matrices are frequently used in computing derivatives of functions of symmetric matrices. Section 9.5 of Lütkepohl (1996) lists many useful properties of duplication matrices.

Conformability

Dmatrix(*n*):
$$\quad n: \quad 1 \times 1$$
$$\quad result: \quad n^2 \times n(n+1)/2$$

Diagnostics

Dmatrix(*n*) aborts with error if *n* is less than 0 or is missing. *n* is interpreted as trunc(*n*).

Reference

Lütkepohl, H. 1996. *Handbook of Matrices*. New York: Wiley.

Also see

[M-5] **Kmatrix( )** — Commutation matrix

[M-5] **Lmatrix( )** — Elimination matrix

[M-5] **vec( )** — Stack matrix columns

[M-4] **standard** — Functions to create standard matrices

Title

[M-5] **dsign( )** — FORTRAN-like DSIGN( ) function

Syntax

real scalar   dsign(*real scalar a*, *real scalar b*)

Description

dsign(a, b) returns a with the sign of b, defined as $|a|$ if $b \geq 0$ and $-|a|$ otherwise.

This function is useful when translating FORTRAN programs.

The in-line construction

$(b >= 0 ? \text{abs}(a) : -\text{abs}(a))$

is clearer. Also, differentiate carefully between what dsign( ) returns (equivalent to the above construction) and signum(b)*abs(a), which is almost equivalent but returns 0 when b is 0 rather than abs(a). (Message: dsign( ) is not one of our favorite functions.)

Conformability

dsign(a, b):

a:	1×1
b:	1×1
result:	1×1

Diagnostics

dsign(., b) returns . for all b.

dsign(a, .) returns abs(a) for all a.

Also see

[M-5] **sign( )** — Sign and complex quadrant functions

[M-4] **scalar** — Scalar mathematical functions

Title

[M-5] e( ) — Unit vectors

Syntax

real rowvector e(*real scalar i, real scalar n*)

Description

e(i, n) returns a $1 \times n$ unit vector, a vector with all elements equal to zero except for the ith, which is set to one.

Conformability

e(i, n):

i:	1×1
n:	1×1
result:	$1 \times n$

Diagnostics

e(i, n) aborts with error if $n < 1$ or if $i < 1$ or if $i > n$. Arguments i and n are interpreted as trunc(i) and trunc(n).

Also see

[M-4] **standard** — Functions to create standard matrices

Title

[M-5] editmissing( ) — Edit matrix for missing values

Syntax

$numeric\ matrix$ editmissing(*numeric matrix A , numeric scalar v*)

$void$ _editmissing(*numeric matrix a , numeric scalar v*)

Description

editmissing(*A , v*) returns *A* with any missing values changed to *v*.

_editmissing(*A , v*) replaces all missing values in *A* with *v*.

Remarks

editmissing( ) and _editmissing( ) are very fast.

If you want to change nonmissing values to other values, including missing, see [M-5] **editvalue( )**.

Conformability

editmissing(*A , v*):

A:	$r \times c$
v:	1×1
result:	$r \times c$

_editmissing(*A , v*):

    input:

A:	$r \times c$
v:	1×1

    output:

A:	$r \times c$

Diagnostics

editmissing(*A , v*) and _editmissing(*A , v*) change all missing elements to *v*, including not only . but also .a, .b, ..., .z.

Also see

[M-5] **editvalue( )** — Edit (change) values in matrix

[M-4] **manipulation** — Matrix manipulation

Title

[M-5] edittoint( ) — Edit matrix for roundoff error (integers)

Syntax

numeric matrix   edittoint(*numeric matrix Z, real scalar amt*)

void   _edittoint(*numeric matrix Z, real scalar amt*)

numeric matrix   edittointtol(*numeric matrix Z, real scalar tol*)

void   _edittointtol(*numeric matrix Z, real scalar tol*)

Description

These edit functions set elements of matrices to integers that are close to integers.

edittoint(Z, *amt*) and _edittoint(Z, *amt*) set

$$Z_{ij} = \text{round}(Z_{ij}) \quad \text{if } \left| Z_{ij} - \text{round}(Z_{ij}) \right| \leq |tol|$$

for Z real and set

$$\text{Re}(Z_{ij}) = \text{round}(\text{Re}(Z_{ij})) \quad \text{if } \left| \text{Re}(Z_{ij}) - \text{round}(\text{Re}(Z_{ij})) \right| \leq |tol|$$

$$\text{Im}(Z_{ij}) = \text{round}(\text{Im}(Z_{ij})) \quad \text{if } \left| \text{Im}(Z_{ij}) - \text{round}(\text{Im}(Z_{ij})) \right| \leq |tol|$$

for Z complex, where in both cases

$$tol = \text{abs}(amt)*\text{epsilon}(\text{sum}(\text{abs}(Z))/(\text{rows}(Z)*\text{cols}(Z)))$$

edittoint() leaves Z unchanged and returns the edited matrix. _edittoint() edits Z in place.

edittointtol(Z, *tol*) and _edittointtol(Z, *tol*) do the same thing, except that *tol* is specified directly.

Remarks

These functions mirror the edittozero() functions documented in [M-5] **edittozero( )**, except that, rather than solely resetting to zero values close to zero, they reset to integer values close to integers.

See [M-5] **edittozero( )**. Whereas use of the functions documented there is recommended, use of the functions documented here generally is not. Although zeros commonly arise in real problems so that there is reason to suspect small numbers would be zero but for roundoff error, integers arise more rarely.

If you have reason to believe that integer values are likely, then by all means use these functions.

Conformability

edittoint(*Z*, *amt*):

	Z:	$r \times c$
	amt:	1×1
	result:	$r \times c$

_edittoint(*Z*, *amt*):

input:

| | *Z*: | $r \times c$ |
| | *amt*: | 1×1 |

output:

| | *Z*: | $r \times c$ |

edittointtol(*Z*, *tol*):

	Z:	$r \times c$
	tol:	1×1
	result:	$r \times c$

_edittointtol(*Z*, *tol*):

input:

| | *Z*: | $r \times c$ |
| | *tol*: | 1×1 |

output:

| | *Z*: | $r \times c$ |

Diagnostics

None.

Also see

[M-5] **edittozero( )** — Edit matrix for roundoff error (zeros)

[M-4] **manipulation** — Matrix manipulation

Title

Syntax

numeric matrix	edittozero(*numeric matrix Z, real scalar amt*)
void	_edittozero(*numeric matrix Z, real scalar amt*)

numeric matrix	edittozerotol(*numeric matrix Z, real scalar tol*)
void	_edittozerotol(*numeric matrix Z, real scalar tol*)

Description

These edit functions set elements of matrices to zero that are close to zero. edittozero(*Z, amt*) and _edittozero(*Z, amt*) set

$$Z_{ij} = 0 \quad \text{if } |Z_{ij}| \le |tol|$$

for Z real and set

$$\mathrm{Re}(Z_{ij}) = 0 \quad \text{if } |\mathrm{Re}(Z_{ij})| \le |tol|$$
$$\mathrm{Im}(Z_{ij}) = 0 \quad \text{if } |\mathrm{Im}(Z_{ij})| \le |tol|$$

for Z complex, where in both cases

        tol = abs(*amt*)*epsilon(sum(abs(Z))/(rows(Z)*cols(Z)))

edittozero() leaves Z unchanged and returns the edited matrix. _edittozero() edits Z in place.

edittozerotol(*Z, tol*) and _edittozerotol(*Z, tol*) do the same thing, except that *tol* is specified directly.

Remarks

Remarks are presented under the following headings:

> *Background*
> *Treatment of complex values*
> *Recommendations*

Background

Numerical roundoff error leads to, among other things, numbers that should be zero being small but not zero, and so it is sometimes desirable to reset those small numbers to zero.

The problem is in identifying those small numbers. Is 1e–14 small? Is 10,000? The answer is that, given some matrix, 1e–14 might not be small because most of the values in the matrix are around 1e–14, and the small values are 1e–28, and given some other matrix, 10,000 might indeed be small because most of the elements are around 1e+18.

edittozero() makes an attempt to determine what is small. edittozerotol() leaves that determination to you. In edittozerotol(Z, tol), you specify *tol* and elements for which $|Z_{ij}| \leq tol$ are set to zero.

Using edittozero(Z, amt), however, you specify *amt* and then *tol* is calculated for you based on the size of the elements in Z and *amt*, using the formula

$$tol = amt * \texttt{epsilon}(\textit{average value of } |Z_{ij}|)$$

epsilon() refers to machine precision, and epsilon(x) is the function that returns machine precision in units of x:

$$\texttt{epsilon}(x) = |x|*\texttt{epsilon}(1)$$

where epsilon(1) is approximately 2.22e–16 on most computers; see [M-5] **epsilon( )**.

Treatment of complex values

The formula

$$tol = amt * \texttt{epsilon}(\textit{average value of } |Z_{ij}|)$$

is used for both real and complex values. For complex, $|Z_{ij}|$ refers to the modulus (length) of the complex element.

However, rather than applying the reset rule

$$Z_{ij} = 0 \qquad \text{if } |Z_{ij}| \leq |tol|$$

as is done when Z is real, the reset rules are

$$\text{Re}(Z_{ij}) = 0 \qquad \text{if } |\text{Re}(Z_{ij})| \leq |tol|$$
$$\text{Im}(Z_{ij}) = 0 \qquad \text{if } |\text{Im}(Z_{ij})| \leq |tol|$$

The first rule, applied even for complex, may seem more appealing, but the use of the second has the advantage of mapping numbers close to being purely real or purely imaginary to purely real or purely imaginary results.

Recommendations

1. Minimal editing is performed by edittozero(Z ,1). Values less than 2.22e–16 times the average would be set to zero.

2. It is often reasonable to code edittozero(Z, 1000), which sets to zero values less than 2.22e–13 times the average.

3. For a given calculation, the amount of roundoff error that arises with complex matrices (matrices with nonzero imaginary part) is greater than the amount that arises with real matrices (matrices with zero imaginary part even if stored as complex). That is because, in addition to all the usual sources of roundoff error, multiplication of complex values involves the addition operator, which introduces additional roundoff error. Hence, whatever is the appropriate value of *amt* or *tol* with real matrices, it is larger for complex matrices.

Conformability

edittozero(*Z*, *amt*):

Z:	$r \times c$
amt:	1×1
result:	$r \times c$

_edittozero(*Z*, *amt*):

input:

Z:	$r \times c$
amt:	1×1

output:

Z:	$r \times c$

edittozerotol(*Z*, *tol*):

Z:	$r \times c$
tol:	1×1
result:	$r \times c$

_edittozerotol(*Z*, *tol*):

input:

Z:	$r \times c$
tol:	1×1

output:

Z:	$r \times c$

Diagnostics

edittozero(*Z*, *amt*) and _edittozero(*Z*, *amt*) leave scalars unchanged because they base their calculation of the likely roundoff error on the average value of $|Z_{ij}|$.

Also see

[M-5] **edittoint( )** — Edit matrix for roundoff error (integers)

[M-4] **manipulation** — Matrix manipulation

Title

Syntax

> *matrix*   editvalue(*matrix A*, *scalar from*, *scalar to*)
>
> *void*   _editvalue(*matrix A*, *scalar from*, *scalar to*)

where *A*, *from*, and *to* may be real, complex, or string.

Description

editvalue(*A*, *from*, *to*) returns *A* with all elements equal to *from* changed to *to*.

_editvalue(*A*, *from*, *to*) does the same thing but modifies *A* itself.

Remarks

editvalue( ) and _editvalue( ) are fast.

If you wish to change missing values to nonmissing values, it is better to use [M-5] **editmissing( )**. editvalue(*A*, ., 1) would change all . missing values to 1 but leave .a, .b, ..., .z unchanged. editmissing(*A*, 1) would change all missing values to 1.

Conformability

editvalue(*A*, *from*, *to*):

A:	$r \times c$
from:	1×1
to:	1×1
result:	$r \times c$

_editvalue(*A*, *from*, *to*):

input:

A:	$r \times c$
from:	1×1
to:	1×1

output:

A:	$r \times c$

Diagnostics

editvalue(*A*, *from*, *to*) returns a matrix of the same type as *A*.

editvalue(*A*, *from*, *to*) and _editvalue(*A*, *from*, *to*) abort with error if *from* and *to* are incompatible with *A*. That is, if *A* is real, *to* and *from* must be real. If *A* is complex, *to* and *from* must each be either real or complex. If *A* is string, *to* and *from* must be string.

_editvalue(*A*, *from*, *to*) aborts with error if *A* is a view.

Also see

[M-5] **editmissing( )** — Edit matrix for missing values

[M-4] **manipulation** — Matrix manipulation

Title

[M-5] eigensystem( ) — Eigenvectors and eigenvalues

Syntax

void	eigensystem(*A*, *X*, *L* [, *rcond* [, *nobalance*]])
void	lefteigensystem(*A*, *X*, *L* [, *rcond* [, *nobalance*]])
complex rowvector	eigenvalues(*A* [, *rcond* [, *nobalance*]])
void	symeigensystem(*A*, *X*, *L*)
real rowvector	symeigenvalues(*A*)
void	_eigensystem(*A*, *X*, *L* [, *rcond* [, *nobalance*]])
void	_lefteigensystem(*A*, *X*, *L* [, *rcond* [, *nobalance*]])
complex rowvector	_eigenvalues(*A* [, *rcond* [, *nobalance*]])
void	_symeigensystem(*A*, *X*, *L*)
real rowvector	_symeigenvalues(*A*)

where inputs are

A:	*numeric matrix*
rcond:	*real scalar* (whether *rcond* desired)
nobalance:	*real scalar* (whether to suppress balancing)

and outputs are

X:	*numeric matrix* of eigenvectors
L:	*numeric vector* of eigenvalues
rcond:	*real vector* of reciprocal condition numbers

The columns of *X* will contain the eigenvectors except when using _lefteigensystem(), in which case the rows of *X* contain the eigenvectors.

The following routines are used in implementing the above routines:

real scalar _eigen_la(*real scalar todo*, *numeric matrix A*, *X*, *L*, *real scalar rcond*, *real scalar nobalance*)

real scalar _symeigen_la(*real scalar todo*, *numeric matrix A*, *X*, *L*)

Description

These routines calculate eigenvectors and eigenvalues of square matrix A.

eigensystem(A, X, L, *rcond*, *nobalance*) calculates eigenvectors and eigenvalues of a general, real or complex, square matrix A. Eigenvectors are returned in X and eigenvalues in L. The remaining arguments are optional:

1. If *rcond* is not specified, then reciprocal condition numbers are not returned in *rcond*.

   If *rcond* is specified and contains a value other than 0 or missing—*rcond*=1 is suggested—in *rcond* will be placed a vector of the reciprocals of the condition numbers for the eigenvalues. Each element of the new *rcond* measures the accuracy to which the corresponding eigenvalue has been calculated; large numbers (numbers close to 1) are better and small numbers (numbers close to 0) indicate inaccuracy; see *Eigenvalue condition* below.

2. If *nobalance* is not specified, balancing is performed to obtain more accurate results.

   If *nobalance* is specified and is not zero nor missing, balancing is not used. Results are calculated more quickly, but perhaps a little less accurately; see *Balancing* below.

lefteigensystem(A, X, L, *rcond*, *nobalance*) mirrors eigensystem( ), the difference being that lefteigensystem( ) solves for left eigenvectors solving $XA = \text{diag}(L)*X$ instead of right eigenvectors solving $AX = X*\text{diag}(L)$.

eigenvalues(A, *rcond*, *nobalance*) returns the eigenvalues of square matrix A; the eigenvectors are not calculated. Arguments *rcond* and *nobalance* are optional.

symeigensystem(A, X, L) and symeigenvalues(A) mirror eigensystem( ) and eigenvalues( ), the difference being that A is assumed to be symmetric (Hermitian). The eigenvalues returned are real. (Arguments *rcond* and *nobalance* are not allowed; *rcond* because symmetric matrices are inherently well conditioned; *nobalance* because it is unnecessary.)

The underscore routines mirror the routines of the same name without underscores, the difference being that A is damaged during the calculation and so the underscore routines use less memory.

_eigen_la( ) and _symeigen_la( ) are the interfaces into the [M-1] **LAPACK** routines used to implement the above functions. Their direct use is not recommended.

Remarks

Remarks are presented under the following headings:

> *Eigenvalues and eigenvectors*
> *Left eigenvectors*
> *Symmetric eigensystems*
> *Normalization and order*
> *Eigenvalue condition*
> *Balancing*
> *eigensystem( ) and eigenvalues( )*
> *lefteigensystem( )*
> *symeigensystem( ) and symeigenvalues( )*

Eigenvalues and eigenvectors

A scalar λ is said to be an eigenvalue of square matrix A: $n \times n$ if there is a nonzero column vector x: $n \times 1$ (called the eigenvector) such that

$$Ax = \lambda x \qquad (1)$$

(1) can also be written

$$(A - \lambda I)x = 0$$

where I is the $n \times n$ identity matrix. A nontrivial solution to this system of n linear homogeneous equations exists if and only if

$$\det(A - \lambda I) = 0 \qquad (2)$$

This nth degree polynomial in λ is called the characteristic polynomial or characteristic equation of A, and the eigenvalues λ are its roots, also known as the characteristic roots.

There are, in fact, n solutions (λ_i, x_i) that satisfy (1)—although some can be repeated—and we can compactly write the full set of solutions as

$$AX = X * \mathrm{diag}(L) \qquad (3)$$

where

$$X = (x_1, x_2, \ldots) \qquad (X : n \times n)$$
$$L = (\lambda_1, \lambda_2, \ldots) \qquad (L : 1 \times n)$$

```
: A = (1, 2 \ 9, 4)
: X = .
: L = .
: eigensystem(A, X, L)
: X
```

	1	2
1	-.316227766	-.554700196
2	-.948683298	.832050294

```
: L
```

	1	2
1	7	-2

The first eigenvalue is 7, and the corresponding eigenvector is $(-.316 \setminus -.949)$. The second eigenvalue is -2, and the corresponding eigenvector is $(-.555 \setminus .832)$.

In general, eigenvalues and vectors can be complex even if A is real.

Left eigenvectors

What we have defined above is properly known as the right-eigensystem problem:

$$Ax = \lambda x \qquad (1)$$

In the above, x is a column vector. The left-eigensystem problem is to find the row vector x satisfying

$$xA = \lambda x \tag{1'}$$

The eigenvalue λ is the same in (1) and (1'), but x can differ.

The n solutions (λ_i, x_i) that satisfy (1') can be compactly written

$$XA = \mathrm{diag}(L) * X \tag{3'}$$

where

$$X = \begin{bmatrix} x_1 \\ x_2 \\ \vdots \\ x_n \end{bmatrix} \qquad L = \begin{bmatrix} \lambda_1 \\ \lambda_2 \\ \vdots \\ \lambda_n \end{bmatrix}$$
$${\scriptstyle n \times n} \qquad\qquad {\scriptstyle n \times 1}$$

For instance,

```
: A = (1, 2 \ 9, 4)
: X = .
: L = .
: lefteigensystem(A, X, L)
: X
```

	1	2
1	-.832050294	-.554700196
2	-.948683298	.316227766

```
: L
```

	1
1	7
2	-2

The first eigenvalue is 7, and the corresponding eigenvector is $(-.832, -.555)$. The second eigenvalue is -2, and the corresponding eigenvector is $(-.949, .316)$.

The eigenvalues are the same as in the previous example; the eigenvectors are different.

Symmetric eigensystems

Below we use the term symmetric to encompass Hermitian matrices, even when we do not emphasize the fact.

Eigensystems of symmetric matrices are conceptually no different from general eigensystems, but symmetry introduces certain simplifications:

1. The eigenvalues associated with symmetric matrices are real, whereas those associated with general matrices may be real or complex.

2. The eigenvectors associated with symmetric matrices—which may be real or complex—are orthogonal.

3. The left and right eigenvectors of symmetric matrices are transposes of each other.

4. The eigenvectors and eigenvalues of symmetric matrices are more easily, and more accurately, computed.

For item 3, let us begin with the right-eigensystem problem:

$$AX = X * \operatorname{diag}(L)$$

Taking the transpose of both sides results in

$$X'A = \operatorname{diag}(L) * X'$$

because $A = A'$ if A is symmetric (Hermitian).

symeigensystem(A, X, L) calculates right eigenvectors. To obtain the left eigenvectors, you simply transpose X.

Normalization and order

If x is a solution to

$$Ax = \lambda x$$

then so is cx, c: 1×1, $c \neq 0$.

The eigenvectors returned by the above routines are scaled to have length (norm) 1.

The eigenvalues are combined and returned in a vector (L) and the eigenvectors in a matrix (X). The eigenvalues are ordered from largest to smallest in absolute value (or, if the eigenvalues are complex, in length). The eigenvectors are ordered to correspond to the eigenvalues.

Eigenvalue condition

Optional argument *rcond* may be specified as a value other than 0 or missing—*rcond* = 1 is suggested— and then *rcond* will be filled in with a vector containing the reciprocals of the condition numbers for the eigenvalues. Each element of *rcond* measures the accuracy with which the corresponding eigenvalue has been calculated; large numbers (numbers close to 1) are better and small numbers (numbers close to 0) indicate inaccuracy.

The reciprocal condition number is calculated as abs($y*x$), where y: $1 \times n$ is the left eigenvector and x: $n \times 1$ is the corresponding right eigenvector. Since y and x each have norm 1, abs($y*x$) = abs(cos(*theta*)), where *theta* is the angle made by the vectors. Thus $0 \leq$ abs($y*x$) ≤ 1. For symmetric matrices, $y*x$ will equal 1. It can be proved that abs($y*x$) is the reciprocal of the condition number for a simple eigenvalue, and so it turns out that the sensitivity of the eigenvalue to a perturbation is a function of how far the matrix is from symmetric on this scale.

Requesting that *rcond* be calculated increases the amount of computation considerably.

Balancing

By default, balancing is performed for general matrices. Optional argument *nobalance* allows you to suppress balancing.

Balancing is related to row-and-column equilibration; see [M-5] **_equilrc( )**. Here, however, a diagonal matrix D is found such that DAD^{-1} is better balanced, the eigenvectors and eigenvalues for DAD^{-1} are extracted, and then the eigenvectors and eigenvalues are adjusted by D so that they reflect those for the original A matrix.

There is no gain from these machinations when A is symmetric, so the symmetric routines do not have a *nobalance* argument.

eigensystem( ) and eigenvalues( )

1. Use $L =$ eigenvalues(A) and eigensystem(A, X, L) for general matrices A.

2. Use $L =$ eigenvalues(A) when you do not need the eigenvectors; it will save both time and memory.

3. The eigenvalues returned by $L =$ eigenvalues(A) and by eigensystem(A, X, L) are of storage type complex even if the eigenvalues are real (i.e., even if Im(L)==0). If the eigenvalues are known to be real, you can save computer memory by subsequently coding

        L = Re(L)

   If you wish to test whether the eigenvalues are real, examine mreldifre(L); see [M-5] **reldif( )**.

4. The eigenvectors returned by eigensystem(A, X, L) are of storage type complex even if the eigenvectors are real (i.e., even if Im(X)==0). If the eigenvectors are known to be real, you can save computer memory by subsequently coding

        X = Re(X)

   If you wish to test whether the eigenvectors are real, examine mreldifre(X); see [M-5] **reldif( )**.

5. If you are using eigensystem(A, X, L) interactively (outside a program), X and L must be predefined. Type

        : eigensystem(A, X=., L=.)

lefteigensystem( )

What was just said about eigensystem( ) applies equally well to lefteigensystem( ).

If you need only the eigenvalues, use $L =$ eigenvalues(A). The eigenvalues are the same for both left and right systems.

symeigensystem( ) and symeigenvalues( )

1. Use $L =$ symeigenvalues(A) and symeigensystem(A, X, L) for symmetric or Hermitian matrices A.

2. Use $L =$ symeigenvalues(A) when you do not need the eigenvectors; it will save both time and memory.

3. The eigenvalues returned by $L =$ symeigenvalues(A) and by symeigensystem(A, X, L) are of storage type real. Eigenvalues of symmetric and Hermitian matrices are always real.

4. The eigenvectors returned by symeigensystem(A, X, L) are of storage type real when A is of storage type real and of storage type complex when A is of storage type complex.

5. If you are using symeigensystem(A, X, L) interactively (outside a program), X and L must be predefined. Type

        : symeigensystem(A, X=., L=.)

Conformability

eigensystem$(A, X, L, rcond, nobalance)$:

   input:

A:	$n \times n$	
rcond:	1×1	(optional, specify as 1 to obtain *rcond*)
nobalance:	1×1	(optional, specify as 1 to prevent balancing)

   output:

X:	$n \times n$	(columns contain eigenvectors)
L:	$1 \times n$	
rcond:	$1 \times n$	(optional)
result:	*void*	

lefteigensystem$(A, X, L, rcond, nobalance)$:

   input:

A:	$n \times n$	
rcond:	1×1	(optional, specify as 1 to obtain *rcond*)
nobalance:	1×1	(optional, specify as 1 to prevent balancing)

   output:

X:	$n \times n$	(rows contain eigenvectors)
L:	$n \times 1$	
rcond:	$n \times 1$	(optional)
result:	*void*	

eigenvalues$(A, rcond, nobalance)$:

   input:

A:	$n \times n$	
rcond:	1×1	(optional, specify as 1 to obtain *rcond*)
nobalance:	1×1	(optional, specify as 1 to prevent balancing)

   output:

rcond:	$1 \times n$	(optional)
result:	$1 \times n$	(contains eigenvalues)

symeigensystem(*A*, *X*, *L*):

    input:

                    A: $n \times n$

    output:

                    X: $n \times n$ (columns contain eigenvectors)
                    L: $1 \times n$
            result: *void*

symeigenvalues(*A*):

    input:

                    A: $n \times n$

    output:

            result: $1 \times n$ (contains eigenvalues)

_eigensystem(*A*, *X*, *L*, *rcond*, *nobalance*):

input:

                    A: $n \times n$
               rcond: 1×1 (optional, specify as 1 to obtain *rcond*)
          nobalance: 1×1 (optional, specify as 1 to prevent balancing)

    output:

                    A: 0×0
                    X: $n \times n$ (columns contain eigenvectors)
                    L: $1 \times n$
               rcond: $1 \times n$ (optional)
            result: *void*

_lefteigensystem(*A*, *X*, *L*, *rcond*, *nobalance*):

    input:

                    A: $n \times n$
                rcond: 1×1 (optional, specify as 1 to obtain *rcond*)
         nobalance: 1×1 (optional, specify as 1 to prevent balancing)

    output:

                    A: 0×0
                    X: $n \times n$ (rows contain eigenvectors)
                    L: $n \times 1$
               rcond: $n \times 1$ (optional)
            result: *void*

_eigenvalues(*A*, *rcond*, *nobalance*):

    input:

                    A: $n \times n$
                rcond: 1×1 (optional, specify as 1 to obtain *rcond*)
         nobalance: 1×1 (optional, specify as 1 to prevent balancing)

    output:

                    A: 0×0
               rcond: $1 \times n$ (optional)
            result: $1 \times n$ (contains eigenvalues)

_symeigensystem(A, X, L):

   input:

                        A: $n \times n$

   output:

                        A: 0×0

                        X: $n \times n$ (columns contain eigenvectors)

                        L: $1 \times n$

              result: *void*

_symeigenvalues(A):

   input:

                        A: $n \times n$

   output:

                        A: 0×0

              result: $1 \times n$ (contains eigenvalues)

_eigen_la(*todo*, A, X, L, *rcond*, *nobalance*):

   input:

              todo: 1×1

                  A: $n \times n$

           rcond: 1×1

      nobalance: 1×1

   output:

                  A: 0×0

                  X: $n \times n$

                  L: $1 \times n$ or $n \times 1$

             rcond: $1 \times n$ or $n \times 1$ (optional)

              result: 1×1 (return code)

_symeigen_la(*todo*, A, X, L):

   input:

              todo: 1×1

                  A: $n \times n$

   output:

                  A: 0×0

                  X: $n \times n$

                  L: $1 \times n$

              result: 1×1 (return code)

Diagnostics

All functions return missing-value results if A has missing values.

symeigensystem(), symeigenvalues(), _symeigensystem(), and _symeigenvalues() use the lower triangle of A without checking for symmetry. When A is complex, only the real part of the diagonal is used.

Also see

[M-5] **matexpsym( )** — Exponentiation and logarithms of symmetric matrices

[M-5] **matpowersym( )** — Powers of a symmetric matrix

[M-4] **matrix** — Matrix functions

Title

> **[M-5] eigensystemselect( )** — Compute selected eigenvectors and eigenvalues

Syntax

void	eigensystemselectr(A, *range*, X, L)
void	lefteigensystemselectr(A, *range*, X, L)
void	eigensystemselecti(A, *index*, X, L)
void	lefteigensystemselecti(A, *index*, X, L)
void	eigensystemselectf(A, f, X, L)
void	lefteigensystemselectf(A, f, X, L)
void	symeigensystemselectr(A, *range*, X, L)
void	symeigensystemselecti(A, *index*, X, L)

where inputs are

A:	*numeric matrix*	
range:	*real vector*	(range of eigenvalues to be selected)
index:	*real vector*	(indices of eigenvalues to be selected)
f:	*pointer scalar*	(points to a function used to select eigenvalue)

and outputs are

X:	*numeric matrix*	of eigenvectors
L:	*numeric vector*	of eigenvalues

The following routines are used in implementing the above routines:

void _eigenselecti_la(*numeric matrix A, XL, XR, L,*
  string scalar side, real vector index)

void _eigenselectr_la(*numeric matrix A, XL, XR, L,*
  string scalar side, real vector range)

void _eigenselectf_la(*numeric matrix A, XL, XR, L,*
  string scalar side, pointer scalar f)

real scalar _eigenselect_la(*numeric matrix A, XL, XR, L, select,*
  string scalar side, real scalar noflopin)

real scalar _symeigenselect_la(*numeric matrix A, X, L, ifail,*
  real scalar type, lower, upper, abstol)

Description

eigensystemselectr(A, *range*, X, L) computes selected right eigenvectors of a square, numeric matrix A along with their corresponding eigenvalues. Only the eigenvectors corresponding to selected eigenvalues are computed. Eigenvalues that lie in a range are selected. The selected eigenvectors are returned in X, and their corresponding eigenvalues are returned in L.

range is a vector of length 2. All eigenvalues with absolute value in the half-open interval (*range*[1], *range*[2]] are selected.

lefteigensystemselectr(*A*, *range*, *X*, *L*) mirrors eigensystemselectr( ), the difference being that it computes selected left eigenvectors instead of selected right eigenvectors.

eigensystemselecti(*A*, *index*, *X*, *L*) computes selected right eigenvectors of a square, numeric matrix, *A*, along with their corresponding eigenvalues. Only the eigenvectors corresponding to selected eigenvalues are computed. Eigenvalues are selected by an index. The selected eigenvectors are returned in *X*, and the selected eigenvalues are returned in *L*.

index is a vector of length 2. The eigenvalues are sorted by their absolute values, in descending order. The eigenvalues whose rank is *index*[1] through *index*[2], inclusive, are selected.

lefteigensystemselecti(*A*, *index*, *X*, *L*) mirrors eigensystemselecti( ), the difference being that it computes selected left eigenvectors instead of selected right eigenvectors.

eigensystemselectf(*A*, *f*, *X*, *L*) computes selected right eigenvectors of a square, numeric matrix, *A*, along with their corresponding eigenvalues. Only the eigenvectors corresponding to selected eigenvalues are computed. Eigenvalues are selected by a user-written function described below. The selected eigenvectors are returned in *X*, and the selected eigenvalues are returned in *L*.

lefteigensystemselectf(*A*, *f*, *X*, *L*) mirrors eigensystemselectf( ), the difference being that it computes selected left eigenvectors instead of selected right eigenvectors.

symeigensystemselectr(*A*, *range*, *X*, *L*) computes selected eigenvectors of a symmetric (Hermitian) matrix, *A*, along with their corresponding eigenvalues. Only the eigenvectors corresponding to selected eigenvalues are computed. Eigenvalues that lie in a range are selected. The selected eigenvectors are returned in *X*, and their corresponding eigenvalues are returned in *L*.

symeigensystemselecti(*A*, *index*, *X*, *L*) computes selected eigenvectors of a symmetric (Hermitian) matrix, *A*, along with their corresponding eigenvalues. Only the eigenvectors corresponding to selected eigenvalues are computed. Eigenvalues are selected by an index. The selected eigenvectors are returned in *X*, and the selected eigenvalues are returned in *L*.

_eigenselectr_la( ), _eigenselecti_la( ), _eigenselectf_la( ), _eigenselect_la( ), and _symeigenselect_la( ) are the interfaces into the [M-1] **LAPACK** routines used to implement the above functions. Their direct use is not recommended.

Remarks

Introduction
Range selection
Index selection
Criterion selection
Other functions

Introduction

These routines compute subsets of the available eigenvectors. This computation can be much faster than computing all the eigenvectors. (See [M-5] **eigensystem**( ) for routines to compute all the eigenvectors and an introduction to the eigensystem problem.)

There are three methods for selecting which eigenvectors to compute; all of them are based on the corresponding eigenvalues. First, we can select only those eigenvectors whose eigenvalues have absolute values that fall in a half-open interval. Second, we can select only those eigenvectors whose eigenvalues have certain indices, after sorting the eigenvalues by their absolute values in descending order. Third, we can select only those eigenvectors whose eigenvalues meet a criterion encoded in a function.

Below we illustrate each of these methods. For comparison purposes, we begin by computing all the eigenvectors of the matrix

```
: A
          1     2     3     4

     1   .31   .69   .13   .56
     2   .31    .5   .72   .42
     3   .68   .37   .71    .8
     4   .09   .16   .83    .9
```

We perform the computation with `eigensystem( )`:

```
: eigensystem(A, X=., L=.)
```

The absolute values of the eigenvalues are

```
: abs(L)
            1              2              3              4

  1   2.10742167    .4658398402    .4005757984    .4005757984
```

The corresponding eigenvectors are

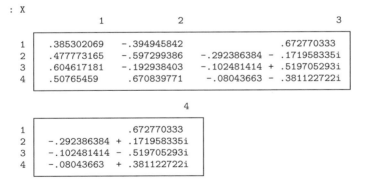

```
: X
              1              2                      3

  1    .385302069   -.394945842            .672770333
  2    .477773165   -.597299386    -.292386384 -  .171958335i
  3    .604617181   -.192938403    -.102481414 +  .519705293i
  4    .50765459     .670839771    -.08043663  -  .381122722i


                      4

  1            .672770333
  2   -.292386384 + .171958335i
  3   -.102481414 - .519705293i
  4   -.08043663  + .381122722i
```

Range selection

In applications, an eigenvalue whose absolute value is greater than 1 frequently corresponds to an explosive solution, whereas an eigenvalue whose absolute value is less than 1 corresponds to a stable solution. We frequently want to use only the eigenvectors from the stable solutions, which we can do using `eigensystemselectr( )`. We begin by specifying

```
: range = (-1, .999999999)
```

which starts from -1 to include 0 and stops at .999999999 to exclude 1. (The half-open interval in range is open on the left and closed on the right.)

Using this range in eigensystemselectr() requests each eigenvector for which the absolute value of the corresponding eigenvalue falls in the interval $(-1, .999999999]$. For the example at hand, we have

```
: eigensystemselectr(A, range, X=., L=.)
: X
```

	1	2	3
1	-.442004357	.201218963 - .875384534i	.201218963 + .875384534i
2	-.668468693	.136296114 + .431873675i	.136296114 - .431873675i
3	-.215927364	-.706872994 - .022093594i	-.706872994 + .022093594i
4	.750771548	.471845361 + .218651289i	.471845361 - .218651289i

```
: L
```

	1	2	3
1	.46583984	-.076630755 + .393177692i	-.076630755 - .393177692i

```
: abs(L)
```

	1	2	3
1	.4658398402	.4005757984	.4005757984

The above output illustrates that eigensystemselectr() has not included the results for the eigenvalue whose absolute value is greater than 1, as desired.

Index selection

In many statistical applications, an eigenvalue measures the importance of an eigenvector factor. In these applications, we want only to compute several of the largest eigenvectors. Here we use eigensystemselecti() to compute the eigenvectors corresponding to the two largest eigenvalues:

```
: index = (1, 2)
: eigensystemselecti(A, index, X=., L=.)
: L
```

	1	2
1	2.10742167	.46583984

```
: X
```

	1	2
1	.385302069	-.442004357
2	.477773165	-.668468693
3	.604617181	-.215927364
4	.50765459	.750771548

Criterion selection

In some applications, we want to compute only those eigenvectors whose corresponding eigenvalues satisfy a more complicated criterion. We can use `eigensystemselectf()` to solve these problems.

We must pass `eigensystemselectf()` a pointer to a function that implements our criterion. The function must accept a complex scalar argument so that it can receive an eigenvalue, and it must return the real value 0 to indicate rejection and a nonzero real value to indicate selection.

In the example below, we consider the common criterion of whether the eigenvalue is real. We want only to compute the eigenvectors corresponding to real eigenvalues. After deciding that anything smaller than 1e–15 is zero, we define our function to be

```
: real scalar onlyreal(complex scalar ev)
> {
>         return( (abs(Im(ev))<1e-15) )
> }
```

We compute only the eigenvectors corresponding to the real eigenvalues by typing

```
: eigensystemselectf(A, &onlyreal(), X=., L=.)
```

The eigenvalues that satisfy this criterion and their corresponding eigenvectors are

```
: L
                  1              2

    1 |  2.10742167      .46583984 |
```

```
: X
                  1              2

    1 |  .385302069    -.442004357 |
    2 |  .477773165    -.668468693 |
    3 |  .604617181    -.215927364 |
    4 |  .50765459      .750771548 |
```

Other functions

`lefteigensystemselectr()` and `symeigensystemselectr()` use a *range* like `eigensystemselectr()`.

`lefteigensystemselecti()` and `symeigensystemselecti()` use an *index* like `eigensystemselecti()`.

`lefteigensystemselectf()` uses a pointer to a function like `eigensystemselectf()`.

(Continued on next page)

Conformability

eigensystemselectr(A, *range*, X, L):

> *input*:
>
> | | |
> |---|---|
> | A: | $n \times n$ |
> | *range*: | 1×2 or 2×1 |
>
> *output*:
>
> | | |
> |---|---|
> | X: | $n \times m$ |
> | L: | $1 \times m$ |

lefteigensystemselectr(A, *range*, X, L):

> *input*:
>
> | | |
> |---|---|
> | A: | $n \times n$ |
> | *range*: | 1×2 or 2×1 |
>
> *output*:
>
> | | |
> |---|---|
> | X: | $m \times n$ |
> | L: | $1 \times m$ |

eigensystemselecti(A, *index*, X, L):

> *input*:
>
> | | |
> |---|---|
> | A: | $n \times n$ |
> | *index*: | 1×2 or 2×1 |
>
> *output*:
>
> | | |
> |---|---|
> | X: | $n \times m$ |
> | L: | $1 \times m$ |

lefteigensystemselecti(A, *index*, X, L):

> *input*:
>
> | | |
> |---|---|
> | A: | $n \times n$ |
> | *index*: | 1×2 or 2×1 |
>
> *output*:
>
> | | |
> |---|---|
> | X: | $m \times n$ |
> | L: | $1 \times m$ |

eigensystemselectf(A, f, X, L):

> *input*:
>
> | | |
> |---|---|
> | A: | $n \times n$ |
> | f: | 1×1 |
>
> *output*:
>
> | | |
> |---|---|
> | X: | $n \times m$ |
> | L: | $1 \times m$ |

```
lefteigensystemselectf(A, f, X, L):
```
   input:

A:	$n \times n$
f:	1×1

   output:

X:	$m \times n$
L:	$1 \times m$

```
symeigensystemselectr(A, range, X, L):
```
   input:

A:	$n \times n$
range:	1×2 or 2×1

   output:

X:	$n \times m$
L:	$1 \times m$

```
symeigensystemselecti(A, index, X, L):
```
   input:

A:	$n \times n$
index:	1×2 or 2×1

   output:

X:	$n \times m$
L:	$1 \times m$

Diagnostics

All functions return missing-value results if A has missing values.

`symeigensystemselectr()` and `symeigensystemselecti()` use the lower triangle of A without checking for symmetry. When A is complex, only the real part of the diagonal is used.

If the ith eigenvector failed to converge, `symeigensystemselectr()` and `symeigensystemselecti()` insert a vector of missing values into the ith column of the returned eigenvector matrix.

Also see

[M-1] **LAPACK** — The LAPACK linear-algebra routines

[M-5] **eigensystem( )** — Eigenvectors and eigenvalues

[M-5] **matexpsym( )** — Exponentiation and logarithms of symmetric matrices

[M-5] **matpowersym( )** — Powers of a symmetric matrix

[M-4] **matrix** — Matrix functions

Title

[M-5] eltype( ) — Element type and organizational type of object

Syntax

> *string scalar*   eltype(*X*)
>
> *string scalar*   orgtype(*X*)

Description

eltype() returns the current *eltype* of the argument.

orgtype() returns the current *orgtype* of the argument.

See [M-6] **Glossary** for a definition of *eltype* and *orgtype*.

Remarks

If *X* is a matrix (syntax 1), returned is

eltype(*X*)	orgtype(*X*)
real	scalar
complex	rowvector
string	colvector
pointer	matrix
struct	

The returned value reflects the current contents of *X*. That is, *X* might be declared a transmorphic matrix, but at any instant, it contains something, and if that something were 5, returned would be "real" and "scalar".

For orgtype(), returned is "scalar" if the object is currently 1×1; "rowvector" if it is $1 \times k$, $k \neq 1$; "colvector" if it is $k \times 1$, $k \neq 1$; and "matrix" otherwise (it is $r \times c$, $r \neq 1$, $c \neq 1$).

X can be a function. Returned is

eltype(*(&func()))	orgtype(*(&func()))
transmorphic	matrix
numeric	vector
real	rowvector
complex	colvector
string	scalar
pointer	void
struct	
structdef	

These types are obtained from the declaration of the function.

Aside: `struct` and `structdef` have to do with structures; see [M-2] **struct**. `structdef` indicates that the function not only returns a structure but is the routine that defines the structure as well.

Conformability

`eltype(`*X*`)`, `orgtype(`*X*`)`:
$$
\begin{array}{ll}
X: & r \times c \\
result: & 1 \times 1
\end{array}
$$

Diagnostics

None.

Also see

[M-5] **isreal( )** — Storage type of matrix

[M-5] **isview( )** — Whether matrix is view

[M-4] **utility** — Matrix utility functions

Title

[M-5] **epsilon( )** — Unit roundoff error (machine precision)

Syntax

> *real scalar*   epsilon(*real scalar x*)

Description

epsilon(*x*) returns the unit roundoff error in quantities of size abs(*x*).

Remarks

On all computers on which Stata and Mata are currently implemented—which are computers following IEEE standards—epsilon(1) is 1.0X–34, or about 2.22045e–16. This is the smallest amount by which a real number can differ from 1.

epsilon(*x*) is abs(*x*)*epsilon(1). This is an approximation of the smallest amount by which a real number can differ from *x*. The approximation is exact at integer powers of 2.

Conformability

epsilon(*x*):

x:	1 × 1
result:	1 × 1

Diagnostics

epsilon(*x*) returns . if *x* is missing.

Also see

[M-5] **mindouble( )** — Minimum and maximum nonmissing value

[M-5] **edittozero( )** — Edit matrix for roundoff error (zeros)

[M-4] **utility** — Matrix utility functions

Title

Syntax

void	_equilrc(*numeric matrix A* , *r* , *c*)
void	_equilr(*numeric matrix A* , *r*)
void	_equilc(*numeric matrix A* , *c*)
real scalar	_perhapsequilrc(*numeric matrix A* , *r* , *c*)
real scalar	_perhapsequilr(*numeric matrix A* , *r*)
real scalar	_perhapsequilc(*numeric matrix A* , *c*)
real colvector	rowscalefactors(*numeric matrix A*)
real rowvector	colscalefactors(*numeric matrix A*)

The types of *r* and *c* are irrelevant because they are overwritten.

Description

_equilrc(*A* , *r* , *c*) performs row and column equilibration (balancing) on matrix *A*, returning the equilibrated matrix in *A*, the row-scaling factors in *r*, and the column-scaling factors in *c*.

_equilr(*A* , *r*) performs row equilibration on matrix *A*, returning the row-equilibrated matrix in *A* and the row-scaling factors in *r*.

_equilc(*A* , *c*) performs column equilibration on matrix *A*, returning the column-equilibrated matrix in *A* and the column-scaling factors in *c*.

_perhapsequilrc(*A* , *r* , *c*) performs row and/or column equilibration on matrix *A*—as is necessary and which decision is made by _perhapsequilrc()—returning the equilibrated matrix in *A*, the row-scaling factors in *r*, the column-scaling factors in *c*, and returning 0 (no equilibration performed), 1 (row equilibration performed), 2 (column equilibration performed), or 3 (row and column equilibration performed).

_perhapsequilr(*A* , *r*) performs row equilibration on matrix *A*—if necessary and which decision is made by _perhapsequilr()—returning the equilibrated matrix in *A*, the row-scaling factors in *r*, and returning 0 (no equilibration performed) or 1 (row equilibration performed).

_perhapsequilc(*A* , *c*) performs column equilibration on matrix *A*—if necessary and which decision is made by _perhapsequilc()—returning the equilibrated matrix in *A*, the column-scaling factors in *c*, and returning 0 (no equilibration performed) or 1 (column equilibration performed).

rowscalefactors(*A*) returns the row-scaling factors of *A*.

colscalefactors(*A*) returns the column-scaling factors of *A*.

Remarks

Remarks are presented under the following headings:

Introduction
Is equilibration necessary?
The _equil( ) family of functions*
The _perhapsequil( ) family of functions*
rowscalefactors( ) and colscalefactors( )

Introduction

Equilibration (also known as balancing) takes a matrix with poorly scaled rows and columns, such as

	1	2
1	1.00000e+10	5.00000e+10
2	2.00000e-10	8.00000e-10

and produces a related matrix, such as

	1	2
1	.2	1
2	.25	1

that will yield improved accuracy in, for instance, the solution to linear systems. The improved matrix above has been row equilibrated. All we did was find the maximum of each row of the original and then divide the row by its maximum. If we were to take the result and repeat the process on the columns—divide each column by the column's maximum—we would obtain

	1	2
1	.8	1
2	1	1

which is the row-and-column equilibrated form of the original matrix.

In terms of matrix notation, equilibration can be thought about in terms of multiplication by diagonal matrices. The row-equilibrated form of A is RA, where R contains the reciprocals of the row maximums on its diagonal. The column-equilibrated form of A is AC, where C contains the reciprocals of the column maximums on its diagonal. The row-and-column equilibrated form of A is RAC, where R contains the reciprocals of the row maximums of A on its diagonal, and C contains the reciprocals of the column maximums of RA on its diagonal.

Say we wished to find the solution x to

$$Ax = b$$

We could compute the solution by solving for y in the equilibrated system

$$(RAC)y = Rb$$

and then setting

$$x = Cy$$

Thus routines that perform equilibration need to return to you, in some fashion, R and C. The routines here do that by returning r and c, the reciprocals of the maximums in vector form. You could obtain R and C from them by coding

$$R = \text{diag}(r)$$

$$C = \text{diag}(c)$$

but that is not in general necessary, and it is wasteful of memory. In code, you will need to multiply by R and C, and you can do that using the :* operator with r and c:

$$RA \leftrightarrow r\!:\!*A$$

$$AC \leftrightarrow A\!:\!*c$$

$$RAC \leftrightarrow r\!:\!*A\!:\!*c$$

Is equilibration necessary?

Equilibration is not a panacea. Equilibration can reduce the condition number of some matrices and thereby improve the accuracy of the solution to linear systems, but equilibration is not guaranteed to reduce the condition number, and counterexamples exist in which equilibration actually decreases the accuracy of the solution. That said, you have to look long and hard to find such examples.

Equilibration is not especially computationally expensive, but neither is it cheap, especially when you consider the extra computational costs of using the equilibrated matrices. In statistical contexts, equilibration may buy you little because matrices are already nearly equilibrated. Data analysts know variables should be on roughly the same scale, and observations are assumed to be draws from an underlying distribution. The computational cost of equilibration is probably better spent somewhere else. For instance, consider obtaining regression estimates from $X'X$ and $X'y$. The gain from equilibrating $X'X$ and $X'y$, or even from equilibrating the original X matrix, is nowhere near that from the gain to be had in removal of the means before $X'X$ and $X'y$ are formed.

In the example in the previous section, we showed you a matrix that assuredly benefited from equilibration. Even so, after row equilibration, column equilibration was unnecessary. It is often the case that solely row or column equilibration is sufficient, and in those cases, although the extra equilibration will do no numerical harm, it will burn computer cycles. And, as we have already argued, some matrices do not need equilibration at all.

Thus programmers who want to use equilibration and obtain the best speed possible examine the matrix and on that basis perform (1) no equilibration, (2) row equilibration, (3) column equilibration, or (4) both. They then write four branches in their subsequent code to handle each case efficiently.

In terms of determining whether equilibration is necessary, measures of the row and column condition can be obtained from $\min(r)/\max(r)$ and $\min(c)/\max(c)$, where r and c are the scaling factors (reciprocals of the row and column maximums). If those measures are near 1 (LAPACK uses $\geq .1$), then equilibration can be skipped.

There is also the issue of the overall scale of the matrix. In theory, the overall scale should not matter, but many packages set tolerances in absolute rather than relative terms, and so overall scale does matter. In most of Mata's other functions, relative scales are used, but provisions are made so that you can specify tolerances in relative or absolute terms. In any case, LAPACK uses the rule that equilibration is necessary if the matrix is too small (its maximum value is less than epsilon(100), approximately 2.22045e–14) or too large (greater than 1/epsilon(100), approximately 4.504e+13).

To summarize,

1. In statistical applications, we believe that equilibration burns too much computer time and adds too much code complexity for the extra accuracy it delivers. This is a judgment call, and we would probably recommend the use of equilibration were it computationally free.

2. If you are going to use equilibration, there is nothing numerically wrong with simply equilibrating matrices in all cases, including those in which equilibration is not necessary. The advantages of this is that you will gain the precision to be had from equilibration while still keeping your code reasonably simple.

3. If you wish to minimize execution time, then you want to perform the minimum amount of equilibration possible and write code to deal efficiently with each case: (1) no equilibration, (2) row equilibration, (3) column equilibration, and (4) row and column equilibration. The defaults used by LAPACK and incorporated in Mata's _perhapsequil*() routines are

   a. Perform row equilibration if $\min(r)/\max(r) < .1$, or if $\min(abs(A)) < ep-silon(100)$, or if $\min(abs(A)) > 1/epsilon(100)$.

   b. After performing row equilibration, perform column equilibration if $\min(c)/\max(c) < .1$, where c is calculated on $r*A$, the row-equilibrated A, if row equilibration was performed.

The _equil*( ) family of functions

The _equil*() family of functions performs equilibration as follows:

_equilrc(A, r, c) performs row equilibration followed by column equilibration; it returns in r and in c the row- and column-scaling factors, and it modifies A to be the fully equilibrated matrix $r:*A:*c$.

_equilr(A, r) performs row equilibration only; it returns in r the row-scaling factors, and it modifies A to be the row-equilibrated matrix $r:*A$.

_equilc(A, c) performs column equilibration only; it returns in c the row-scaling factors, and it modifies A to be the row-equilibrated matrix $A:*c$.

Here is code to solve $Ax = b$ using the fully equilibrated form, which damages A in the process:

```
_equilrc(A, r, c)
x = c:*lusolve(A, r:*b)
```

The _perhapsequil*( ) family of functions

The _perhapsequil*() family of functions mirrors _equil*(), except that these functions apply the rules mentioned above for whether equilibration is necessary.

Here is code to solve $Ax = b$, which may damage A in the process:

```
result = _perhapsequilrc(A, r, c)
if (result==0)         x = lusolve(A, b)
else if (result==1)    x = lusolve(A, r:*b)
else if (result==2)    x = c:*lusolve(A, b)
else if (result==3)    x = c:*lusolve(A, r:*b)
```

As a matter of fact, the _perhapsequil*() family returns a vector of 1s when equilibration is not performed, so you could code

```
(void) _perhapsequilrc(A, r, c)
x = c:*lusolve(A, r:*b)
```

but that would be inefficient.

rowscalefactors( ) and colscalefactors( )

rowscalefactors(A) and colscalefactors(A) return the scale factors (reciprocals of row and column maximums) to perform row and column equilibration. These functions are used by the other functions above and are provided for those who wish to write their own equilibration routines.

Conformability

_equilrc(A, r, c):
  input:
              A: $m \times n$
  output:
              A: $m \times n$
              r: $m \times 1$
              c: $1 \times n$

_equilr(A, r):
  input:
              A: $m \times n$
  output:
              A: $m \times n$
              r: $m \times 1$

_equilc(A, c):
  input:
              A: $m \times n$
  output:
              A: $m \times n$
              c: $1 \times n$

_perhapsequilrc(A, r, c):
  input:
              A: $m \times n$
  output:
              A: $m \times n$ (unmodified if *result* = 0)
              r: $m \times 1$
              c: $1 \times n$
         result: 1×1

_perhapsequilr(A, r):

   input:

         A: $m \times n$

   output:

         A: $m \times n$ (unmodified if *result* = 0)

         r: $m \times 1$

     result: 1×1

_perhapsequilc(A, c):

   input:

         A: $m \times n$

   output:

         A: $m \times n$ (unmodified if *result* = 0)

         c: $1 \times n$

     result: 1×1

rowscalefactors(A):

         A: $m \times n$

     result: $m \times 1$

colscalefactors(A):

         A: $m \times n$

     result: $1 \times n$

Diagnostics

Scale factors used and returned by all functions are calculated by rowscalefactors(A) and colscalefactors(A). The functions are defined as 1:/rowmaxabs(A) and 1:/colmaxabs(A), with missing values changed to 1. Thus rows or columns that contain missing or are entirely zero are defined to have scale factors of 1.

Equilibration functions do not equilibrate rows or columns that contain missing or all zeros.

The _equil*() functions convert A to an array if A was a view. The Stata dataset is not changed.

The _perhapsequil*() functions convert A to an array if A was a view and returned is nonzero. The Stata dataset is not changed.

Also see

[M-4] **matrix** — Matrix functions

Title

[M-5] error( ) — Issue error message

Syntax

> *real scalar*   error(*real scalar rc*)
>
> *void*   _error(*real scalar errnum*)
>
> *void*   _error(*string scalar errtxt*)
>
> *void*   _error(*real scalar errnum*, *string scalar errtxt*)

Description

error(*rc*) displays the standard Stata error message associated with return code *rc* and returns *rc*; see [P] **error** for a listing of return codes. error( ) does not abort execution; standard usage is exit(error(*rc*)).

_error( ) aborts execution and produces a traceback log.

_error(*errnum*) produces a traceback log with standard Mata error message *errnum*; see [M-2] **errors** for a listing of the standard Mata error codes.

_error(*errtxt*) produces a traceback log with error number 3498 and custom text *errtext*.

_error(*errnum*, *errtxt*) produces a traceback log with error number *errnum* and custom text *errtext*.

If *errtxt* is specified, it should contain straight text; SMCL codes are not interpreted.

Remarks

Remarks are presented under the following headings:

> *Use of _error( )*
> *Use of error( )*

Use of _error( )

_error( ) aborts execution and produces a traceback log:

```
        : myfunction(A,B)
                    mysub():   3200  conformability error
                myfunction():    -   function returned error
                    <istmt>:     -   function returned error
        r(3200);
```

The above output was created because function mysub( ) contained the line

```
        _error(3200)
```

445

and 3200 is the error number associated with the standard message "conformability error"; see [M-2] **errors**. Possibly, the code read

```
if (rows(A)!=rows(B) | cols(A)!=cols(B)) {
        _error(3200)
}
```

Another kind of mistake might produce

```
: myfunction(A,B)
            mysub():  3498  zeros on diagonal not allowed
        myfunction():    -  function returned error
            <istmt>:    -  function returned error
    r(3498);
```

and that could be produced by the code

```
if (diag0cnt(A)>0) {
        _error("zeros on diagonal not allowed")
}
```

If we wanted to produce the same text but change the error number to 3300, we could have coded

```
if (diag0cnt(A)>0) {
        _error(3300, "zeros on diagonal not allowed")
}
```

Coding _error() is not always necessary. In our conformability-error example, imagine that more of the code read

```
...
if (rows(A)!=rows(B) | cols(A)!=cols(B)) {
        _error(3200)
}
C = A + B
...
```

If we simplified the code to read

```
...
C = A + B
...
```

the conformability error would still be detected because + requires p-conformability:

```
: myfunction(A,B)
                +:  3200  conformability error
            mysub():    -  conformability error
        myfunction():    -  function returned error
            <istmt>:    -  function returned error
    r(3200);
```

Sometimes, however, you must detect the error yourself. For instance,

```
...
if (rows(A)!=rows(B) | cols(A)!=cols(B) | rows(A)!=2*cols(A)) {
        _error(3200)
}
C = A + B
...
```

We assume we have some good reason to require that A has twice as many rows as columns. +, however, will not require that, and perhaps no other calculation we will make will require that, either. Or perhaps it will be subsequently detected, but in a way that leads to a confusing error message for the caller.

Use of error( )

error(*rc*) does not cause the program to terminate. Standard usage is

    exit(error(*rc*))

such as

    exit(error(503))

In any case, error( ) does not produce a traceback log:

```
: myfunction(A,B)
conformability error
r(503);
```

error( ) is intended to be used in functions that are subroutines of ado-files:

—————————————————————————————— begin example.ado ———————

```
program example
        version 11
        ...
        mata: myfunction("`mat1'", "`mat2'")
        ...
end
version 11
mata:
void myfunction(string scalar matname1, string scalar matname2)
{
        ...
        A = st_matrix(matname1)
        B = st_matrix(matname2)
        if (rows(A)!=rows(B) | cols(A)!=cols(B)) {
                exit(error(503))
        }
        C = A + B
        ...
}
end
```

——————————————————————————————— end example.ado ———————

This way, when the user uses our example command incorrectly, he or she will see

```
. example ...
conformability error
r(503);
```

rather than the traceback log that would have been produced had we omitted the test and
exit(error(503)):

```
                    . example ...
                                    +:   3200   conformability error
                        myfunction():    -    function returned error
                              <istmt>:    -    function returned error
                    r(3200);
```

Conformability

error(*rc*):

rc:	1×1
result:	1×1

_error(*errnum*):

errnum:	1×1
result:	*void*

_error(*errtxt*):

errtxt:	1×1
result:	*void*

_error(*errnum*, *errtxt*):

errnum:	1×1
errtxt:	1×1
result:	*void*

Diagnostics

error(*rc*) does not abort execution; code exit(error(*rc*)) if that is your desire; see [M-5] **exit( )**.

The code error(*rc*) returns can differ from *rc* if *rc* is not a standard code or if there is a better
code associated with it.

error(*rc*) with *rc* = 0 produces no output and returns 0.

_error(*errnum*), _error(*errtxt*), and _error(*errnum*, *errtxt*) always abort with error. _error( )
will abort with error because you called it wrong if you specify an *errnum* less than 1 or greater than
2,147,483,647 or if you specify an *errtxt* longer than 100 characters. If you specify an *errnum* that
is not a standard code, the text of the error messages will read "Stata returned error".

Also see

[M-2] **errors** — Error codes

[M-5] **exit( )** — Terminate execution

[M-4] **programming** — Programming functions

Title

> **[M-5] errprintf( )** — Format output and display as error message

Syntax

$$\textit{void} \quad \texttt{errprintf}(\textit{string scalar fmt}, r_1, r_2, \ldots, r_N)$$

Description

`errprintf()` is a convenience tool for displaying an error message.

`errprintf(...)` is equivalent to `printf(...)` except that it executes `displayas("error")` before the `printf()` is executed; see [M-5] **printf( )** and [M-5] **displayas( )**.

Remarks

You have written a program. At one point in the code, you have variable `fn` that should contain the name of an existing file:

```
if (!fileexists(fn)) {
        // you wish to display the error message
        // file ____ not found
        exit(601)
}
```

One solution is

```
if (!fileexists(fn)) {
        displayas("error")
        printf("file %s not found\n", fn)
        exit(601)
}
```

Equivalent is

```
if (!fileexists(fn)) {
        errprintf("file %s not found\n", fn)
        exit(601)
}
```

It is important that you either `displayas("error")` before using `printf()` or that you use `errprintf()`, to ensure that your error message is displayed (is not suppressed by a `quietly`) and that it is displayed in red; see [M-5] **displayas( )**.

Conformability

errprintf(*fmt*, r_1, r_2, ..., r_N)

fmt:	1×1
r_1:	1×1
r_2:	1×1
...	
r_N:	1×1
result:	▪ *void*

Diagnostics

errprintf( ) aborts with error if a %*fmt* is misspecified, if a numeric %*fmt* corresponds to a string result or a string %*fmt* corresponds to a numeric result, or there are too few or too many %*fmts* in *fmt* relative to the number of *results* specified.

Also see

[M-5] **printf( )** — Format output

[M-5] **displayas( )** — Set display level

[M-4] **io** — I/O functions

Title

Syntax

exit(*real scalar rc*)

exit()

Description

exit(*rc*) terminates execution and sets the overall return code to *rc*.

exit() with no argument specified is equivalent to exit(0).

Remarks

Do not confuse exit() and return. return stops execution of the current function and returns to the caller, whereupon execution continues. exit() terminates execution. For instance, consider

```
function first()
{
        "begin execution"
        second()
        "this message will never be seen"
}
function second()
{
        "hello from second()"
        exit(0)
}
```

The result of running this would be

```
: first
  begin execution
  hello from second()
```

If we changed the exit(0) to be exit(198) in second(), the result would be

```
: first
  begin execution
  hello from second()
r(198);
```

No error message is presented. If you want to present an error message and exit, you should code exit(error(198)); see [M-5] **error( )**.

Conformability

exit(*rc*):
        rc: 1×1 (optional)

Diagnostics

exit(*rc*) and exit() do not return.

Also see

[M-5] **error( )** — Issue error message

[M-4] **programming** — Programming functions

Title

Syntax

> *numeric matrix*   exp(*numeric matrix Z*)
>
> *numeric matrix*   ln(*numeric matrix Z*)
>
> *numeric matrix*   log(*numeric matrix Z*)
>
> *numeric matrix*   log10(*numeric matrix Z*)

Description

exp(Z) returns the elementwise exponentiation of Z. exp( ) returns real if Z is real and complex if Z is complex.

ln(Z) and log(Z) return the elementwise natural logarithm of Z. The functions are synonyms. ln( ) and log( ) return real if Z is real and complex if Z is complex.

ln(x), x real, returns the natural logarithm of x or returns missing (.) if $x \leq 0$.

ln(z), z complex, returns the complex natural logarithm of z. Im(ln( )) is chosen to be in the interval $[-pi, pi]$.

log10(Z) returns the elementwise log base 10 of Z. log10( ) returns real if Z is real and complex if Z is complex. log10(Z) is defined mathematically and operationally as ln(Z)/ln(10).

Conformability

exp(Z), ln(Z), log(Z), log10(Z):
        Z: $r \times c$
    result: $r \times c$

Diagnostics

exp(Z) returns missing when Re(Z) > 709.

ln(Z), log(Z), and log10(Z) return missing when Z is real and $Z \leq 0$. In addition, the functions return missing (.) for real arguments when the result would be complex. For instance, ln(-1) = ., whereas ln(-1+0i) = 3.14159265i.

Also see

[M-4] **scalar** — Scalar mathematical functions

Title

Syntax

real matrix	factorial(*real matrix R*)
real matrix	lnfactorial(*real matrix R*)
numeric matrix	lngamma(*numeric matrix Z*)
numeric matrix	gamma(*numeric matrix Z*)
real matrix	digamma(*real matrix R*)
real matrix	trigamma(*real matrix R*)

Description

factorial(*R*) returns the elementwise factorial of *R*.

lnfactorial(*R*) returns the elementwise ln(factorial(*R*)), calculated differently. Very large values of *R* may be evaluated.

lngamma(*Z*), for *Z* real, returns the elementwise real result ln(abs(gamma(*Z*))), but calculated differently. lngamma(*Z*), for *Z* complex, returns the elementwise ln(gamma(*Z*)), calculated differently. Thus, lngamma(-2.5) = −.056244, whereas lngamma(-2.5+0i) = −.056244 + 3.1416i. In both cases, very large values of *Z* may be evaluated.

gamma(*Z*) returns exp(lngamma(*Z*)) for complex arguments and Re(exp(lngamma(C(*Z*)))) for real arguments. Thus gamma() can correctly calculate, say, gamma(-2.5) even for real arguments.

digamma(*R*) returns the derivative of lngamma() for *R* > 0, sometimes called the psi function. digamma() requires a real argument.

trigamma(*R*) returns the second derivative of lngamma() for *R* > 0. trigamma() requires a real argument.

Conformability

All functions return a matrix of the same dimension as input, containing element-by-element calculated results.

Diagnostics

factorial() returns missing for noninteger arguments, negative arguments, and arguments > 167.

lnfactorial() returns missing for noninteger arguments, negative arguments, and arguments > 2,147,483,647.

lngamma() returns missing for 0, negative integer arguments, negative arguments $< -2{,}147{,}483{,}647$, and arguments $> 1\mathrm{e}{+}307$.

gamma() returns missing for real arguments > 171, for negative integer arguments, and for arguments $< -2{,}147{,}483{,}647$.

digamma() returns missing for 0 and negative integer arguments, and for arguments $< -10{,}000{,}000$.

trigamma() returns missing for 0 and negative integer arguments, and for arguments $< -10{,}000{,}000$.

Also see

[M-4] **scalar** — Scalar mathematical functions

[M-4] **statistical** — Statistical functions

Title

[M-5] **favorspeed( )** — Whether speed or space is to be favored

Syntax

> *real scalar* favorspeed()

Description

favorspeed() returns 1 if the user has mata set matafavor speed and 0 if the user has mata set matafavor space or has not set matafavor at all; see [M-3] **mata set**.

Remarks

Sometimes in programming you can choose between writing code that runs faster but consumes more memory or writing code that conserves memory at the cost of execution speed. favorspeed() tells you the user's preference:

```
if (favorspeed()) {
        /* code structured for speed over memory */
}
else {
        /* code structured for memory over speed */
}
```

Conformability

```
favorspeed():
        result:     1 × 1
```

Diagnostics

None.

Also see

[M-3] **mata set** — Set and display Mata system parameters

[M-4] **programming** — Programming functions

Title

> **[M-5] ferrortext( )** — Text and return code of file error code

Syntax

> *string scalar* ferrortext(*real scalar ec*)
>
> *real scalar*   freturncode(*real scalar ec*)

Description

ferrortext(*ec*) returns the text associated with a file error code returned by, for instance, _fopen(),
_fwrite(), fstatus(), or any other file-processing functions that return an error code. See [M-5]
fopen( ).

freturncode(*ec*) returns the Stata return code associated with the file error code.

Remarks

Most file-processing functions abort with error if something goes wrong. You attempt to read a
nonexisting file, or attempt to write a read-only file, and the file functions you use to do that, usually
documented in [M-5] **fopen( )**, abort with error. Abort with error means not only that the file function
you called stops when the error arises but also that the calling function is aborted. The user is
presented with a traceback log documenting what went wrong.

Sometimes you will want to write code to handle such errors for itself. For instance, you are writing
a subroutine for a command to be used in Stata and, if the file does not exist, you want the subroutine
to present an informative error message and exit without a traceback log but with a nonzero return
code. Or in another application, if the file does not exist, that is not an error at all; your code will
create the file.

Most file-processing functions have a corresponding underscore function that, rather than aborting,
returns an error code when things go wrong. fopen() opens a file or aborts with error. _fopen()
opens a file or returns an error code. The error code is sufficient for your program to take the
appropriate action. One uses the underscore functions when the calling program will deal with any
errors that might arise.

Let's take the example of simply avoiding traceback logs. If you code

```
fh = fopen(filename, "r")
```

and the file does not exist, execution aborts and you see a traceback log. If you code

```
if ((fh = _fopen(filename, "r"))<0) {
        errprintf("%s\n", ferrortext(fh))
        exit(freturncode(fh))
}
```

execution still stops if the file does not exist, but this time, it stops because you coded exit(). You
still see an error message, but this time, you see the message because you coded errprintf(). No
traceback log is produced because you did not insert code to produce one. You could have coded
_exit() if you wanted one.

The file error codes and the messages associated with them are

Negative code	Meaning
0	all is well
−1	end of file (this code is usually not an error)
−601	file not found
−602	file already exists
−603	file could not be opened
−608	file is read-only
−610	file not Stata format
−612	unexpected end of file
−630	web files not supported in this version of Stata
−631	host not found
−632	web file not allowed in this context
−633	may not write to web file
−639	file transmission error—checksums do not match
−660	proxy host not found
−662	proxy server refused request to send
−663	remote connection to proxy failed
−665	could not set socket nonblocking
−667	wrong version of winsock.dll
−668	could not find valid winsock.dll or astsys0.lib
−669	invalid URL
−670	invalid network port number
−671	unknown network protocol
−672	server refused to send file
−673	authorization required by server
−674	unexpected response from server
−675	server reported error
−676	server refused request to send
−677	remote connection failed—see r(677) for troubleshooting
−678	could not open local network socket
−679	unexpected web error
−680	could not find valid odbc32.dll
−681	too many open files
−682	could not connect to ODBC data source name
−683	could not fetch variable in ODBC table
−684	could not find valid dlxabi32.dll
−691	I/O error
−699	insufficient disk space

−3601	invalid file handle
−3602	invalid filename
−3611	too many open files
−3621	attempt to write read-only file
−3622	attempt to read write-only file
−3623	attempt to seek append-only file
−3698	file seek error

File error codes are usually negative, but neither `ferrortext`(ec) nor `freturncode`(ec) cares whether ec is of the positive or negative variety.

Conformability

`ferrortext`(ec), `freturncode`(ec):
　　　　ec: 1×1
　　result: 1×1

Diagnostics

`ferrortext`(ec) and `freturncode`(ec) treat $ec = -1$ (end of file) the same as $ec = 612$ (unexpected end of file). Code -1 usually is not an error; it just denotes end of file. It is assumed that you will not call `ferrortext`() and `freturncode`() in such cases. If you do call the functions here, it is assumed that you mean that the end of file was unexpected.

Also see

[M-5] **fopen**( ) — File I/O

[M-4] **io** — I/O functions

Title

Syntax

complex vector	`fft`(*numeric vector h*)
numeric vector	`invfft`(*numeric vector H*)
void	`_fft`(*complex vector h*)
void	`_invfft`(*complex vector H*)
numeric vector	`convolve`(*numeric vector r*, *numeric vector s*)
numeric vector	`deconvolve`(*numeric vector r*, *numeric vector sm*)
numeric vector	`Corr`(*numeric vector g*, *numeric vector h*, *real scalar k*)
real vector	`ftperiodogram`(*numeric vector H*)
numeric vector	`ftpad`(*numeric vector h*)
numeric vector	`ftwrap`(*numeric vector r*, *real scalar n*)
numeric vector	`ftunwrap`(*numeric vector H*)
numeric vector	`ftretime`(*numeric vector r*, *numeric vector s*)
real vector	`ftfreqs`(*numeric vector H*, *real scalar delta*)

Description

H=`fft`(h) and h=`invfft`(H) calculate the Fourier transform and inverse Fourier transform. The length of h (H) must be a power of 2.

`_fft`(h) and `_invfft`(H) do the same thing, but they perform the calculation in place, replacing the contents of h and H.

`convolve`(r, s) returns the convolution of the signal s with the response function r. `deconvolve`(r, sm) deconvolves the smeared signal sm with the response function r and is thus the inverse of `convolve`().

`Corr`(g, h, k) returns a $2k + 1$ element vector containing the correlations of g and h for lags and leads as large as k.

`ftperiodogram`(H) returns a real vector containing the one-sided periodogram of H.

`ftpad`(h) returns h padded with 0s to have a length that is a power of 2.

`ftwrap`(r, n) converts the symmetrically stored response function r into wraparound format of length n, $n \geq$ `rows`(r)*`cols`(r) and `rows`(r)*`cols`(r) odd.

ftunwrap(*H*) unwraps frequency-wraparound order such as returned by fft(). You may find this useful when graphing or listing results, but it is otherwise unnecessary.

ftretime(*r*, *s*) retimes the signal *s* to be on the same time scale as convolve(*r*, *s*). This is useful in graphing data and listing results but is otherwise not required.

ftfreqs(*H*, *delta*) returns a vector containing the frequencies associated with the elements of *H*; *delta* is the sampling interval and is often specified as 1.

Remarks

Remarks are presented under the following headings:

> *Definitions, notation, and conventions*
> *Fourier transform*
> *Convolution and deconvolution*
> *Correlation*
> *Utility routines*
> *Warnings*

Definitions, notation, and conventions

A signal *h* is a row or column vector containing real or complex elements. The length of the signal is defined as the number of elements of the vector. It is occasionally necessary to pad a signal to a given length. This is done by forming a new vector equal to the original and with zeros added to the end.

The Fourier transform of a signal *h*, typically denoted by capital letter *H* of *h*, is stored in frequency-wraparound order. That is, if there are *n* elements in *H*:

$H[1]$	frequency 0
$H[2]$	frequency 1
$H[3]$	frequency 2
$\vdots$	
$H[n/2]$	frequency $n/2-1$
$H[n/2 + 1]$	frequency $n/2$ $(-n/2,$ aliased)
$H[n/2 + 2]$	frequency $-(n/2-1)$
$\vdots$	
$H[n - 1]$	frequency -2
$H[n]$	frequency -1

All routines expect and use this order, but see ftunwrap() below.

A response function *r* is a row or column vector containing $m = 2k + 1$ real or complex elements. *m* is called the duration of the response function. Response functions are generally stored symmetrically, although the response function itself need not be symmetric. The response vector contains

$r[1]$	response at lag $-k$
$r[2]$	response at lag $-k+1$
$\vdots$	
$r[k]$	response at lag -1
$r[k+1]$	contemporaneous response
$r[k+2]$	response at lead 1
$r[k+3]$	response at lead 2
$\vdots$	
$r[2k+1]$	response at lead k

Response functions always have odd lengths. Response vectors are never padded.

You may occasionally find it convenient to store a response vector in "wraparound" order (similar to frequency-wraparound order), although none of the routines here require this. In wraparound order:

$\text{wrap}[1]$	contemporaneous response
$\text{wrap}[2]$	response at lead 1
$\text{wrap}[3]$	response at lead 2
$\vdots$	
$\text{wrap}[k+1]$	response at lead k
$\text{wrap}[k+2]$	response at lag $-k$
$\text{wrap}[k+3]$	response at lag $-k+1$
$\vdots$	
$\text{wrap}[2k+1]$	response at lag -1

Response vectors stored in wraparound order may be internally padded (as opposed to merely padded) to a given length by the insertion of zeros between $\text{wrap}[k+1]$ and $\text{wrap}[k+2]$.

Fourier transform

$\texttt{fft}(h)$ returns the discrete Fourier transform of h. h may be either real or complex, but its length must be a power of 2, so one typically codes $\texttt{fft(ftpad}(h))$; see $\texttt{ftpad()}$, below. The returned result is p-conformable with h. The calculation is performed by $\texttt{\_fft()}$.

$\texttt{invfft}(H)$ returns the discrete inverse Fourier transform of H. H may be either real or complex, but its length must be a power of 2. The returned result is p-conformable with H. The calculation is performed by $\texttt{\_invfft()}$.

$\texttt{invfft}(H)$ may return a real or complex. This should be viewed as a feature, but if you wish to ensure the complex interpretation, code $\texttt{C(invfft}(H))$.

$\texttt{\_fft}(h)$ is the built-in procedure that performs the fast Fourier transform in place. h must be complex, and its length must be a power of 2.

$\texttt{\_invfft}(H)$ is the built-in procedure that performs the inverse fast Fourier transform in place. H must be complex, and its length must be a power of 2.

Convolution and deconvolution

convolve(*r*, *s*) returns the convolution of the signal *s* with the response function *r*. Calculation is performed by taking the fft() of the elements, multiplying, and using invfft() to transform the results back. Nevertheless, it is not necessary that the length of *s* be a power of 2. convolve() handles all paddings necessary, including paddings to *s* required to prevent the result from being contaminated by erroneous wrapping around of *s*. Although one thinks of the convolution operator as being commutative, convolve() is not commutative since required zero-padding of the response and signal differ.

If *n* is the length of the signal and $2k + 1$ is the length of the response function, the returned result has length $n + 2k$. The first *k* elements are the convoluted signal before the true signal begins, and the last *k* elements are the convoluted signal after the true signal ends. See ftretime(), below. In any case, you may be interested only in the elements convolve()[|*k*+1*n*−*k*|], the part contemporaneous with *s*.

The returned vector is a row vector if *s* is a row vector and a column vector otherwise. The result is guaranteed to be real if both *r* and *s* are real; the result may be complex or real, otherwise.

It is not required that the response function be shorter than the signal, although this will typically be the case.

deconvolve(*r*, *sm*) deconvolves the smeared signal *sm* with the response function *r* and is thus the inverse of convolve(). In particular,

deconvolve(*r*, convolve(*r*,*s*)) = *s*   (up to roundoff error)

Everything said about convolve() applies equally to deconvolve().

Correlation

Here we refer to correlation in the signal-processing sense, not the statistical sense.

Corr(*g*, *h*, *k*) returns a $2k + 1$ element vector containing the correlations of *g* and *h* for lags and leads as large as *k*. For instance, Corr(*g*, *h*, 2) returns a five-element vector, the first element of which contains the correlation for lag 2, the second element lag 1, the third (middle) element the contemporaneous correlation, the fourth element lead 1, and the fifth element lead 2. *k* must be greater than or equal to 1. The returned vector is a row or column vector depending on whether *g* is a row or column vector. *g* and *h* must have the same number of elements but need not be p-conformable.

The result is obtained by padding with zeros to avoid contamination, taking the Fourier transform, multiplying $G \times \text{conj}(H)$, and rearranging the inverse transformed result. Nevertheless, it is not required that the number of elements of *g* and *h* be powers of 2 because the program pads internally.

Utility routines

ftpad(*h*) returns *h* padded with 0s to have a length that is a power of 2. For instance,

```
: h = (1,2,3,4,5)
: ftpad(h)
```

	1	2	3	4	5	6	7	8
1	1	2	3	4	5	0	0	0

If *h* is a row vector, a row vector is returned. If *h* is a column vector, a column vector is returned.

ftwrap(*r*, *n*) converts the symmetrically stored response function *r* into wraparound format of length *n*, $n \geq$ rows(*r*)*cols(*r*) and rows(*r*)*cols(*r*) odd. A symmetrically stored response function is a vector of odd length, for example:

(.1, .5, 1, .2, .4)

The middle element of the vector represents the response function at lag 0. Elements before the middle represent lags while elements after the middle represent leads. Here .1 is the response for lag 2 and .5 for lag 1, 1 the contemporaneous response, .2 the response for lead 1, and .4 the response for lead 2. The wraparound format of a response function records the response at times 0, 1, and so on in the first positions, has some number of zeros, and then records the most negative time value of the response function, and so on.

For instance,

```
: r
            1    2    3    4    5

       1   .1   .5    1   .2   .4

: ftwrap(r, 5)
            1    2    3    4    5

       1    1   .2   .4   .1   .5

: ftwrap(r, 6)
            1    2    3    4    5    6

       1    1   .2   .4    0   .1   .5

: ftwrap(r, 8)
            1    2    3    4    5    6    7    8

       1    1   .2   .4    0    0    0   .1   .5
```

ftunwrap(*H*) unwraps frequency-wraparound order such as returned by fft(). You may find this useful when graphing or listing results, but it is otherwise unnecessary. Frequency-unwrapped order is defined as

unwrap[1]	frequency $-(n/2) + 1$
unwrap[2]	frequency $-(n/2) + 2$
$\vdots$	
unwrap[$n/2 - 1$]	frequency -1
unwrap[$n/2$]	frequency 0
unwrap[$n/2 + 1$]	frequency 1
$\vdots$	
unwrap[$n - 1$]	frequency $n/2 - 1$
unwrap[n]	frequency $n/2$

Here we assume that *n* is even, as will usually be true. The aliased (highest) frequency is assigned the positive sign.

Also see ftperiodogram(), below.

ftretime(r, s) retimes the signal s to be on the same time scale as convolve(r, s). This is useful in graphing and listing results but is otherwise not required. ftretime() uses only the length of r, and not its contents, to perform the retiming. If the response vector is of length $2k + 1$, a vector containing k zeros, s, and k more zeros is returned. Thus the result of ftretime(r, s) is p-conformable with convolve(r, s).

ftfreqs(H, *delta*) returns a p-conformable-with-H vector containing the frequencies associated with the elements of H. *delta* is the sampling interval and is often specified as 1.

ftperiodogram(H) returns a real vector of length $n/2$ containing the one-sided periodogram of H (length n), calculated as

$$|H(f)|^2 + |H(-f)|^2$$

excluding frequency 0. Thus ftperiodogram(H)[1] corresponds to frequency 1 (-1), ftperiodogram(H)[2] to frequency 2 (-2), and so on.

Warnings

invfft(H) will cast the result down to real if possible. Code C(invfft(H)) if you want to be assured of the result being stored as complex.

convolve(r, s) is not the same as convolve(s, r).

convolve(r, s) will cast the result down to real if possible. Code C(convolve(r, s)) if you want to be assured of the result being stored as complex.

For convolve(r, s), the response function r must have odd length.

Conformability

fft(h):

h:	$1 \times n$	or	$n \times 1$,	n a power of 2
result:	$1 \times n$	or	$n \times 1$	

invfft(H):

H:	$1 \times n$	or	$n \times 1$,	n a power of 2
result:	$1 \times n$	or	$n \times 1$	

_fft(h):

h:	$1 \times n$	or	$n \times 1$,	n a power of 2
result:	*void*			

_invfft(H):

H:	$1 \times n$	or	$n \times 1$,	n a power of 2
result:	*void*			

convolve(r, s):

r:	$1 \times n$	or	$n \times 1$,	$n > 0$, n odd
s:	$1 \times 2k + 1$	or	$2k + 1 \times 1$,	i.e., s of odd length
result:	$1 \times 2k + n$	or	$2k + n \times 1$	

deconvolve(r, sm):

r:	$1 \times n$	or	$n \times 1$,	$n > 0$, n odd
sm:	$1 \times 2k + n$	or	$2k + n \times 1$	
result:	$1 \times 2k + 1$	or	$2k + 1 \times 1$	

Corr(g, h, k):

g:	$1 \times n$	or	$n \times 1$,	$n > 0$
h:	$1 \times n$	or	$n \times 1$	
k:	1×1	or	1×1,	$k > 0$
result:	$1 \times 2k + 1$	or	$2k + 1 \times 1$	

ftperiodogram(H):

H:	$1 \times n$	or	$n \times 1$,	n even
result:	$n/2 \times 1$	or	$1 \times n/2$	

ftpad(h):

h:	$1 \times n$	or	$n \times 1$	
result:	$1 \times N$	or	$N \times 1$,	$N = n$ rounded up to power of 2

ftwrap(r, n):

r:	$1 \times m$	or	$m \times 1$,	$m > 0$, m odd
n:	1×1	or	1×1,	$n \geq m$
result:	$1 \times n$	or	$n \times 1$	

ftunwrap(H):

H:	$1 \times n$	or	$n \times 1$	
result:	$1 \times n$	or	$n \times 1$	

ftretime(r, s):

r:	$1 \times n$	or	$n \times 1$,	$n > 0$, n odd
s:	$1 \times 2k + 1$	or	$2k + 1 \times 1$,	i.e., s of odd length
result:	$1 \times 2k + n$	or	$2k + n \times 1$	

ftfreqs(H, $delta$):

H:	$1 \times n$	or	$n \times 1$,	n even
delta:	1×1			
result:	$1 \times n$	or	$n \times 1$	

Diagnostics

All functions abort with error if the conformability requirements are not met. This is always true, of course, but pay particular attention to the requirements outlined under *Conformability* directly above.

fft(h), _fft(h), invfft(H), _invfft(H), convolve(r, s), deconvolve(r, sm), and Corr(g, h, k) return missing results if any argument contains missing values.

ftwrap(r, n) aborts with error if n contains missing value.

Also see

[M-4] **mathematical** — Important mathematical functions

Title

[M-5] fileexists( ) — Whether file exists

Syntax

> *real scalar* fileexists(*string scalar fn*)

Description

fileexists(*fn*) returns 1 if file *fn* exists and is readable and returns 0 otherwise.

Conformability

fileexists(*fn*):

fn:	1×1	
result:	1×1	

Diagnostics

None.

Also see

[M-4] **io** — I/O functions

Title

Syntax

> *void*   _fillmissing(*transmorphic matrix A*)

Description

_fillmissing(*transmorphic matrix A*) changes the contents of *A* to missing values.

Remarks

The definition of missing depends on the storage type of *A*:

Storage type	Contents
real	.
complex	C(.)
string	" "
pointer	NULL

Conformability

_fillmissing(*A*):

    input:

          A: *r* × *c*

    output:

          A: *r* × *c*

Diagnostics

None.

Also see

[M-4] **manipulation** — Matrix manipulation

Title

Syntax

    pointer( )   *scalar*   findexternal(*string scalar name*)

    pointer( )   *scalar*   crexternal(*string scalar name*)

    void   rmexternal(*string scalar name*)

    string   *scalar*   nameexternal(*pointer( ) scalar p*)

Description

findexternal(*name*) returns a pointer (see [M-2] **pointers**) to the external global matrix, vector, or scalar whose name is specified by *name*, or to the external global function if the contents of *name* end in (). findexternal() returns NULL if the external global is not found.

crexternal(*name*) creates a new external global 0×0 real matrix with the specified name and returns a pointer to it; it returns NULL if an external global of that name already exists.

rmexternal(*name*) removes (deletes) the specified external global or does nothing if no such external global exists.

nameexternal(*p*) returns the name of *$*p$*.

Remarks

Remarks are presented under the following headings:

       Definition of a global
       Use of globals

Also see *Linking to external globals* in [M-2] **declarations**.

Definition of a global

When you use Mata interactively, any variables you create are known, equivalently, as externals, globals, or external globals.

```
: myvar = x
```

Such variables can be used by subsequent functions that you run, and there are two ways that can happen:

```
function example1(...)
{
        external real myvar

        ... myvar ...
}
```

469

and

```
function example2(...)
{
        pointer(real) p

        p = findexternal("myvar")
        ... *p ...
}
```

Using the first method, you must know the name of the global at the time you write the source code, and when you run your program, if the global does not exist, it will refuse to run (abort with myvar not found). With the second method, the name of the global can be specified at run time and what is to happen when the global is not found is up to you.

In the second example, although we declared p as a pointer to a real, myvar will not be required to contain a real. After p = findexternal("myvar"), if p!=NULL, p will point to whatever myvar contains, whether it be real, complex, string, or another pointer. (You can diagnose the contents of *p using eltype(*p) and orgtype(*p).)

Use of globals

Globals are useful when a function must remember something from one call to the next:

```
function example3(real scalar x)
{
        pointer() scalar p

        if ( (p = findexternal("myprivatevar")) == NULL) {
                printf("you haven't called me previously")
                p = crexternal("myprivatevar")
        }
        else {
                printf("last time, you said "%g", *p)
        }
        *p = x
}
```

```
: example3(2)
you haven't called me previously
: example3(31)
last time, you said 2
: example3(9)
last time, you said 31
```

Note our use of the name myprivatevar. It actually is not a private variable; it is global, and you would see the variable listed if you described the contents of Mata's memory. Because global variables are so exposed, it is best that you give them long and unlikely names.

In general, programs do not need global variables. The exception is when a program must remember something from one invocation to the next, and especially if that something must be remembered from one invocation of Mata to the next.

When you do need globals, you probably will have more than one thing you will need to recall. There are two ways to proceed. One way is simply to create separate global variables for each thing you need to remember. The other way is to create one global pointer vector and store everything in that. In the following example, we remember one scalar and one matrix:

```
function example4()
{
        pointer(pointer() vector) scalar   p
        scalar                             s
        real matrix                        X
        pointer() scalar                   ps, pX

        if ( (p = findexternal("mycollection")) == NULL) {
                ... calculate scalar s and X from nothing ...
                ... and save them:
                p = crexternal("mycollection")
                *p = (&s, &X)
        }
        else {
                ps = (*p)[1]
                pX = (*p)[2]
                ... calculate using *ps and *pX ...
        }
}
```

In the above example, even though `crexternal()` created a 0×0 real global, we morphed it into a 1×2 pointer vector:

```
p = crexternal("mycollection")        *p is 0 × 0 real
*p = (&s, &X)                         *p is 1 × 2 vector
```

just as we could with any nonpointer object.

In the else part of our program, where we use the previous values, we do not use variables s and X, but ps and pX. Actually, we did not really need them, we could just as well have used `*((*p)[1])` and `*((*p)[2])`, but the code is more easily understood by introducing `*ps` and `*pX`.

Actually, we could have used the variables s and X by changing the else part of our program to read

```
else {
        s = *(*p)[1]
        X = *(*p)[2]
        ... calculate using s and X ...
        *p = (&s, &X)                 ← remember to put them back
}
```

Doing that is inefficient because s and X contain copies of the global values. Obviously, the amount of inefficiency depends on the sizes of the elements being copied. For s, there is really no inefficiency at all because s is just a scalar. For X, the amount of inefficiency depends on the dimensions of X. Making a copy of a small X matrix would introduce just a little inefficiency.

The best balance between efficiency and readability is achieved by introducing a subroutine:

```
function example5()
{
        pointer(pointer() vector) scalar   p
        scalar                             s
        real matrix                        X

        if ( (p = findexternal("mycollection")) == NULL) {
                example5_sub(1, s=., X=J(0,0,.))
                p = crexternal("mycollection")
                *p = (&s, &X)
        }
        else {
                example5_sub(0, (*p)[1], (*p)[2])
        }
}

function example5_sub(scalar firstcall, scalar x, matrix X)
{
        ...
}
```

The last two lines in the not-found case

```
p = crexternal("mycollection")
*p = (&s, &X)
```

could also be coded

```
*crexternal("mycollection") = (&s, &X)
```

Conformability

findexternal(*name*), crexternal(*name*):
>   *name*:   1 × 1
>   *result*:   1 × 1

rmexternal(*name*):
>   *name*:   1 × 1
>   *result*:   *void*

nameexternal(*p*):
>   *p*:   1 × 1
>   *result*:   1 × 1

Diagnostics

findexternal(*name*), crexternal(*name*), and rmexternal(*name*) abort with error if *name* contains an invalid name.

findexternal(*name*) returns NULL if *name* does not exist.

crexternal(*name*) returns NULL if *name* already exists.

nameexternal(*p*) returns "" if p = NULL. Also, nameexternal() may be used not just with pointers to globals but pointers to locals as well. For example, you can code nameexternal(&myx), where myx is declared in the same program or a calling program. nameexternal() will usually return the expected local name, such as "myx". In such cases, however, it is also possible that "" will be returned. That can occur because, in the compilation/optimization process, the identity of local variables can be lost.

Also see

[M-5] **valofexternal( )** — Obtain value of external global

[M-4] **programming** — Programming functions

Title

[M-5] **findfile( )** — Find file

Syntax

>   *string scalar*   findfile(*string scalar fn*, *string scalar dirlist*)

>   *string scalar*   findfile(*string scalar fn*)

Description

findfile(*fn*, *dirlist*) looks for file *fn* along the semicolon-separated list of directories *dirlist* and returns the fully qualified path and filename if *fn* is found. findfile() returns "" if the file is not found.

findfile(*fn*) is equivalent to findfile(*fn*, c("adopath")). findfile() with one argument looks along the official Stata ado-path for file *fn*.

Remarks

For instance,

>   findfile("kappa.ado")

might return C:\Program Files\Stata11\ado\updates\k\kappa.ado.

Conformability

findfile(*fn*, *dirlist*):
>   *fn*: 1×1
>   *dirlist*: 1×1 (optional)
>   *result*: 1×1

Diagnostics

findfile(*fn*, *dirlist*) and findfile(*fn*) return "" if the file is not found. If the file is found, the returned fully qualified path and filename is guaranteed to exist and be readable at the instant findfile() concluded.

Also see

[M-4] **io** — I/O functions

Title

Syntax

real matrix floatround(*real matrix x*)

Description

floatround(*x*) returns *x* rounded to IEEE 4-byte real (float) precision. floatround() is the element-by-element equivalent of Stata's float() function. The Mata function could not be named float() because the word float is reserved in Mata.

Remarks

```
: printf("  %21x\n", .1)
  +1.999999999999aX-004
: printf("  %21x\n", floatround(.1))
  +1.99999a0000000X-004
```

Conformability

floatround(*x*):
        x: $r \times c$
  result: $r \times c$

Diagnostics

floatround(*x*) returns missing (.) if $x < -1.\text{fffffeX+7e}$ (approximately $-1.70141173319e+38$) or $x > 1.\text{fffffeX+7e}$ (approximately $1.70141173319e+38$).

By contrast with most functions, floatround(*x*) returns the same kind of missing value if *x* contains missing; . if $x == .$, .a if $x == .a$, .b if $x == .b$, ..., and .z if $x == .z$.

Also see

[M-4] **utility** — Matrix utility functions

Title

[M-5] **fmtwidth( )** — Width of %fmt

Syntax

> *real matrix* fmtwidth(*string matrix f*)

Description

fmtwidth(*f*) returns the width of the %*fmt* contained in *f*.

Remarks

fmtwidth("%9.2f") returns 9.

fmtwidth("%20s") returns 20.

fmtwidth("%tc") returns 18.

fmtwidth("%tcDay_Mon_DD_hh:mm:ss_!C!D!T_CCYY") returns 28.

fmtwidth("not a format") returns . (missing).

Conformability

fmtwidth(*f*):
        f: $r \times c$
   result: $r \times c$

Diagnostics

fmtwidth(*f*) returns . (missing) when *f* does not contain a valid Stata format.

Also see

[M-5] **strlen( )** — Length of string

[M-4] **string** — String manipulation functions

Title

Syntax

> *real scalar*   fopen(*string scalar fn*, *mode*)
>
> *real scalar*   fopen(*string scalar fn*, *mode*, *public*)
>
> *real scalar*   _fopen(*string scalar fn*, *mode*)
>
> *real scalar*   _fopen(*string scalar fn*, *mode*, *public*)

where

> *mode*:   *string scalar* containing "r", "w", "rw", or "a"
>
> *public*: optional *real scalar* containing zero or nonzero

In what follows, *fh* is the value returned by fopen( ) or _fopen( ):

> *void*   fclose(*fh*)
> *real scalar*   _fclose(*fh*)
>
> *string scalar* fget(*fh*)
> *string scalar* _fget(*fh*)
> *string scalar* fgetnl(*fh*)
> *string scalar* _fgetnl(*fh*)
>
> *string scalar* fread(*fh*, *real scalar len*)
> *string scalar* _fread(*fh*, *real scalar len*)
>
> *void*   fput(*fh*, *string scalar s*)
> *real scalar*   _fput(*fh*, *string scalar s*)
>
> *void*   fwrite(*fh*, *string scalar s*)
> *real scalar*   _fwrite(*fh*, *string scalar s*)
>
> *matrix*   fgetmatrix(*fh*)
> *matrix*   _fgetmatrix(*fh*)
> *void*   fputmatrix(*fh*, *transmorphic matrix X*)
> *real scalar*   _fputmatrix(*fh*, *transmorphic matrix X*)

477

real scalar   fstatus(*fh*)

real scalar   ftell(*fh*)

real scalar   _ftell(*fh*)

void   fseek(*fh*, *real scalar offset*, *real scalar whence*)

real scalar   _fseek(*fh*, *real scalar offset*, *real scalar whence*)

(*whence* is coded −1, 0, or 1, meaning from start of file, from current position, or from end of file; *offset* is signed: positive values mean after *whence* and negative values mean before)

void   ftruncate(*fh*)

real scalar   _ftruncate(*fh*)

Description

These functions read and write files. First, open the file and get back a file handle (*fh*). The file handle, which is nothing more than an integer, is how you refer to the file in the calls to other file I/O functions. When you are finished, close the file.

Most file I/O functions come in two varieties: without and with an underscore in front of the name, such as fopen() and _fopen(), and fwrite() and _fwrite().

In functions without a leading underscore, errors cause execution to be aborted. For instance, you attempt to open a file for read and the file does not exist. Execution stops. Or, having successfully opened a file, you attempt to write into it and the disk is full. Execution stops. When execution stops, the appropriate error message is presented.

In functions with the leading underscore, execution continues and no error message is displayed; it is your responsibility (1) to verify that things went well and (2) to take the appropriate action if they did not. Concerning (1), some underscore functions return a status value; others require that you call fstatus() to obtain the status information.

You can mix and match use of underscore and nonunderscore functions, using, say, _fopen() to open a file and fread() to read it, or fopen() to open and _fwrite() to write.

Remarks

Remarks are presented under the following headings:

> *Opening and closing files*
> *Reading from a file*
> *Writing to a file*
> *Reading and writing in the same file*
> *Reading and writing matrices*
> *Repositioning in a file*
> *Truncating a file*
> *Error codes*

Opening and closing files

Functions

> fopen(*string scalar fn, string scalar mode*)
>
> _fopen(*string scalar fn, string scalar mode*)
>
> fopen(*string scalar fn, string scalar mode, real scalar public*)
>
> _fopen(*string scalar fn, string scalar mode, real scalar public*)

open a file. The file may be on a local disk, a network disk, or even on the web (such as http://www.stata.com/index.html). *fn* specifies the filename, and *mode* specifies how the file is to opened:

mode	meaning
"r"	Open for reading; file must exist and be readable.
	File may be "http://..." file.
	File will be positioned at the beginning.
"w"	Open for writing; file must not exist and the directory be writable.
	File may not be "http://..." file.
	File will be positioned at the beginning.
"rw"	Open for reading and writing; file must either exist and be writable or not exist and directory be writable.
	File may not be "http://..." file.
	File will be positioned at the beginning (new file) or at the end (existing file).
"a"	Open for appending; file must either exist and be writable or not exist and directory be writable.
	File may not be "http://..." file.
	File will be positioned at the end.

Other values for *mode* cause fopen() and _fopen() to abort with an invalid-mode error.

Optional third argument *public* specifies whether the file, if it is being created, should be given permissions so that everyone can read it, or if it instead should be given the normal permissions. Not specifying *public*, or specifying *public* as 0, gives the file the normal permissions. Specifying *public* as nonzero makes the file publicly readable. *public* is relevant only when the file is being created, i.e., is being opened "w", or being opened "rw" and not previously existing.

fopen() returns a file handle; the file is opened or execution is aborted.

_fopen() returns a file handle or returns a negative number. If a negative number is returned, the file is not open, and the number indicates the reason. For _fopen(), there are a few likely possibilities

negative value	meaning
−601	file not found
−602	file already exists
−603	file could not be opened
−608	file is read-only
−691	I/O error

and there are many other possibilities. For instance, perhaps you attempted to open a file on the web (say, *http://www.newurl.org/upinfo.doc*) and the URL was not found, or the server refused to send back the file, etc. See *Error codes* below for a complete list of codes.

After opening the file, you use the other file I/O commands to read and write it, and then you close the file with fclose() or _fclose(). fclose() returns nothing; if the file cannot be closed, execution is aborted. _fclose returns 0 if successful, or a negative number otherwise. For _fclose(), the likely possibilities are

negative value	meaning
−691	filesystem I/O error

Reading from a file

You may read from a file opened "r" or "rw". The commands to read are

    fget(*fh*)

    fgetnl(*fh*)

    fread(*fh*, *real scalar len*)

and, of course,

    _fget(*fh*)

    _fgetnl(*fh*)

    _fread(*fh*, *real scalar len*)

All functions, with or without an underscore, require a file handle be specified, and all the functions return a string scalar or they return J(0,0,""), a 0 × 0 string matrix. They return J(0,0,"") on end of file and, for the underscore functions, when the read was not successful for other reasons. When using the underscore functions, you use fstatus() to obtain the status code; see *Error codes* below. The underscore read functions are rarely used because the only reason a read can fail is I/O error, and there is not much that can be done about that except abort, which is exactly what the nonunderscore functions do.

fget(*fh*) is for reading ASCII files; the next line from the file is returned, without end-of-line characters. (If the line is longer then 32,768 characters, the first 32,768 characters are returned.)

fgetnl(*fh*) is much the same as fget(), except that the new-line characters are not removed from the returned result. (If the line is longer then 32,768 characters, the first 32,768 characters are returned.)

fread(*fh*, *len*) is usually used for reading binary files and returns the next *len* characters (bytes) from the file or, if there are fewer than that remaining to be read, however many remain. (*len* may not exceed 2,147,483,647 on 32-bit computers and 9,007,199,254,740,991 [*sic*] on 64-bit computers; memory shortages for storing the result will arise long before these limits are reached on most computers.)

The following code reads and displays a file:

```
fh = fopen(filename, "r")
while ((line=fget(fh))!=J(0,0,"")) {
      printf("%s\n", line)
}
fclose(fh)
```

Writing to a file

You may write to a file opened "w", "rw", or "a". The functions are

> fput(*fh*, *string scalar s*)
>
> fwrite(*fh*, *string scalar s*)

and, of course,

> _fput(*fh*, *string scalar s*)
>
> _fwrite(*fh*, *string scalar s*)

fh specifies the file handle, and *s* specifies the string to be written. fput() writes *s* followed by the new-line characters appropriate for your operating system. fwrite() writes *s* alone.

fput() and fwrite() return nothing; _fput() and _fwrite() return a real scalar equal to 0 if all went well or a negative error code; see *Error codes* below.

The following code copies text from one file to another:

```
fh_in  = fopen(inputname, "r")
fh_out = fopen(outputname, "w")
while ((line=fget(fh_in))!=J(0,0,"")) {
      fput(fh_out, line)
}
fclose(fh_out)
fclose(fh_in)
```

The following code reads a file (binary or text) and changes every occurrence of "a" to "b":

```
fh_in  = fopen(inputname, "r")
fh_out = fopen(outputname, "w")
while ((c=fread(fh_in, 1))!=J(0,0,"")) {
    fwrite(fh_out, (c=="a" ? "b" : c))
}
fclose(fh_out)
fclose(fh_in)
```

Reading and writing in the same file

You may read and write from a file opened "rw", using any of the read or write functions described above. When reading and writing in the same file, one often uses file repositioning functions, too; see *Repositioning in a file* below.

Reading and writing matrices

Functions

        fputmatrix(*fh*, *transmorphic matrix X*)

and

        _fputmatrix(*fh*, *transmorphic matrix X*)

will write a matrix to a file. In the usual fashion, fputmatrix() returns nothing (it aborts if there is an I/O error) and _fputmatrix() returns a scalar equal to 0 if all went well and a negative error code otherwise.

Functions

        fgetmatrix(*fh*)

and

        _fgetmatrix(*fh*)

will read a matrix written by fputmatrix() or _fputmatrix(). Both functions return the matrix read or return J(0,0,.) on end of file (both functions) or error (_fgetmatrix() only). Since J(0,0,.) could be the matrix that was written, distinguishing between that and end of file requires subsequent use of fstatus(). fstatus() will return 0 if fgetmatrix() or _fgetmatrix() returned a written matrix, −1 if end of file, or (after _fgetmatrix()) a negative error code.

fputmatrix() writes matrices in a compact, efficient, and portable format; a matrix written on a Windows computer can be read back on a Mac or Unix computer and vice versa.

The following code creates a file containing three matrices

```
fh = fopen(filename, "w")
fputmatrix(fh, a)
fputmatrix(fh, b)
fputmatrix(fh, c)
fclose(fh)
```

and the following code reads them back:

```
fh = fopen(filename, "r")
a = fgetmatrix(fh)
b = fgetmatrix(fh)
c = fgetmatrix(fh)
fclose(fh)
```

❑ Technical note

You may even write pointer matrices

```
mats = (&a, &b, &c, NULL)
fh = fopen(filename,"w")
fputmatrix(fh, mats)
fclose(fh)
```

and read them back:

```
fh = fopen(filename, "r")
mats = fgetmatrix(fh)
fclose(fh)
```

When writing pointer matrices, fputmatrix() writes NULL for any pointer-to-function value. It is also recommended that you do not write self-referential matrices (matrices that point to themselves, either directly or indirectly), although the elements of the matrix may be cross linked and even recursive themselves. If you are writing pointer matrix p, no element of p, *p, **p, etc., should contain &p. That one address cannot be preserved because in the assignment associated with reading back the matrix (the "*result=*" part of *result*=fgetmatrix(*fh*), a new matrix with a different address is associated with the contents.

❑

Repositioning in a file

The function

```
ftell(fh)
```

returns a real scalar reporting where you are in a file, and function

```
fseek(fh, real scalar offset, real scalar whence)
```

changes where you are in the file to be *offset* bytes from the beginning of the file (*whence* = −1), *offset* bytes from the current position (*whence* = 0), or *offset* bytes from the end of the file (*whence* = 1).

Functions _ftell() and _fseek() do the same thing as ftell() and fseek(), the difference being that, rather than aborting on error, the underscore functions return negative error codes. _ftell() is pretty well useless as the only error that can arise is I/O error, and what else are you going to do other than abort? _fseek(), however, has a use, because it allows you to try out a repositioning and check whether it was successful. With fseek(), if the repositioning is not successful, execution is aborted.

Say you open a file for read:

```
fh = fopen(filename, "r")
```

After opening the file in mode r, you are positioned at the beginning of the file or, in the jargon of file processing, at position 0. Now say that you read 10 bytes from the file:

```
part1 = fread(fh, 10)
```

Assuming that was successful, you are now at position 10 of the file. Say that you next read a line from the file

```
line = fget(fh)
```

and assume that fget() returns "abc". You are now at position 14 or 15. (No, not 13: fget() read the line and the new-line characters and returned the line. abc was followed by carriage return and line feed (two characters) if the file was written under Windows and by a carriage return or line feed alone (one character) if the file was written under Mac or Unix).

ftell(*fh*) and _ftell(*fh*) tell you where you are in the file. Coding

```
pos = ftell(fh)
```

would store 14 or 15 in pos. Later in your code, after reading more of the file, you could return to this position by coding

```
fseek(fh, pos, -1)
```

You could return to the beginning of the file by coding

```
fseek(fh, 0, -1)
```

and you could move to the end of the file by coding

```
fseek(fh, 0, 1)
```

ftell(*fh*) is equivalent to _fseek(*fh*, 0, 0).

Repositioning functions cannot be used when the file has been opened "a".

Truncating a file

Truncation refers to making a longer file shorter. If a file was opened "w" or "rw", you may truncate it at its current position by using

```
ftruncate(fh)
```

or

```
_ftruncate(fh)
```

ftruncate() returns nothing; _ftruncate() returns 0 on success and otherwise returns a negative error code.

The following code shortens a file to its first 100 bytes:

```
fh = fopen(filename, "rw")
fseek(fh, 100, -1)
ftruncate(fh)
fclose(fh)
```

Error codes

If you use the underscore I/O functions, if there is an error, they will return a negative code. Those codes are

negative code	meaning
0	all is well
−1	end of file
−601	file not found
−602	file already exists
−603	file could not be opened
−608	file is read-only
−610	file format error
−630	web files not supported in this version of Stata
−631	host not found
−632	web file not allowed in this context
−633	web file not allowed in this context
−660	proxy host not found
−662	proxy server refused request to send
−663	remote connection to proxy failed
−665	could not set socket nonblocking
−669	invalid URL
−670	invalid network port number
−671	unknown network protocol
−672	server refused to send file
−673	authorization required by server
−674	unexpected response from server
−678	could not open local network socket
−679	unexpected web error
−691	I/O error
−699	insufficient disk space
−3601	invalid file handle
−3602	invalid filename
−3611	too many open files
−3621	attempt to write read-only file
−3622	attempt to read write-only file
−3623	attempt to seek append-only file
−3698	file seek error

Other codes in the −600 to −699 range are possible. The codes in this range correspond to the negative of the corresponding Stata return code; see [P] **error**.

Underscore functions that return a real scalar will return one of these codes if there is an error.

If an underscore function does not return a real scalar, then you obtain the outcome status using fstatus(). For instance, the read-string functions return a string scalar or J(0,0,"") on end of file. The underscore variants do the same, and they also return J(0,0,"") on error, meaning error looks like end of file. You can determine the error code using the function

> fstatus(*fh*)

fstatus() returns 0 (no previous error) or one of the negative codes above.

fstatus() may be used after any underscore I/O command to obtain the current error status.

fstatus() may also be used after the nonunderscore I/O commands; then fstatus() will return −1 or 0 because all other problems would have stopped execution.

Conformability

fopen(*fn*, *mode*, *public*), _fopen(*fn*, *mode*, *public*):

fn:	1 × 1	
mode:	1 × 1	
public:	1 × 1	(optional)
result:	1 × 1	

fclose(*fh*):

fh:	1 × 1
result:	*void*

_fclose(*fh*):

fh:	1 × 1
result:	1 × 1

fget(*fh*), _fget(*fh*), fgetnl(*fh*), _fgetnl(*fh*):

fh:	1 × 1	
result:	1 × 1	or 0 × 0 if end of file

fread(*fh*, *len*), _fread(*fh*, *len*):

fh:	1 × 1	
len:	1 × 1	
result:	1 × 1	or 0 × 0 if end of file

fput(*fh*, *s*), fwrite(*fh*, *s*):

fh:	1 × 1
s:	1 × 1
result:	*void*

_fput(*fh*, *s*), _fwrite(*fh*, *s*):

fh:	1 × 1
s:	1 × 1
result:	1 × 1

fgetmatrix(*fh*), _fgetmatrix(*fh*):

fh:	1 × 1	
result:	*r* × *c*	or 0 × 0 if end of file

fputmatrix(*fh*, *X*):

fh:	1 × 1
X:	*r* × *c*
result:	*void*

_fputmatrix(*fh*, *X*):

fh:	1 × 1
X:	*r* × *c*
result:	1 × 1

fstatus(*fh*):

fh:	1 × 1
result:	1 × 1

ftell(*fh*), _ftell(*fh*):

fh:	1 × 1
result:	1 × 1

fseek(*fh*, *offset*, *whence*):

fh:	1 × 1
offset:	1 × 1
whence:	1 × 1
result:	*void*

_fseek(*fh*, *offset*, *whence*):

fh:	1 × 1
offset:	1 × 1
whence:	1 × 1
result:	1 × 1

ftruncate(*fh*):

fh:	1 × 1
result:	*void*

_ftruncate(*fh*):

fh:	1 × 1
result:	1 × 1

Diagnostics

fopen(*fn*, *mode*) aborts with error if *mode* is invalid or if *fn* cannot be opened or if an attempt is made to open too many files simultaneously.

_fopen(*fn*, *mode*) aborts with error if *mode* is invalid or if an attempt is made to open too many files simultaneously. _fopen() returns the appropriate negative error code if *fn* cannot be opened.

All remaining I/O functions—even functions with leading underscore—abort with error if *fh* is not a handle to a currently open file.

Also, the functions that do not begin with an underscore abort with error when a file was opened read-only and a request requiring write access is made, when a file is opened write-only and a request requiring read access is made, etc. See *Error codes* above; all problems except code −1 (end of file) cause the nonunderscore functions to abort with error.

Finally, the following functions will also abort with error for the following specific reasons:

fseek(*fh*, *offset*, *whence*) and _fseek(*fh*, *offset*, *whence*) abort with error if *offset* is outside the range ±2,147,483,647 on 32-bit computers; if *offset* is outside the range ±9,007,199,254,740,991 on 64-bit computers; or, on all computers, if *whence* is not −1, 0, or 1.

Also see

[M-5] **cat( )** — Load file into string matrix sprintf( ) in

[M-5] **printf( )** — Format output

[M-5] **bufio( )** — Buffered (binary) I/O

[M-4] **io** — I/O functions

Title

[M-5] **fullsvd( )** — Full singular value decomposition

Syntax

void	fullsvd(*numeric matrix A*, *U*, *s*, *Vt*)
numeric matrix	fullsdiag(*numeric colvector s*, *real scalar k*)
void	_fullsvd(*numeric matrix A*, *U*, *s*, *Vt*)
real scalar	_svd_la(*numeric matrix A*, *U*, *s*, *Vt*)

Description

fullsvd(*A*, *U*, *s*, *Vt*) calculates the singular value decomposition of $m \times n$ matrix *A*, returning the result in *U*, *s*, and *Vt*. Singular values in *s* are sorted from largest to smallest.

fullsdiag(*s*, *k*) converts column vector *s* returned by fullsvd() into matrix *S*. In all cases, the appropriate call for this function is

$$S = \text{fullsdiag}(s, \text{rows}(A)\text{-cols}(A))$$

_fullsvd(*A*, *U*, *s*, *Vt*) does the same as fullsvd(), except that, in the process, it destroys *A*. Use of _fullsvd() in place of fullsvd() conserves memory.

_svd_la() is the interface into the [M-1] **LAPACK** SVD routines and is used in the implementation of the previous functions. There is no reason you should want to use it. _svd_la() is similar to _fullsvd(). It differs in that it returns a real scalar equal to 1 if the numerical routines fail to converge, and it returns 0 otherwise. The previous SVD routines set *s* to contain missing values in this unlikely case.

Remarks

Remarks are presented under the following headings:

> *Introduction*
> *Relationship between the full and thin SVDs*
> *The contents of s*
> *Possibility of convergence problems*

Documented here is the full SVD, appropriate in all cases, but of interest mainly when $A: m \times n$, $m < n$. There is a thin SVD that conserves memory when $m \geq n$; see [M-5] **svd( )**. The relationship between the two is discussed in *Relationship between the full and thin SVDs* below.

Introduction

The SVD is used to compute accurate solutions to linear systems and least-squares problems, to compute the 2-norm, and to determine the numerical rank of a matrix.

The singular value decomposition (SVD) of A: $m \times n$ is given by

$$A = USV'$$

where

     U: $m \times m$ and orthogonal (unitary)
     S: $m \times n$ and diagonal
     V: $n \times n$ and orthogonal (unitary)

When A is complex, the transpose operator $'$ is understood to mean the conjugate transpose operator.

Diagonal matrix S contains the singular values and those singular values are real even when A is complex. It is usual (but not required) that S is arranged so that the largest singular value appears first, then the next largest, and so on. The SVD routines documented here do this.

The full SVD routines return U and $Vt = V'$. S is returned as a column vector s, and S can be obtained by

$$S = \texttt{fullsdiag}(s, \texttt{rows}(A)\texttt{-cols}(A))$$

so we will write the SVD as

$$A = U * \texttt{fullsdiag}(s, \texttt{rows}(A)\texttt{-cols}(A)) * Vt$$

Function $\texttt{fullsvd}(A, U, s, Vt)$ returns the U, s, and Vt corresponding to A.

Relationship between the full and thin SVDs

A popular variant of the SVD is known as the thin SVD and is suitable for use when $m \geq n$. Both SVDs have the same formula,

$$A = USV'$$

but U and S have reduced dimensions in the thin version:

Matrix	Full SVD	Thin SVD
U:	$m \times m$	$m \times n$
S:	$m \times n$	$n \times n$
V:	$n \times n$	$n \times n$

When $m = n$, the two variants are identical.

The thin SVD is of use when $m > n$, because then only the first m diagonal elements of S are nonzero, and therefore only the first m columns of U are relevant in $A = USV'$. There are considerable memory savings to be had in calculating the thin SVD when $m \gg n$.

As a result, many people call the thin SVD the SVD and ignore the full SVD altogether. If the matrices you deal with have $m \geq n$, you will want to do the same. To obtain the thin SVD, see [M-5] **svd( )**.

Regardless of the dimension of your matrix, you may wish to obtain only the singular values. In this case, see $\texttt{svdsv}()$ documented in [M-5] **svd( )**. That function is appropriate in all cases.

The contents of s

Given A: $m \times n$, the singular values are returned in s: $\min(m, n) \times 1$.

Let's consider the $m = n$ case first. A is $m \times m$ and the m singular values are returned in s, an $m \times 1$ column vector. If A were 3×3, perhaps we would get back

```
: s
          1

1     13.47
2      5.8
3      2.63
```

If we needed it, we could obtain S from s simply by creating a diagonal matrix from s

```
: S = diag(s)
: S
[symmetric]
          1       2       3

1     13.47
2         0     5.8
3         0       0    2.63
```

although the official way we are supposed to do this is

```
: S = fullsdiag(s, rows(A)-cols(A))
```

and that will return the same result.

Now let's consider $m < n$. Let's pretend that A is 3×4. The singular values will be returned in 3×1 vector s. For instance, s might still contain

```
: s
          1

1     13.47
2      5.8
3      2.63
```

The S matrix here needs to be 3×4, and `fullsdiag()` will form it:

```
: fullsdiag(s, rows(A)-cols(A))
          1       2       3       4

1     13.47       0       0       0
2         0     5.8       0       0
3         0       0    2.63       0
```

The final case is $m > n$. We will pretend that A is 4×3. The s vector we get back will look the same

```
: s
              1

    1 ┌ 13.47 ┐
    2 │  5.8  │
    3 └  2.63 ┘
```

but this time, we need a 4 × 3 rather than a 3 × 4 matrix formed from it.

```
: fullsdiag(s, rows(A)-cols(A))
        1       2       3

    1 ┌ 13.47   0       0   ┐
    2 │   0    5.8      0   │
    3 │   0     0     2.63  │
    4 └   0     0       0   ┘
```

Possibility of convergence problems

See *Possibility of convergence problems* in [M-5] **svd( )**; what is said there applies equally here.

Conformability

fullsvd(A, U, s, Vt):

> *input*:

>> A: $m \times n$

> *output*:

>> U: $m \times m$
>> s: $\min(m,n) \times 1$
>> Vt: $n \times n$
>> *result*: *void*

fullsdiag(s, k):

> *input*:

>> s: $r \times 1$
>> k: 1×1

> *output*:

>> *result*: $r + k \times r$, if k $\geq$ 0
>> $r \times r - k$, otherwise

_fullsvd(A, U, s, Vt):

> *input*:

>> A: $m \times n$

> *output*:

>> A: 0×0
>> U: $m \times m$
>> s: $\min(m,n) \times 1$
>> Vt: $n \times n$
>> *result*: *void*

_svd_la(A, U, s, Vt):

> *input*:
>
> | | |
> |---|---|
> | A: | $m \times n$ |
>
> *output*:
>
> | | |
> |---|---|
> | A: | $m \times n$,   but contents changed |
> | U: | $m \times m$ |
> | s: | $\min(m, n) \times 1$ |
> | Vt: | $n \times n$ |
> | *result*: | 1×1 |

Diagnostics

fullsvd(A, U, s, Vt) and _fullsvd(A, s, Vt) return missing results if A contains missing. In all other cases, the routines should work, but there is the unlikely possibility of convergence problems, in which case missing results will also be returned; see *Possibility of convergence problems* in [M-5] **svd( )**.

_fullsvd( ) aborts with error if A is a view.

Direct use of _svd_la( ) is not recommended.

Also see

[M-5] **svd( )** — Singular value decomposition

[M-5] **svsolve( )** — Solve AX=B for X using singular value decomposition

[M-5] **pinv( )** — Moore–Penrose pseudoinverse

[M-5] **norm( )** — Matrix and vector norms

[M-5] **rank( )** — Rank of matrix

[M-4] **matrix** — Matrix functions

Title

Syntax

void   geigensystem(A, B, X, w, b)

void   leftgeigensystem(A, B, X, w, b)

void   geigensystemselectr(A, B, *range*, X, w, b)

void leftgeigensystemselectr(A, B, *range*, X, w, b)

void   geigensystemselecti(A, B, *index*, X, w, b)

void leftgeigensystemselecti(A, B, *index*, X, w, b)

void   geigensystemselectf(A, B, f, X, w, b)

void   leftgeigensystemselectf(A, B, f, X, w, b)

where inputs are

A:	*numeric matrix*	
B:	*numeric matrix*	
range:	*real vector*	(range of generalized eigenvalues to be selected)
index:	*real vector*	(indices of generalized eigenvalues to be selected)
f:	*pointer scalar*	(points to a function used to select generalized eigenvalues)

and outputs are

X:	*numeric matrix* of generalized eigenvectors	
w:	*numeric vector* (numerators of generalized eigenvalues)	
b:	*numeric vector* (denominators of generalized eigenvalues)	

The following routines are used in implementing the above routines:

void _geigensystem_la(*numeric matrix H, R, XL, XR, w, b,*
                string scalar side)

void _geigenselectr_la(*numeric matrix H, R, XL, XR, w, b,*
                range, string scalar side)

void _geigenselecti_la(*numeric matrix H, R, XL, XR, w, b,*
                index, string scalar side)

void _geigenselectf_la(*numeric matrix H, R, XL, XR, w, b,*
                pointer scalar f, string scalar side)

real scalar _geigen_la(*numeric matrix H, R, XL, XR, w, select,*
                string scalar side, string scalar howmany)

Description

geigensystem(A, B, X, w, b) computes generalized eigenvectors of two general, real or complex, square matrices, A and B, along with their corresponding generalized eigenvalues.

- A and B are two general, real or complex, square matrices with the same dimensions.

- X contains generalized eigenvectors.

- w contains numerators of generalized eigenvalues.

- b contains denominators of generalized eigenvalues.

leftgeigensystem(A, B, X, w, b) mirrors geigensystem(), the difference being that leftgeigensystem() computes left, generalized eigenvectors.

geigensystemselectr(A, B, *range*, X, w, b) computes selected generalized eigenvectors of two general, real or complex, square matrices, A and B, along with their corresponding generalized eigenvalues. Only the generalized eigenvectors corresponding to selected generalized eigenvalues are computed. Generalized eigenvalues that lie in a range are selected. The selected generalized eigenvectors are returned in X, and their corresponding generalized eigenvalues are returned in (w, b).

> *range* is a vector of length 2. All finite, generalized eigenvalues with absolute value in the half-open interval (*range*[1], *range*[2]] are selected.

leftgeigensystemselectr(A, B, *range*, X, w, b) mirrors geigensystemselectr(), the difference being that leftgeigensystemr() computes left, generalized eigenvectors.

geigensystemselecti(A, B, *index*, X, w, b) computes selected right, generalized eigenvectors of two general, real or complex, square matrices, A and B, along with their corresponding generalized eigenvalues. Only the generalized eigenvectors corresponding to selected generalized eigenvalues are computed. Generalized eigenvalues are selected by an index. The selected generalized eigenvectors are returned in X, and the selected generalized eigenvalues are returned in (w, b).

> The finite, generalized eigenvalues are sorted by their absolute values, in descending order, followed by the infinite, generalized eigenvalues. There is no particular order among infinite, generalized eigenvalues.

> *index* is a vector of length 2. The generalized eigenvalues in elements *index*[1] through *index*[2], inclusive, are selected.

leftgeigensystemselecti(A, B, *index*, X, w, b) mirrors geigensystemselecti(), the difference being that leftgeigensystemi() computes left, generalized eigenvectors.

geigensystemselectf(A, B, f, X, w, b) computes selected generalized eigenvectors of two general, real or complex, square matrices A and B along with their corresponding generalized eigenvalues. Only the generalized eigenvectors corresponding to selected generalized eigenvalues are computed. Generalized eigenvalues are selected by a user-written function described below. The selected generalized eigenvectors are returned in X, and the selected generalized eigenvalues are returned in (w, b).

leftgeigensystemselectf(A, B, f, X, w, b) mirrors geigensystemselectf(), the difference being that leftgeigensystemselectf() computes selected left, generalized eigenvectors.

_geigen_la(), _geigensystem_la(), _geigenselectr_la(), _geigenselecti_la(), and _geigenselectf_la() are the interfaces into the [M-1] **LAPACK** routines used to implement the above functions. Their direct use is not recommended.

Remarks

Remarks are presented under the following headings:

>*Generalized eigenvalues*
>*Generalized eigenvectors*
>*Criterion selection*
>*Range selection*
>*Index selection*

Generalized eigenvalues

A scalar, l (usually denoted by *lambda*), is said to be a generalized eigenvalue of a pair of $n \times n$ square, numeric matrices $(\mathbf{A}, \mathbf{B})$ if there is a nonzero column vector $\mathbf{x}$: $n \times 1$ (called the generalized eigenvector) such that

$$\mathbf{A}\mathbf{x} = l\mathbf{B}\mathbf{x} \tag{1}$$

(1) can also be written

$$(\mathbf{A} - l\mathbf{B})\mathbf{x} = 0$$

A nontrivial solution to this system of n linear homogeneous equations exists if and only if

$$\det(\mathbf{A} - l\mathbf{B}) = 0 \tag{2}$$

In practice, the generalized eigenvalue problem for the matrix pair $(\mathbf{A}, \mathbf{B})$ is usually formulated as finding a pair of scalars (w, b) and a nonzero column vector $\mathbf{x}$ such that

$$w\mathbf{A}\mathbf{x} = b\mathbf{B}\mathbf{x}$$

The scalar w/b is a finite, generalized eigenvalue if b is not zero. The pair (w, b) represents an infinite, generalized eigenvalue if b is zero or numerically close to zero. This situation may arise if $\mathbf{B}$ is singular.

The Mata functions that compute generalized eigenvalues return them in two complex vectors, $\mathbf{w}$ and $\mathbf{b}$, of length n. If b[i]=0, the ith generalized eigenvalue is infinite; otherwise, the ith generalized eigenvalue is w[i]/b[i].

Generalized eigenvectors

A column vector, $\mathbf{x}$, is a right, generalized eigenvector or simply a generalized eigenvector of a generalized eigenvalue (w, b) for a pair of matrices, $\mathbf{A}$ and $\mathbf{B}$, if

$$w\mathbf{A}\mathbf{x} = b\mathbf{B}\mathbf{x}$$

A row vector, $\mathbf{v}$, is a left, generalized eigenvector of a generalized eigenvalue (w, b) for a pair of matrices, $\mathbf{A}$ and $\mathbf{B}$, if

$$w\mathbf{v}\mathbf{A} = b\mathbf{v}\mathbf{B}$$

For instance, let's consider the linear system

$$dx/dt = \text{A1} \times x + \text{A2} \times u$$

$$dy/dt = \text{A3} \times x + \text{A4} \times u$$

where

```
: A1 = (-4, -3 \ 2, 1)
: A2 = (3 \ 1)
: A3 = (1, 2)
```

and

```
: A4 = 0
```

The finite solutions of zeros for the transfer function

$$g(s) = \text{A3} \times (sI - \text{A1})^{-1} \times \text{A2} + \text{A4} \tag{3}$$

of this linear time-invariant state-space model is given by the finite, generalized eigenvalues of A and B where

```
: A = (A1, A2 \ A3, A4)
```

and

```
: B = (1, 0, 0 \ 0, 1, 0 \ 0, 0, 0)
```

We obtain generalized eigenvectors in X and generalized eigenvalues in w and b by using

```
: geigensystem(A, B, X=., w=., b=.)
: X
```

	1	2	3
1	-1	0	2.9790e-16
2	.5	0	9.9301e-17
3	.1	1	1

```
: w
```

	1	2	3
1	-1.97989899	3.16227766	2.23606798

```
: b
```

	1	2	3
1	.7071067812	0	0

The only finite, generalized eigenvalue of A and B is

```
: w[1,1]/b[1,1]
  -2.8
```

In this simple example, (3) can be explicitly written out as

$$g(s) = (5s + 14)/(s^2 + 3s + 2)$$

which clearly has the solution of zero at -2.8.

Criterion selection

We sometimes want to compute only those generalized eigenvectors whose corresponding generalized eigenvalues satisfy certain criterion. We can use `geigensystemselectf()` to solve these problems.

We must pass `geigensystemselectf()` a pointer to a function that implements our conditions. The function must accept two numeric scalar arguments so that it can receive the numerator `w` and the denominator `b` of a generalized eigenvalue, and it must return the real value 0 to indicate rejection and a nonzero real value to indicate selection.

In this example, we want to compute only finite, generalized eigenvalues for each of which `b` is not zero. After deciding that anything smaller than 1e–15 is zero, we define our function to be

```
: real scalar finiteonly(numeric scalar w, numeric scalar b)
> {
>            return((abs(b)>=1e-15))
> }
```

By using

```
: geigensystemselectf(A, B, &finiteonly(), X=., w=., b=.)
```

we get the only finite, generalized eigenvalue of A and B in (w, b) and its corresponding eigenvector in X:

```
: X
                    1

    1    -.894427191
    2     .447213595
    3     .089442719
```

```
: w
 -1.97989899
: b
  .7071067812
: w:/b
 -2.8
```

Range selection

We can use `geigensystemselectr()` to compute only those generalized eigenvectors whose generalized eigenvalues have absolute values that fall in a half-open interval.

For instance,

```
: A = (-132, -88, 84, 104 \ -158.4, -79.2, 76.8, 129.6 \
> 129.6, 81.6, -79.2, -100.8 \ 160, 84, -80, -132)
: B = (-60, -50, 40, 50 \ -69, -46.4, 38, 58.2 \ 58.8, 46, -37.6, -48 \
> 70, 50, -40, -60)
: range = (0.99, 2.1)
```

We obtain generalized eigenvectors in X and generalized eigenvalues in w and b by using

```
: geigensystemselectr(A, B, range, X=., w=., b=.)
: X
```

	1	2
1	.089442719	.02236068
2	.04472136	.067082039
3	.04472136	.067082039
4	.089442719	.02236068

```
: w
```

	1	2
1	.02820603	.170176379

```
: b
```

	1	2
1	.0141030148	.1701763791

The generalized eigenvalues have absolute values in the half-open interval (0.99, 2.1].

```
: abs(w:/b)
```

	1	2
1	2	1

Index selection

geigensystemselect( ) sorts the finite, generalized eigenvalues using their absolute values, in descending order, placing the infinite, generalized eigenvalues after the finite, generalized eigenvalues. There is no particular order among infinite, generalized eigenvalues.

If we want to compute only generalized eigenvalues whose ranks are *index*[1] through *index*[2] in the list of generalized eigenvalues obtained by geigensystemselect( ), we can use geigensystemseleci( ).

To compute the first two generalized eigenvalues and generalized eigenvectors in this example, we can specify

```
: index = (1, 2)
: geigensystemselecti(A, B, index, X=., w=., b=.)
```

The results are

```
: X
```

	1	2
1	.02981424	−.059628479
2	.04472136	−.059628479
3	.089442719	−.02981424
4	.01490712	−.119256959

```
: w
```

	1	2
1	.012649111	.379473319

```
: b
```

	1	2
1	.0031622777	.1264911064

```
: w:/b
```

	1	2
1	4	3

Conformability

geigensystem(A, B, X, w, b):

> *input*:
>
> > A: $n \times n$
> > B: $n \times n$
>
> *output*:
>
> > X: $n \times n$
> > w: $1 \times n$
> > b: $1 \times n$

leftgeigensystem(A, B, X, w, b):

> *input*:
>
> > A: $n \times n$
> > B: $n \times n$
>
> *output*:
>
> > X: $n \times n$
> > w: $1 \times n$
> > b: $1 \times n$

geigensystemselectr(A, B, *range*, X, w, b):

> *input*:
>
> > A: $n \times n$
> > B: $n \times n$
> > *range*: 1×2 or 2×1
>
> *output*:
>
> > X: $n \times m$
> > w: $1 \times m$
> > b: $1 \times m$

`leftgeigensystemselectr(`A, B, *range*, X, w, b`)`:

    input:

	A:	$n \times n$
	B:	$n \times n$
	range:	1×2 or 2×1

    output:

	X:	$m \times n$
	w:	$1 \times m$
	b:	$1 \times m$

`geigensystemselecti(`A, B, *index*, X, w, b`)`:

    input:

	A:	$n \times n$
	B:	$n \times n$
	index:	1×2 or 2×1

    output:

	X:	$n \times m$
	w:	$1 \times m$
	b:	$1 \times m$

`leftgeigensystemselecti(`A, B, *index*, X, w, b`)`:

    input:

	A:	$n \times n$
	B:	$n \times n$
	index:	1×2 or 2×1

    output:

	X:	$m \times n$
	w:	$1 \times m$
	b:	$1 \times m$

`geigensystemselectf(`A, B, f, X, w, b`)`:

    input:

	A:	$n \times n$
	B:	$n \times n$
	f:	1×1

    output:

	X:	$n \times m$
	w:	$1 \times m$
	b:	$1 \times m$

`leftgeigensystemselectf(`A, B, f, X, w, b`)`:

    input:

	A:	$n \times n$
	B:	$n \times n$
	f:	1×1

    output:

	X:	$m \times n$
	w:	$1 \times m$
	b:	$1 \times m$

Diagnostics

All functions return missing-value results if A or B has missing values.

Also see

[M-1] **LAPACK** — The LAPACK linear-algebra routines

[M-5] **geigensystem( )** — Generalized eigenvectors and eigenvalues

[M-5] **ghessenbergd( )** — Generalized Hessenberg decomposition

[M-5] **gschurd( )** — Generalized Schur decomposition

[M-4] **matrix** — Matrix functions

Title

[M-5] **ghessenbergd( )** — Generalized Hessenberg decomposition

Syntax

void ghessenbergd(*numeric matrix A , B , H , R , U , V*)

void _ghessenbergd(*numeric matrix A , B ,* $\quad\quad$ *U , V*)

Description

ghessenbergd(*A , B , H , R , U , V*) computes the generalized Hessenberg decomposition of two general, real or complex, square matrices, *A* and *B*, returning the upper Hessenberg form matrix in *H*, the upper triangular matrix in *R*, and the orthogonal (unitary) matrices in *U* and *V*.

_ghessenbergd(*A , B , U , V*) mirrors ghessenbergd(), the difference being that it returns *H* in *A* and *R* in *B*.

_ghessenbergd_la() is the interface into the LAPACK routines used to implement the above function; see [M-1] **LAPACK**. Its direct use is not recommended.

Remarks

The generalized Hessenberg decomposition of two square, numeric matrices (**A** and **B**) can be written as

$$\mathbf{U}' \times \mathbf{A} \times \mathbf{V} = \mathbf{H}$$

$$\mathbf{U}' \times \mathbf{B} \times \mathbf{V} = \mathbf{R}$$

where **H** is in upper Hessenberg form, **R** is upper triangular, and **U** and **V** are orthogonal matrices if **A** and **B** are real or are unitary matrices otherwise.

In the example below, we define A and B, obtain the generalized Hessenberg decomposition, and list H and Q.

```
: A = (6, 2, 8, -1\-3, -4, -6, 4\0, 8, 4, 1\-8, -7, -3, 5)
: B = (8, 0, -8, -1\-6, -2, -6, -1\-7, -6, 2, -6\1, -7, 9, 2)
: ghessenbergd(A, B, H=., R=., U=., V=.)
: H
```

	1	2	3	4
1	-4.735680169	1.363736029	5.097381347	3.889763589
2	9.304479208	-8.594240253	-7.993282943	4.803411217
3	0	4.553169015	3.236266637	-2.147709419
4	0	0	6.997043028	-3.524816722

```
: R
```

	1	2	3	4
1	-12.24744871	-1.089095534	-1.848528639	-5.398470103
2	0	-5.872766311	8.891361089	3.86967647
3	0	0	9.056748937	1.366322731
4	0	0	0	8.357135399

Conformability

ghessenbergd(A, B, H, R, U, V):

> *input*:
>> A: $n \times n$
>> B: $n \times n$
>
> *output*:
>> H: $n \times n$
>> R: $n \times n$
>> U: $n \times n$
>> V: $n \times n$

_ghessenbergd(A, B, U, V):

> *input*:
>> A: $n \times n$
>> B: $n \times n$
>
> *output*:
>> A: $n \times n$
>> B: $n \times n$
>> U: $n \times n$
>> V: $n \times n$

Diagnostics

_ghessenbergd( ) aborts with error if A or B is a view.

ghessenbergd( ) and _ghessenbergd( ) return missing results if A or B contains missing values.

Also see

[M-1] **LAPACK** — The LAPACK linear-algebra routines

[M-5] **gschurd( )** — Generalized Schur decomposition

[M-4] **matrix** — Matrix functions

Title

[M-5] **ghk( )** — Geweke–Hajivassiliou–Keane (GHK) multivariate normal simulator

Syntax

S = ghk_init(*real scalar npts*)

(varies)	ghk_init_method(S [, *string scalar method*])
(varies)	ghk_init_start(S [, *real scalar start*])
(varies)	ghk_init_pivot(S [, *real scalar pivot*])
(varies)	ghk_init_antithetics(S [, *real scalar anti*])

real scalar ghk_query_npts(S)

real scalar ghk(S, *real vector x*, V)

real scalar ghk(S, *real vector x*, V, *real rowvector dfdx, dfdv*)

where S, if declared, should be declared

transmorphic S

and where *method*, optionally specified in ghk_init_method(), is

method	Description
"halton"	Halton sequences
"hammersley"	Hammersley's variation of the Halton set
"random"	pseudorandom uniforms

Description

S = ghk_init(*npts*) constructs the transmorphic object S. Calls to ghk_init_method(S, *method*), ghk_init_start(S, *start*), ghk_init_pivot(S, *pivot*), and ghk_init_antithetics(S, *anti*) prior to calling ghk(S, ...) allow you to modify the simulation algorithm through the object S.

ghk(S, x, V) returns a real scalar containing the simulated value of the multivariate normal (MVN) distribution with variance–covariance V at the point x. First, code S = ghk_init(*npts*) and then use ghk(S, ...) to obtain the simulated value based on *npts* simulation points.

ghk(S, x, V, *dfdx*, *dfdv*) does the same thing but also returns the first-order derivatives of the simulated probability with respect to x in *dfdx* and the simulated probability derivatives with respect to vech(V) in *dfdv*. See vech() in [M-5] **vec( )** for details of the half-vectorized operator.

The ghk_query_npts(S) function returns the number of simulation points, the same value given in the construction of the transmorphic object S.

505

Remarks

Halton and Hammersley point sets are composed of deterministic sequences on [0,1] and, for sets of dimension less than 10, generally have better coverage than that of the uniform pseudorandom sequences.

Antithetic draws effectively double the number of points and reduce the variability of the simulated probability. For draw u, the antithetic draw is $1 - u$. To use antithetic draws, call ghk_init_antithetic(S, 1) prior to executing ghk( ).

By default, ghk( ) will pivot the wider intervals of integration (and associated rows/columns of the covariance matrix) to the interior of the multivariate integration. This improves the accuracy of the quadrature estimate. When ghk( ) is used in a likelihood evaluator for [R] **ml** or [M-5] **optimize( )**, discontinuities may result in the computation of numerical second-order derivatives using finite differencing (for the Newton–Raphson optimize technique) when few simulation points are used, resulting in a nonpositive-definite Hessian. To turn off the interval pivoting, call ghk_init_pivot(S, 0) prior to executing ghk( ).

If you are using ghk( ) in a likelihood evaluator, be sure to use the same sequence with each call to the likelihood evaluator. For a uniform pseudorandom sequence, ghk_init_method("random"), set the seed of the uniform random-variate generator—see rseed( ) in [M-5] **runiform( )**—to the same value with each call to the likelihood evaluator.

If you are using the Halton or Hammersley point sets, you will want to keep the sequences going with each call to ghk( ) within one likelihood evaluation. This can be done in one expression executed after each call to ghk( ): ghk_init_start(S, ghk_init_start(S) + ghk_query_npts(S)). With each call to the likelihood evaluator, you will need to reset the starting index to 1. This last point assumes that the transmorphic object S is not recreated on each call to the likelihood evaluator.

Unlike ghkfast_init( ), the transmorphic object S created by ghk_init( ) is inexpensive to create, so it is possible to re-create it with each call to your likelihood evaluator instead of storing it as external global and reusing the object with each likelihood evaluation. Alternatively, the initialization function for optimize( ), optimize_init_arguments( ), can be used.

Conformability

All initialization functions have 1×1 inputs and have 1×1 or *void* outputs except

ghk_init(*npts*):

input:

	npts:	1×1

output:

	S:	transmorphic

ghk_query_npts(*S*):

input:

	S:	transmorphic

output:

	result:	1×1

ghk(*S*, *x*, *V*):

input:

	S:	transmorphic
	x:	$1 \times m$ or $m \times 1$
	V:	$m \times m$ (symmetric, positive definite)

output:

	result:	1×1

ghk(*S*, *x*, *V*, *dfdx*, *dfdv*):

input:

	S:	transmorphic
	x:	$1 \times m$ or $m \times 1$
	V:	$m \times m$ (symmetric, positive definite)

output:

	result:	1×1
	dfdx:	$1 \times m$
	dfdv:	$1 \times m(m+1)/2$

Diagnostics

The maximum dimension, m, is 20.

V must be symmetric and positive definite. ghk() will return a missing value when V is not positive definite. When ghk() is used in an ml (or optimize()) likelihood evaluator, return a missing likelihood to ml and let ml take the appropriate action (i.e., step halving).

Also see

[M-5] **ghkfast( )** — GHK multivariate normal simulator using pregenerated points

[M-5] **halton( )** — Generate a Halton or Hammersley set

[M-4] **statistical** — Statistical functions

Title

[M-5] **ghkfast( )** — GHK multivariate normal simulator using pregenerated points

Syntax

S = ghkfast_init(*real scalar n*, *npts*, *dim*, *string scalar method*)

(varies)	ghkfast_init_pivot(S [, *real scalar pivot*])
(varies)	ghkfast_init_antithetics(S [, *real scalar anti*])
real scalar	ghkfast_query_n(S)
real scalar	ghkfast_query_npts(S)
real scalar	ghkfast_query_dim(S)
string scalar	ghkfast_query_method(S)
string scalar	ghkfast_query_rseed(S)
real matrix	ghkfast_query_pointset_i(S, *i*)
real colvector	ghkfast(S, *real matrix X*, *V*)
real colvector	ghkfast(S, *real matrix X*, *V*, *dfdx*, *dfdv*)
real scalar	ghkfast_i(S, *real matrix X*, *V*, *i*)
real scalar	ghkfast_i(S, *real matrix X*, *V*, *i*, *dfdx*, *dfdv*)

where S, if it is declared, should be declared

transmorphic S

and where *method* specified in ghkfast_init() is

method	Description
"halton"	Halton sequences
"hammersley"	Hammersley's variation of the Halton set
"random"	pseudorandom uniforms
"ghalton"	generalized Halton sequences

Description

Please see [M-5] **ghk( )**. The routines documented here do the same thing, but ghkfast() can be faster at the expense of using more memory. First, code S = ghkfast_init(...) and then use ghkfast(S, ...) to obtain the simulated values. There is a time savings because the simulation points are generated once in ghkfast_init(), whereas for ghk() the points are generated on each call to ghk(). Also, ghkfast() can generate simulated probabilities from the generalized Halton sequence; see [M-5] **halton( )**.

508

ghkfast_init(n, *npts*, *dim*, *method*) computes the simulation points to be used by ghkfast( ). Inputs n, *npts*, and *dim* are the number of observations, the number of repetitions for the simulation, and the maximum dimension of the multivariate normal (MVN) distribution, respectively. Input *method* specifies the type of points to generate and can be one of "halton", "hammersley", "random", or "ghalton".

ghkfast(S, X, V) returns an $n \times 1$ real vector containing the simulated values of the MVN distribution with *dim* $\times$ *dim* variance–covariance matrix V at the points stored in the rows of the $n \times$ *dim* matrix X.

ghkfast(S, X, V, *dfdx*, *dfdv*) does the same thing as ghkfast(S, X, V) but also returns the first-order derivatives of the simulated probability with respect to the rows of X in *dfdx* and the simulated probability derivatives with respect to vech(V) in *dfdv*. See vech( ) in [M-5] **vec( )** for details of the half-vectorized operator.

The ghk_query_n(S), ghk_query_npts(S), ghk_query_dim(S), and ghk_query_method(S) functions extract the number of observations, number of simulation points, maximum dimension, and method of point-set generation that is specified in the construction of the transmorphic object S. Use ghk_query_rseed(S) to retrieve the uniform random-variate seed used to generate the "random" or "ghalton" point sets. The ghkfast_query_pointset_i(S, i) function will retrieve the ith point set used to simulate the MVN probability for the ith observation.

The ghkfast_i(S, X, V, i, ...) function computes the probability and derivatives for the ith observation, $i = 1, \ldots, n$.

Remarks

For problems where repetitive calls to the GHK algorithm are required, ghkfast( ) might be a preferred alternative to ghk( ). Generating the points once at the outset of a program produces a speed increase. For problems with many observations or many simulation points per observation, ghkfast( ) will be faster than ghk( ) at the cost of requiring more memory.

If ghkfast( ) is used within a likelihood evaluator for ml or optimize( ), you will need to store the transmorphic object S as an external global and reuse the object with each likelihood evaluation. Alternatively, the initialization function for optimize( ), optimize_init_arguments( ), can be used.

Prior to calling ghkfast( ), call ghkfast_init_npivot(S, 1) to turn off the integration interval pivoting that takes place in ghkfast( ). By default, ghkfast( ) pivots the wider intervals of integration (and associated rows/columns of the covariance matrix) to the interior of the multivariate integration to improve quadrature accuracy. This option may be useful when ghkfast( ) is used in a likelihood evaluator for [R] **ml** or [M-5] **optimize( )** and few simulation points are used for each observation. Here the pivoting may cause discontinuities when computing numerical second-order derivatives using finite differencing (for the Newton–Raphson technique), resulting in a nonpositive-definite Hessian.

Also the sequences "halton", "hammersley", and "random", ghkfast( ) will use the generalized Halton sequence, "ghalton". Generalized Halton sequences have the same uniform coverage (low discrepancy) as the Halton sequences with the addition of a pseudorandom uniform component. Therefore, "ghalton" sequences are like "random" sequences in that you should set the random-number seed before using them if you wish to replicate the same point set; see [M-5] **runiform( )**.

Conformability

All initialization functions have 1×1 inputs and have 1×1 or *void* outputs, and all query functions have the *transmorphic* input and 1×1 outputs except

ghkfast_init(n, *npts*, *dim*, *method*):

> *input*:

n:	1×1
npts:	1×1
dim:	1×1
method:	1×1

> *output*:

result:	*transmorphic*

ghkfast_query_pointset_i(S, i):

> *input*:

S:	*transmorphic*
i:	1×1

> *output*:

result:	*npts* $\times$ *dim*

ghkfast(S, X, V):

> *input*:

S:	*transmorphic*
X:	$n \times dim$
V:	$dim \times dim$ (symmetric, positive definite)

> *output*:

result:	$n \times 1$

ghkfast(S, X, V, *dfdx*, *dfdv*):

> *input*:

S:	*transmorphic*
X:	$n \times dim$
V:	$dim \times dim$ (symmetric, positive definite)

> *output*:

result:	$n \times 1$
dfdx:	$n \times dim$
dfdv:	$n \times dim(dim + 1)/2$

ghkfast_i(S, X, V, i, *dfdx*, *dfdv*):

> *input*:

S:	*transmorphic*
X:	$n \times dim$ or $1 \times dim$
V:	$dim \times dim$ (symmetric, positive definite)
i:	1×1 $(1 \leq i \leq n)$

> *output*:

result:	$n \times 1$
dfdx:	$1 \times dim$
dfdv:	$1 \times dim(dim + 1)/2$

Diagnostics

ghkfast_init(n, *npts*, *dim*, *method*) aborts with error if the dimension, *dim*, is greater than 20.

ghkfast(S, X, V, ...) and ghkfast_i(S, X, V, i, ...) require that V be symmetric and positive definite. If V is not positive definite, then the returned vector (scalar) is filled with missings.

Also see

[M-5] **ghk( )** — Geweke–Hajivassiliou–Keane (GHK) multivariate normal simulator

[M-5] **halton( )** — Generate a Halton or Hammersley set

[M-4] **statistical** — Statistical functions

Title

[M-5] gschurd( ) — Generalized Schur decomposition

Syntax

void   gschurd(A, B, T, R, U, V, w, b)

void   _gschurd(A, B, U, V, w, b)

void   gschurdgroupby(A, B, f, T, R, U, V, w, b, m)

void _gschurdgroupby(A, B, f, U, V, w, b, m)

Description

gschurd(A, B, T, R, U, V, w, b) computes the generalized Schur decomposition of two square, numeric matrices, A and B, and the generalized eigenvalues. The decomposition is returned in the Schur-form matrix, T; the upper-triangular matrix, R; and the orthogonal (unitary) matrices, U and V. The generalized eigenvalues are returned in the complex vectors w and b.

gschurdgroupby(A, B, f, T, R, U, V, w, b, m) computes the generalized Schur decomposition of two square, numeric matrices, A and B, and the generalized eigenvalues, and groups the results according to whether a condition on each generalized eigenvalue is satisfied. f is a pointer to the function that implements the condition on each generalized eigenvalue, as discussed below. The number of generalized eigenvalues for which the condition is true is returned in m.

_gschurd() mirrors gschurd(), the difference being that it returns T in A and R in B.

_gschurdgroupby() mirrors gschurdgroupby(), the difference being that it returns T in A and R in B.

_gschurd_la() and _gschurdgroupby_la() are the interfaces into the LAPACK routines used to implement the above functions; see [M-1] **LAPACK**. Their direct use is not recommended.

Remarks

Remarks are presented under the following headings:

> *Generalized Schur decomposition*
> *Grouping the results*

Generalized Schur decomposition

The generalized Schur decomposition of a pair of square, numeric matrices, $\mathbf{A}$ and $\mathbf{B}$, can be written as

$$\mathbf{U}' \times \mathbf{A} \times \mathbf{V} = \mathbf{T}$$

$$\mathbf{U}' \times \mathbf{B} \times \mathbf{V} = \mathbf{R}$$

where $\mathbf{T}$ is in Schur form, $\mathbf{R}$ is upper triangular, and $\mathbf{U}$ and $\mathbf{V}$ are orthogonal if $\mathbf{A}$ and $\mathbf{B}$ are real and are unitary if $\mathbf{A}$ or $\mathbf{B}$ is complex. The complex vectors $\mathbf{w}$ and $\mathbf{b}$ contain the generalized eigenvalues.

512

If **A** and **B** are real, **T** is in real Schur form and **R** is a real upper-triangular matrix. If **A** or **B** is complex, **T** is in complex Schur form and **R** is a complex upper-triangular matrix.

In the example below, we define A and B, obtain the generalized Schur decomposition, and list T and R.

```
: A = (6, 2, 8, -1\-3, -4, -6, 4\0, 8, 4, 1\-8, -7, -3, 5)
: B = (8, 0, -8, -1\-6, -2, -6, -1\-7, -6, 2, -6\1, -7, 9, 2)
: gschurd(A, B, T=., R=., U=., V=., w=., b=.)
```

```
: T
```

	1	2	3	4
1	12.99313938	1.746927947	3.931212285	-10.91622337
2	0	.014016016	6.153566902	1.908835695
3	0	-4.362999645	1.849905717	-2.998194791
4	0	0	0	-5.527285433

```
: R
```

	1	2	3	4
1	4.406836593	6.869534063	-1.840892081	1.740906311
2	0	13.88730687	0	-.6995556735
3	0	0	9.409495218	-4.659386723
4	0	0	0	9.453808732

```
: w
```

	1	2	3	4
1	12.9931394	.409611804+1.83488354i	.024799819-.111092453i	-5.52728543

```
: b
```

	1	2	3	4
1	4.406836593	4.145676341	.2509986829	9.453808732

Generalized eigenvalues can be obtained by typing

```
: w:/b
```

	1	2	3	4
1	2.94840508	.098804579+.442601735i	.098804579-.442601735i	-.584662287

Grouping the results

gschurdgroupby( ) reorders the generalized Schur decomposition so that a selected group of generalized eigenvalues appears in the leading block of the pair w and b. It also reorders the generalized Schur form T, R, and orthogonal (unitary) matrices, U and V, correspondingly.

We must pass gschurdgroupby( ) a pointer to a function that implements our criterion. The function must accept two arguments, a complex scalar and a real scalar, so that it can receive a generalized eigenvalue, and it must return the real value 0 to indicate rejection and a nonzero real value to indicate selection.

In the following example, we use gschurdgroupby( ) to put the finite, real, generalized eigenvalues first. One of the arguments to schurdgroupby( ) is a pointer to the function onlyreal( ) which accepts two arguments, a complex scalar and a real scalar that define a generalized eigenvalue. onlyreal() returns 1 if the generalized eigenvalue is finite and real; it returns zero otherwise.

```
: real scalar onlyreal(complex scalar w, real scalar b)
> {
>         if(b==0) return(0)
>         if(Im(w/b)==0) return(1)
>         return(0)
> }
: gschurdgroupby(A, B, &onlyreal(), T=., R=., U=., V=., w=., b=., m=.)
```

We obtain

```
: T
                      1              2              3              4

         1     12.99313938     8.19798168     6.285710813     5.563547054
         2               0    -5.952366071    -1.473533834     2.750066482
         3               0              0     -.2015830885     3.882051743
         4               0              0     6.337230739     1.752690714
```

```
: R
                      1              2              3              4

         1     4.406836593     2.267479575    -6.745927817     1.720793701
         2               0    10.18086202     -2.253089622     5.74882307
         3               0              0     -12.5704981              0
         4               0              0              0     9.652818299
```

```
: w
               1             2               3                        4

    1     12.9931394    -5.95236607    .36499234+1.63500766i    .36499234-1.63500766i
```

```
: b
                 1              2              3              4

    1      4.406836593    10.18086202    3.694083258    3.694083258
```

```
: w:/b
               1              2               3                        4

    1      2.94840508    -.584662287    .098804579+.442601735i    .098804579-.442601735i
```

m contains the number of real, generalized eigenvalues

```
: m
  2
```

Conformability

gschurd(A, B, T, R, U, V, w, b):

   input:

A:	$n \times n$
B:	$n \times n$

   output:

T:	$n \times n$
R:	$n \times n$
U:	$n \times n$
V:	$n \times n$
w:	$1 \times n$
b:	$1 \times n$

_gschurd(A, B, U, V, w, b):

   input:

A:	$n \times n$
B:	$n \times n$

   output:

A:	$n \times n$
B:	$n \times n$
U:	$n \times n$
V:	$n \times n$
w:	$1 \times n$
b:	$1 \times n$

gschurdgroupby(A, B, f, T, R, U, V, w, b, m):

   input:

A:	$n \times n$
B:	$n \times n$
f:	1×1

   output:

T:	$n \times n$
R:	$n \times n$
U:	$n \times n$
V:	$n \times n$
w:	$1 \times n$
b:	$1 \times n$
m:	1×1

(Continued on next page)

_gschurdgroupby(A, B, f, U, V, w, b, m):

   input:

A:	$n \times n$
B:	$n \times n$
f:	$.1 \times 1$

   output:

A:	$n \times n$
B:	$n \times n$
U:	$n \times n$
V:	$n \times n$
w:	$1 \times n$
b:	$1 \times n$
m:	1×1

Diagnostics

_gschurd( ) and _gschurdgroupby( ) abort with error if A or B is a view.

gschurd( ), _gschurd( ), gschurdgroupby( ), and _gschurdgroupby( ) return missing results if A or B contains missing values.

Also see

[M-1] **LAPACK** — The LAPACK linear-algebra routines

[M-5] **ghessenbergd( )** — Generalized Hessenberg decomposition

[M-5] **geigensystem( )** — Generalized eigenvectors and eigenvalues

[M-4] **matrix** — Matrix functions

Title

[M-5] halton( ) — Generate a Halton or Hammersley set

Syntax

real matrix	`halton`(*real scalar n*, *real scalar d*)
real matrix	`halton`(*real scalar n*, *real scalar d*, *real scalar start*)
real matrix	`halton`(*real scalar n*, *real scalar d*, *real scalar start*, *real scalar hammersley*)
void	`_halton`(*real matrix x*)
void	`_halton`(*real matrix x*, *real scalar start*)
void	`_halton`(*real matrix x*, *real scalar start*, *real scalar hammersley*)
real colvector	`ghalton`(*real scalar n*, *real scalar base*, *real scalar u*)

Description

`halton`(*n*, *d*) returns an $n \times d$ matrix containing a Halton set of length *n* and dimension *d*.

`halton`(*n*, *d*, *start*) does the same thing, but the first row of the returned matrix contains the sequences starting at index *start*. The default is *start* = 1.

`halton`(*n*, *d*, *start*, *hammersley*), with *hammersley* $\neq$ 0, returns a Hammersley set of length *n* and dimension *d* with the first row of the returned matrix containing the sequences starting at index *start*.

`_halton`(*x*) modifies the $n \times d$ matrix *x* so that it contains a Halton set of dimension *d* of length *n*.

`_halton`(*x*, *start*) does the same thing, but the first row of the returned matrix contains the sequences starting at index *start*. The default is *start* = 1.

`_halton`(*x*, *start*, *hammersley*), with *hammersley* $\neq$ 0, returns a Hammersley set of length *n* and dimension *d* with the first row of the returned matrix containing the sequences starting at index *start*.

`ghalton`(*n*, *base*, *u*) returns an $n \times 1$ vector containing a (generalized) Halton sequence using base *base* and starting from scalar $0 \leq u < 1$. For *u* = 0, the standard Halton sequence is generated. If *u* is uniform $(0, 1)$, a randomized Halton sequence is generated.

Remarks

The Halton sequences are generated from the first *d* primes and generally have more uniform coverage over the unit cube of dimension *d* than that of sequences generated from pseudouniform random numbers. However, Halton sequences based on large primes ($d > 10$) can be highly correlated, and their coverage can be worse than that of the pseudorandom uniform sequences.

The Hammersley set contains the sequence $(2 * i - 1)/(2 * n)$, $i = 1, \ldots, n$, in the first dimension and Halton sequences for dimensions 2, ..., *d*.

_halton() modifies *x* and can be used when repeated calls are made to generate long sequences in blocks. Here update the *start* index between calls by using *start* = *start* + rows(*x*).

ghalton() uses the base *base*, preferably a prime, and generates a Halton sequence using $0 \leq u < 1$ as a starting value. If *u* is uniform $(0, 1)$, the sequence is a randomized Halton sequence. For $u = 0$, the sequence is the standard Halton sequence. Blocks of sequences can be generated by ghalton() by using the last value in the vector returned from a previous call as *u*. For example,

```
x = J(n,1,0)
for (i=1; i<=k; i++) {
    x[.] = ghalton(n, base, x[n])
    ...
}
```

Conformability

halton(*n*, *d*, *start*, *hammersley*):

   input:

n:	1×1	
d:	1×1	
start:	1×1	(optional)
hammersley:	1×1	(optional)

   output:

result:	$n \times d$

_halton(*x*, *start*, *hammersley*):

   input:

x:	$n \times d$	
start:	1×1	(optional)
hammersley:	1×1	(optional)

   output:

x:	$n \times d$

ghalton(*n*, *base*, *u*):

   input:

n:	1×1
base:	1×1
u:	1×1

   output:

result:	$n \times 1$

Diagnostics

The maximum dimension, *d*, is 20. The scalar index *start* must be a positive integer, and the scalar *u* must be such that $0 \leq u < 1$.

Also see

[M-4] **mathematical** — Important mathematical functions

Title

[M-5] **hash1( )** — Jenkins' one-at-a-time hash function

Syntax

real scalar hash1(x [, n [, *byteorder*]])

where

x:	of any type except struct and of any dimension.
n:	*real scalar*; $1 \le n \le 2{,}147{,}483{,}647$ or . (missing). Optional; default . (missing).
byteorder:	*real scalar*; 1 (HILO), 2 (LOHI), . (missing, natural byte order). Optional; default . (missing).

Description

hash1(x) returns Jenkins' one-at-a-time hash calculated over the bytes of x; $0 \le$ hash1(x) $\le$ 4,294,967,295.

hash1(x, n) returns Jenkins' one-at-a-time hash scaled to $1 \le$ hash1(x, n) $\le n$, assuming $n <$. (missing). hash1(x, .) is equivalent to hash1(x).

hash1(x, n, *byteorder*) returns hash1(x, n) performed on the bytes of x ordered as they would be on a HILO computer (*byteorder* $= 1$), or as they would be on a LOHI computer (*byteorder* $= 2$), or as they are on this computer (*byteorder* $\ge$.). See [M-5] **byteorder( )** for a definition of byte order.

In all cases, the values returned by hash1( ) are integers.

Remarks

Calculation is significantly faster using the natural byte order of the computer. Argument *byteorder* is included for those rare cases when it is important to calculate the same hash value across different computers, which in the case of hash1( ) is mainly for testing. hash1( ), being a one-at-a-time method, is not sufficient for constructing digital signatures. It is sufficient for constructing hash tables; see [M-5] **asarray( )**, in which case, byte order is irrelevant. Also note that because strings occur in the same order on all computers, the value of *byteorder* is irrelevant when x is a string.

For instance,

```
: hash1("this"), hash1("this",.,1), hash1("this",.,2)
            1              2              3

  1     2385389520     2385389520     2385389520


: hash1(15), hash1(15,.,1), hash1(15,.,2)
           1              2              3

  1      463405819     3338064604      463405819
```

519

The computer on which this example was run is evidently *byteorder* $= 2$, meaning LOHI, or least-significant byte first.

In a Mata context, it is the two-argument form of hash1() that is most useful. In that form, the full result is mapped onto $[1, n]$:

$$\text{hash}(x, n) = \text{floor}((\text{hash}(x)/4294967295)\text{*n}) + 1$$

For instance,

```
: hash1("this", 10)
  6
: hash1(15, 10)
  2
```

The result of hash$(x, 10)$ could be used directly to index a 10×1 array.

Conformability

hash1(x, n, *byteorder*):

x:	$r \times c$	
n:	1×1	(optional)
byteorder:	1×1	(optional)
result:	1×1	

Diagnostics

None.

Note that hash1$(x[, \ldots])$ never returns a missing result, even if x is or contains a missing value. In the missing case, the hash value is calculated of the missing value. Also note that x can be a vector or a matrix, in which case the result is calculated over the elements aligned row-wise as if they were a single element. Thus hash1(("a", "b")) == hash1("ab").

References

Jenkins, B. 1997. *Dr. Dobb's Journal.* Algorithm alley: Hash functions. http://www.ddj.com/184410284.

——. unknown. A hash function for hash table lookup. http://www.burtleburtle.net/bob/hash/doobs.html.

Also see

[M-5] **asarray()** — Associative arrays

[M-4] **programming** — Programming functions

Title

Syntax

void hessenbergd(*numeric matrix A*, *H*, *Q*)

void _hessenbergd(*numeric matrix A*, *Q*)

Description

hessenbergd(*A*, *H*, *Q*) calculates the Hessenberg decomposition of a square, numeric matrix, *A*, returning the upper Hessenberg form matrix in *H* and the orthogonal (unitary) matrix in *Q*. *Q* is orthogonal if *A* is real and unitary if *A* is complex.

_hessenbergd(*A*, *Q*) does the same as hessenbergd() except that it returns *H* in *A*.

_hessenbergd_la() is the interface into the LAPACK routines used to implement the above function; see [M-1] **LAPACK**. Its direct use is not recommended.

Remarks

The Hessenberg decomposition of a matrix, **A**, can be written as

$$\mathbf{Q}' \times \mathbf{A} \times \mathbf{Q} = \mathbf{H}$$

where **H** is upper Hessenberg; **Q** is orthogonal if **A** is real or unitary if **A** is complex.

A matrix **H** is in upper Hessenberg form if all entries below its first subdiagonal are zero. For example, a 5 × 5 upper Hessenberg matrix looks like

```
        1   2   3   4   5
    1   x   x   x   x   x
    2   x   x   x   x   x
    3   0   x   x   x   x
    4   0   0   x   x   x
    5   0   0   0   x   x
```

For instance,

```
: A
        1   2   3    4    5
    1   3   2   1   -2   -5
    2   4   2   1    0    3
    3   4   4   0    1   -1
    4   5   6   7   -2    4
    5   6   7   1    2   -1
```

```
: hessenbergd(A, H=., Q=.)
```

```
: H
```

521

	1	2	3	4	5
1	3	2.903464745	-.552977683	-4.78764119	-1.530555451
2	-9.643650761	7.806451613	2.878001755	5.1085876	5.580422694
3	0	-3.454023879	-6.119229633	-.2347200215	1.467932097
4	0	0	1.404136249	-1.715823624	-.9870601994
5	0	0	0	-2.668128952	-.971398356

: Q

	1	2	3	4	5
1	1	0	0	0	0
2	0	-.4147806779	-.0368006164	-.4047768558	-.8140997488
3	0	-.4147806779	-.4871239484	-.5692309155	.5163752637
4	0	-.5184758474	.8096135604	-.0748449196	.2647771074
5	0	-.6221710168	-.3253949238	.7117092805	-.0221645995

Many algorithms use a Hessenberg decomposition in the process of finding another decomposition with more structure.

Conformability

hessenbergd(A, H, Q):

   input:

      A: $n \times n$

   output:

      H: $n \times n$

      Q: $n \times n$

_hessenbergd(A, Q):

   input:

      A: $n \times n$

   output:

      A: $n \times n$

      Q: $n \times n$

Diagnostics

_hessenbergd( ) aborts with error if A is a view.

hessenbergd( ) and _hessenbergd( ) return missing results if A contains missing values.

Also see

[M-1] **LAPACK** — The LAPACK linear-algebra routines

[M-5] **schurd( )** — Schur decomposition

[M-4] **matrix** — Matrix functions

Title

Syntax

>   *real matrix*   Hilbert(*real scalar n*)
>
>   *real matrix*   invHilbert(*real scalar n*)

Description

Hilbert(*n*) returns the $n \times n$ Hilbert matrix, defined as matrix H with elements $H_{ij} = 1/(i+j-1)$.

invHilbert(*n*) returns the inverse of the $n \times n$ Hilbert matrix, defined as the matrix with elements $-1^{i+j}\,(i+j-1) \times \text{comb}(n+i-1, n-j) \times \text{comb}(n+j-1, n-i) \times \text{comb}(i+j-2, i-1)^2$.

Remarks

Hilbert(*n*) and invHilbert(*n*) are used in testing Mata. Hilbert matrices are notoriously ill conditioned. The determinants of the first five Hilbert matrices are 1, 1/12, 1/2,160, 1/6,048,000, and 1/266,716,800,000.

Conformability

Hilbert(*n*), invHilbert(*n*):
>   *n*: 1×1
>   *result*: $\text{trunc}(n) \times \text{trunc}(n)$

Diagnostics

None.

David Hilbert (1862–1943) was born near Königsberg, Prussia (now Kaliningrad, Russia), and studied mathematics at the university there. He joined the staff from 1886 to 1895, when he moved to Göttingen, where he stayed despite tempting offers to move. Hilbert was one of the outstanding mathematicians of his time, producing major work in several fields, including invariant theory, algebraic number theory, the foundations of geometry, functional analysis, integral equations, and the calculus of variations. In 1900 he identified 23 key problems in an address to the Second International Congress of Mathematicians in Paris that continues to influence directions in research (Hilbert 1902). Hilbert's name is perhaps best remembered through the idea of Hilbert space. His work on what are now known as Hilbert matrices was published in 1894.

References

Choi, M.-D. 1983. Tricks or treats with the Hilbert matrix. *American Mathematical Monthly* 90: 301–312.

Hilbert, D. 1894. Ein Beitrag zur Theorie des Legendreschen Polynoms. *Acta Mathematica* 18: 155–159.

——. 1902. Mathematical problems. *Bulletin of the American Mathematical Society* 8: 437–479.

Reid, C. 1970. *Hilbert.* Berlin: Springer.

Also see

[M-4] **standard** — Functions to create standard matrices

Title

[M-5] I( ) — Identity matrix

Syntax

real matrix   I(*real scalar n*)

real matrix   I(*real scalar m*, *real scalar n*)

Description

I(n) returns the $n \times n$ identity matrix.

I(m, n) returns an $m \times n$ matrix with 1s down its principal diagonal and 0s elsewhere.

Remarks

I() must be typed in uppercase.

Conformability

I(n):

n:	1×1	
result:	$n \times n$	

I(m, n):

m:	1×1	
n:	1×1	
result:	$m \times n$	

Diagnostics

I(n) aborts with error if n is less than 0 or is missing. n is interpreted as trunc(n).

I(m, n) aborts with error if m or n are less than 0 or if they are missing. m and n are interpreted as trunc(m) and trunc(n).

Also see

[M-4] **standard** — Functions to create standard matrices

Title

[M-5] **inbase( )** — Base conversion

Syntax

string matrix inbase(*real scalar base*, *real matrix x* [, *real scalar fdigits* [, *err*]])

real matrix   frombase(*real scalar base*, *string matrix s*)

Description

inbase(*base*, *x*) returns a string matrix containing the values of *x* in base *base*.

inbase(*base*, *x*, *fdigits*) does the same; *fdigits* specifies the maximum number of digits to the right of the base point to appear in the returned result when *x* has a fractional part. inbase(*base*, *x*) is equivalent to inbase(*base*, *x*, 8).

inbase(*base*, *x*, *fdigits*, *err*) is the same as inbase(*base*, *x*, *fdigits*), except that it returns in *err* the difference between *x* and the converted result.

x = frombase(*base*, *s*) is the inverse of *s* = inbase(*base*, *x*). It returns base *base* number *s* as a number. We are tempted to say, "as a number in base 10", but that is not exactly true. It returns the result as a *real*, that is, as an IEEE base-2 double-precision float that, when you display it, is displayed in base 10.

Remarks

Remarks are presented under the following headings:

> *Positive integers*
> *Negative integers*
> *Numbers with nonzero fractional parts*
> *Use of the functions*

Positive integers

inbase(2, 1691) is 11010011011; that is, 1691 base 10 equals 11010011011 base 2. frombase(2, "11010011011") is 1691.

inbase(3, 1691) is 2022122; that is, 1691 base 10 equals 2022122 base 3. frombase(3, "2022122") is 1691.

inbase(16, 1691) is 69b; that is, 1691 base 10 equals 1691 base 16. frombase(16, "69b") is 1691. (The base-16 digits are 0, 1, 2, 3, 4, 5, 6, 7, 8, 9, a, b, c, d, e, f.)

inbase(62, 1691) is rh; that is, 1691 base 10 equals rh base 62. frombase(62, "rh") is 1691. (The base-62 digits are 0, 1, 2, 3, 4, 5, 6, 7, 8, 9, a, b, . . . , z, A, B, . . . , Z.)

There is a one-to-one correspondence between the integers in different bases. The error of the conversion is always zero.

Negative integers

Negative integers are no different from positive integers. For instance, inbase(2, -1691) is -11010011011; that is, -1691 base 10 equals -11010011011 base 2. frombase(2, "-11010011011") is -1691.

The error of the conversion is always zero.

Numbers with nonzero fractional parts

inbase(2, 3.5) is 11.1; that is, 3.5 base 10 equals 11.1 base 2. frombase(2, "11.1") is 3.5.

inbase(3, 3.5) is 10.11111111.

inbase(3, 3.5, 20) is 10.11111111111111111111.

inbase(3, 3.5, 30) is 10.111111111111111111111111111111.

Therefore, 3.5 base 10 equals 1.1111... in base 3. There is no exact representation of one-half in base 3. The errors of the above three conversions are .0000762079, 1.433399e–10, and 2.45650e–15. Those are the values that would be returned in *err* if inbase(3, 3.5, *fdigits*, *err*) were coded.

frombase(3, "10.11111111") is 3.499923792.

frombase(3, "10.11111111111111111111") is 3.4999999998566.

frombase(3, "10.111111111111111111111111111111") is 3.49999999999999734.

inbase(16, 3.5) is 3.8; that is, 3.5 base 10 equals 3.8 base 16. The error is zero. frombase(16, "3.8") is 3.5.

inbase(62, 3.5) is 3.v; that is, 3.5 base 10 equals 3.v base 62. frombase(62, "3.v") is 3.5. The error is zero.

In inbase(*base*, *x*, *fdigits*), *fdigits* specifies the maximum number of digits to appear to the right of the base point. *fdigits* is required to be greater than or equal to 1. inbase(16, 3.5, *fdigits*) will be 3.8 regardless of the value of *fdigits* because more digits are unnecessary.

The error that is returned in inbase(*base*, *x*, *fdigits*, *err*) can be an understatement. For instance, inbase(16, .1, 14, *err*) is 0.1999999999999a and returned in *err* is 0 even though there is no finite-digit representation of 0.1 base 10 in base 16. That is because the .1 you specified in the call was not actually 0.1 base 10. The computer that you are using is binary, and it converted the .1 you typed to

$$0.0001100110011001100110011001100110011001100110011010 \text{ base } 2$$

before inbase() was ever called. 0.1999999999999a base 16 is an exact representation of that number.

Use of the functions

These functions are used mainly for teaching, especially on the sources and avoidance of roundoff error; see Gould (2006).

The functions can have a use in data processing, however, when used with integer arguments. You have a dataset with 10-digit identification numbers. You wish to record the 10-digit number, but more densely. You could convert the number to base 62. The largest 10-digit ID number possible is 9999999999, or aUKYOz base 62. You can record the ID numbers in a six-character string by using inbase(). If you needed the original numbers back, you could use frombase().

In a similar way, Stata internally uses base 36 for naming temporary files, and that was important when filenames were limited to eight characters. Base 36 allows Stata to generate up to 2,821,109,907,455 filenames before wrapping of filenames occurs.

Conformability

inbase(*base*, *x*, *fdigits*, *err*):

   input:

base:	1×1		
x:	$r \times c$		
fdigits:	1×1	(optional)	

   output:

err:	$r \times c$	(optional)
result:	$r \times c$	

frombase(*base*, *s*):

base:	1×1
s:	$r \times c$
result:	$r \times c$

Diagnostics

The digits used by inbase()/frombase() to encode/decode results are

0 0	*10* a	*20* k	*30* u	*40* E	*50* O	*60* Y
1 1	*11* b	*21* l	*31* v	*41* F	*51* P	*61* Z
2 2	*12* c	*22* m	*32* w	*42* G	*52* Q	
3 3	*13* d	*23* n	*33* x	*43* H	*53* R	
4 4	*14* e	*24* o	*34* y	*44* I	*54* S	
5 5	*15* f	*25* p	*35* z	*45* J	*55* T	
6 6	*16* g	*26* q	*36* A	*46* K	*56* U	
7 7	*17* h	*27* r	*37* B	*47* L	*57* V	
8 8	*18* i	*28* s	*38* C	*48* M	*58* W	
9 9	*19* j	*29* t	*39* D	*49* N	*59* X	

When *base* $\leq$ 36, frombase() treats A, B, C, . . ., as if they were a, b, c, ...

inbase(*base*, *x*, *fdigits*, *err*) returns . (missing) if *base* < 2, *base* > 62, *base* is not an integer, or *x* is missing. If *fdigits* is less than 1 or *fdigits* is missing, results are as if *fdigits* = 8 were specified.

frombase(*base*, *s*) returns . (missing) if *base* < 2, *base* > 62, *base* is not an integer, or *s* is missing; if *s* is not a valid base *base* number; or if the converted value of *s* is greater than 8.988e+307 in absolute value.

Reference

Gould, W. W. 2006. Mata Matters: Precision. *Stata Journal* 6: 550–560.

Also see

[M-4] **mathematical** — Important mathematical functions

Title

[M-5] indexnot( ) — Find character not in list

Syntax

> *real matrix*   indexnot(*string matrix* s_1, *string matrix* s_2)

Description

indexnot(s_1, s_2) returns the position of the first character of s_1 not found in s_2, or it returns 0 if all characters of s_1 are found in s_2.

Conformability

indexnot(s_1, s_2):
      s_1: $r_1 \times c_1$
      s_2: $r_2 \times c_2$, s_1 and s_2 r-conformable
   result: $\max(r_1, r_2) \times \max(c_1, c_2)$

Diagnostics

indexnot(s_1, s_2) returns 0 if all characters of s_1 are found in s_2.

Also see

[M-4] **string** — String manipulation functions

Title

Syntax

real vector   invorder(*real vector p*)

real vector   revorder(*real vector p*)

where p is assumed to be a permutation vector.

Description

invorder(p) returns the permutation vector that undoes the permutation performed by p.

revorder(p) returns the permutation vector that is the reverse of the permutation performed by p.

Remarks

See [M-1] **permutation** for a description of permutation vectors. To summarize,

1. Permutation vectors p are used to permute the rows or columns of a matrix X: $r \times c$.

   If p is intended to permute the rows of X, the permuted X is obtained via $Y = X[p, \cdot]$.

   If p is intended to permute the columns of X, the permuted X is obtained via $Y = X[\cdot, p]$.

2. If p is intended to permute the rows of X, it is called a row-permutation vector. Row-permutation vectors are $r \times 1$ column vectors.

3. If p is intended to permute the columns of X, it is called a column-permutation vector. Column-permutation vectors are $1 \times c$ row vectors.

4. Row-permutation vectors contain a permutation of the integers 1 to r.

5. Column-permutation vectors contain a permutation of the integers 1 to c.

Let us assume that p is a row-permutation vector, so that

$$Y = X[p, \cdot]$$

invorder(p) returns the row-permutation vector that undoes p:

$$X = Y[\text{invorder}(p), \cdot]$$

That is, using the matrix notation of [M-1] **permutation**,

$$Y = PX \qquad \text{implies} \qquad X = P^{-1}Y$$

If p is the permutation vector corresponding to permutation matrix P, invorder(p) is the permutation vector corresponding to permutation matrix P^{-1}.

revorder(p) returns the permutation vector that reverses the order of p. For instance, say that row-permutation vector p permutes the rows of X so that the diagonal elements are in ascending order. Then revorder(p) would permute the rows of X so that the diagonal elements would be in descending order.

Conformability

invorder(p), revorder(p):

$$
\begin{array}{rccc}
p: & r \times 1 & \text{or} & 1 \times c \\
\textit{result}: & r \times 1 & \text{or} & 1 \times c
\end{array}
$$

Diagnostics

invorder(p) and revorder(p) can abort with error or can produce meaningless results when p is not a permutation vector.

Also see

[M-1] **permutation** — An aside on permutation matrices and vectors

[M-4] **manipulation** — Matrix manipulation

Title

[M-5] invsym( ) — Symmetric real matrix inversion

Syntax

real matrix	invsym(*real matrix A*)
real matrix	invsym(*real matrix A*, *real vector order*)
void	_invsym(*real matrix A*)
void	_invsym(*real matrix A*, *real vector order*)

Description

invsym(*A*) returns a generalized inverse of real, symmetric, positive-definite matrix *A*.

invsym(*A*, *order*) does the same but allows you to specify which columns are to be swept first.

_invsym(*A*) and _invsym(*A*, *order*) do the same thing as invsym(*A*) and invsym(*A*, *order*) except that *A* is replaced with the generalized inverse result rather than the result being returned. _invsym( ) uses less memory than invsym( ).

invsym( ) and _invsym( ) are the routines Stata uses for calculating inverses of symmetric matrices.

Also see [M-5] **luinv( )**, [M-5] **qrinv( )**, and [M-5] **pinv( )** for general matrix inversion.

Remarks

Remarks are presented under the following headings:

> *Definition of generalized inverse*
> *Specifying the order in which columns are dropped*
> *Determining the rank, or counting the number of dropped columns*
> *Extracting linear dependencies*

Definition of generalized inverse

When the matrix is of full rank and positive definite, the generalized inverse equals the inverse, i.e., assuming *A* is $n \times n$,

$$\text{invsym}(A)*A = A*\text{invsym}(A) = \text{I}(n)$$

or, at least the above restriction is true up to roundoff error. When *A* is not full rank, the generalized inverse invsym( ) satisfies (ignoring roundoff error)

$$A*\text{invsym}(A)*A = A$$

$$\text{invsym}(A)*A*\text{invsym}(A) = \text{invsym}(A)$$

In the generalized case, there are an infinite number of inverse matrices that can satisfy the above restrictions. The one `invsym()` chooses is one that sets entire columns (and therefore rows) to 0, thus treating A as if it were of reduced dimension. Which columns (rows) are selected is determined on the basis of minimizing roundoff error.

In the above we talk as if determining whether a matrix is of full rank is an easy calculation. That is not true. Because of the roundoff error in the manufacturing and recording of A itself, columns that ought to be perfectly collinear will not be and yet you will still want `invsym()` to behave as if they were. `invsym()` tolerates a little deviation from collinearity in making the perfectly collinear determination.

Specifying the order in which columns are dropped

Left to make the decision itself, `invsym()` will choose which columns to drop (to set to 0) to minimize the overall roundoff error of the generalized inverse calculation. If column 1 and column 3 are collinear, then `invsym()` will choose to drop column 1 or column 3.

There are occasions, however, when you would like to ensure that a particular column or set of columns are not dropped. Perhaps column 1 corresponds to the intercept of a regression model and you would much rather, if one of columns 1 and 3 has to be dropped, that it be column 3.

Order allows you to specify the columns of the matrix that you would prefer not be dropped in the generalized inverse calculation. In the above example, to prevent column 1 from being dropped, you could code

>   invsym(A, 1)

If you would like to keep columns 1, 5, and 10 from being dropped, you can code

>   invsym(A, (1,5,10))

Specifying columns not to be dropped does not guarantee that they will not be dropped because they still might be collinear with each other or they might equal constants. However, if any other column can be dropped to satisfy your desire, it will be.

Determining the rank, or counting the number of dropped columns

If a column is dropped, 0 will appear on the corresponding diagonal entry. Hence, the rank of the original matrix can be extracted after inversion by `invsym()`:

```
: Ainv = invsym(A)
: rank = rows(Ainv)-diag0cnt(Ainv)
```

See [M-5] **diag0cnt( )**.

Extracting linear dependencies

The linear dependencies can be read from the rows of $A*$invsym(A):

```
: A*invsym(A)
```

	1	2	3
1	1	0	-5.20417e-17
2	-1	0	1
3	1.34441e-16	0	1

The above is interpreted to mean

$$x_1 = x_1$$
$$x_2 = -x_1 + x_3$$
$$x_3 = x_3$$

ignoring roundoff error.

Conformability

invsym(A), invsym(A, *order*):

A:	$n \times n$		
order:	$1 \times k$	or $k \times 1, k \le n$	(optional)
result:	$n \times n$		

_invsym(A), _invsym(A, *order*):

A:	$n \times n$		
order:	$1 \times k$	or $k \times 1, k \le n$	(optional)
output:			
A:	$n \times n$		

Diagnostics

invsym(A), invsym(A, *order*), _invsym(A), and _invsym(A, *order*) assume that A is symmetric; they do not check. If A is nonsymmetric, they treat it as if it were symmetric and equal to its upper triangle.

invsym( ) and _invsym( ) return a result containing missing values if A contains missing values.

_invsym( ) aborts with error if A is a view. Both functions abort with argument-out-of-range error if *order* is specified and contains values less than 1, greater than rows(A), or the same value more than once.

invsym( ) and _invsym( ) return a matrix of zeros if A is not positive definite.

Also see

[M-5] **cholinv( )** — Symmetric, positive-definite matrix inversion

[M-5] **luinv( )** — Square matrix inversion

[M-5] **qrinv( )** — Generalized inverse of matrix via QR decomposition

[M-5] **pinv( )** — Moore–Penrose pseudoinverse

[M-5] **diag0cnt( )** — Count zeros on diagonal

[M-4] **matrix** — Matrix functions

[M-4] **solvers** — Functions to solve AX=B and to obtain A inverse

Title

[M-5] **invtokens( )** — Concatenate string rowvector into string scalar

Syntax

>*string scalar* invtokens(*string rowvector s*)
>
>*string scalar* invtokens(*string rowvector s*, *string scalar c*)

Description

invtokens(*s*) returns the elements of *s*, concatenated into a string scalar with the elements separated by spaces. invtokens(*s*) is equivalent to invtokens(*s*, " ").

invtokens(*s*, *c*) returns the elements of *s*, concatenated into a string scalar with the elements separated by *c*.

Remarks

invtokens(*s*) is the inverse of tokens( ) (see [M-5] **tokens( )**); invtokens( ) returns the string obtained by concatenating the elements of *s* into a space-separated list.

invtokens(*s*, *c*) places *c* between the elements of *s* even when the elements of *s* are equal to "". For instance,

```
: s = ("alpha", "", "gamma", "")
: invtokens(s, ";")
  alpha;;gamma;
```

To remove separators between empty elements, use select( ) (see [M-5] **select( )**) to remove the empty elements from *s* beforehand:

```
: s2 = select(s, strlen(s):>0)
: s2
```

	1	2
1	alpha	gamma

```
: invtokens(s2, ";")
  alpha;gamma
```

Conformability

invtokens(*s*, *c*):

s:	$1 \times p$	
c:	1×1	(optional)
result:	1×1	

Diagnostics

If s is 1×0, invtokens(s,c) returns "".

Also see

[M-5] **tokens( )** — Obtain tokens from string

[M-4] **string** — String manipulation functions

Title

Syntax

> *real scalar*   isdiagonal(*numeric matrix A*)

Description

isdiagonal(*A*) returns 1 if *A* has only zeros off the principal diagonal and returns 0 otherwise. isdiagonal( ) may be used with either real or complex matrices.

Remarks

See [M-5] **diag( )** for making diagonal matrices out of vectors or out of nondiagonal matrices; see [M-5] **diagonal( )** for extracting the diagonal of a matrix into a vector.

Conformability

isdiagonal(*A*):
> *A*: $r \times c$
> *result*: 1×1

Diagnostics

isdiagonal(*A*) returns 1 if *A* is void.

Also see

[M-5] **diag( )** — Create diagonal matrix

[M-5] **diagonal( )** — Extract diagonal into column vector

[M-4] **utility** — Matrix utility functions

Title

> **[M-5] isfleeting( )** — Whether argument is temporary

Syntax

> *real scalar*   isfleeting(*polymorphic matrix A*)

where *A* is an argument passed to your function.

Description

isfleeting(*A*) returns 1 if *A* was constructed for the sole purpose of passing to your function, and returns 0 otherwise. If an argument is fleeting, then you may change its contents and be assured that the caller will not care or even know.

Remarks

Let us assume that you have written function myfunc(*A*) that takes *A*: $r \times c$ and returns an $r \times c$ matrix. Just to fix ideas, we will pretend that the code for myfunc( ) reads

```
real matrix myfunc(real matrix A)
{
        real scalar     i
        real matrix     B

        B=A
        for (i=1; i<=rows(B); i++) B[i,i] = 1
        return(B)
}
```

Function myfunc(*A*) returns a matrix equal to *A*, but with ones along the diagonal. Now let's imagine myfunc( ) in use. A snippet of the code might read

```
...
C = A*myfunc(D)*C
...
```

Here D is passed to myfunc( ), and the argument D is said not to be fleeting. Now consider another code snippet:

```
...
D = A*myfunc(D+E)*D
...
```

In this code snippet, the argument passed to myfunc( ) is D+E and that argument is fleeting. It is fleeting because it was constructed for the sole purpose of being passed to myfunc( ), and once myfunc( ) concludes, the matrix containing D+E will be discarded.

Arguments that are fleeting can be reused to save memory.

Look carefully at the code for myfunc(). It makes a copy of the matrix that it was passed. It did that to avoid damaging the matrix that it was passed. Making that copy, however, is unnecessary if the argument received was fleeting, because damaging something that would have been discarded anyway does not matter. Had we not made the copy, we would have saved not only computer time but also memory. Function myfunc() could be recoded to read

```
real matrix myfunc(real matrix A)
{
        real scalar     i
        real matrix     B

        if (isfleeting(A)) {
                for (i=1; i<=rows(A); i++) A[i,i] = 1
                return(A)
        }
        B=A
        for (i=1; i<=rows(B); i++) B[i,i] = 1
        return(B)
}
```

Here we wrote separate code for the fleeting and nonfleeting cases. That is not always necessary. We could use a pointer here to combine the two code blocks:

```
real matrix myfunc(real matrix A)
 {
        real scalar     i
        real matrix     B
        pointer scalar  p

        if (isfleeting(A)) p = &A
        else {
                B = A
                p = &B
        }
        for (i=1; i<=rows(*p); i++) (*p)[i,i] = 1
        return(*p)
}
```

Many official library functions come in two varieties: _foo(A, ...), which replaces A with the calculated result, and foo(A, ...), which returns the result leaving A unmodified. Invariably, the code for foo() reads

```
function foo(A, ...)
{
        matrix B

        if (isfleeting(A)) {
                _foo(A, ...)
                return(A)
        }
        _foo(B=A, ...)
        return(B)
}
```

This makes function foo() whoppingly efficient. If foo() is called with a temporary argument—an argument that could be modified without the caller being aware of it—then no extra copy of the matrix is ever made.

Conformability

isfleeting(A):
$$A: \quad r \times c$$
$$result: \quad 1 \times 1$$

Diagnostics

isfleeting(A) returns 1 if A is fleeting and not a view. The value returned is indeterminate if A is not an argument of the function in which it appears, and therefore the value of isfleeting() is also indeterminate when used interactively.

Also see

[M-4] **programming** — Programming functions

Title

Syntax

    real scalar isreal(*transmorphic matrix X*)

    real scalar iscomplex(*transmorphic matrix X*)

    real scalar isstring(*transmorphic matrix X*)

    real scalar ispointer(*transmorphic matrix X*)

Description

isreal(X) returns 1 if X is a real and returns 0 otherwise.

iscomplex(X) returns 1 if X is a complex and returns 0 otherwise.

isstring(X) returns 1 if X is a string and returns 0 otherwise.

ispointer(X) returns 1 if X is a pointer and returns 0 otherwise.

Remarks

These functions base their results on storage type. isreal() is not the way to check whether a number is real, since it might be stored as a complex and yet still be a real number, such as $2 + 0i$. To determine whether x is real, you want to use isrealvalues(X); see [M-5] **isrealvalues( )**.

Conformability

isreal(X), iscomplex(X), isstring(X), ispointer(X):
        X: $r \times c$
   result: 1×1

Diagnostics

These functions return 1 or 0; they cannot fail.

Also see

[M-5] **isrealvalues( )** — Whether matrix contains only real values

[M-5] **eltype( )** — Element type and organizational type of object

[M-4] **utility** — Matrix utility functions

Title

Syntax

real scalar   isrealvalues(*numeric matrix X*)

Description

isrealvalues(*X*) returns 1 if *X* is of storage type real or if *X* is of storage type complex and all elements are real or missing; 0 is returned otherwise.

Remarks

isrealvalues(*X*) is logically equivalent to *X*==Re(*X*) but is significantly faster and requires less memory.

Conformability

isrealvalues(*X*):
$$\begin{array}{rl} X: & r \times c \\ result: & 1 \times 1 \end{array}$$

Diagnostics

isrealvalues(*X*) returns 1 if *X* is void and complex.

Also see

[M-5] **isreal( )** — Storage type of matrix

[M-4] **utility** — Matrix utility functions

Title

Syntax

> *real scalar*   issymmetric(*transmorphic matrix A*)
>
> *real scalar*   issymmetriconly(*transmorphic matrix A*)

Description

issymmetric(*A*) returns 1 if $A==A'$ and returns 0 otherwise. (Also see mreldifsym() in [M-5] **reldif( )**.)

issymmetriconly(*A*) returns 1 if A==transposeonly(*A*) and returns 0 otherwise.

Remarks

issymmetric(*A*) and issymmetriconly(*A*) return the same result except when A is complex.

In the complex case, issymmetric(*A*) returns 1 if A is equal to its conjugate transpose, i.e., if A is Hermitian, which is the complex analog of symmetric. A is symmetric (Hermitian) if its off-diagonal elements are conjugates of each other and its diagonal elements are real.

issymmetriconly(*A*), on the other hand, uses the mechanical definition of symmetry: A is symmetriconly [*sic*] if its off-diagonal elements are equal. issymmetriconly() is uninteresting, mathematically speaking, but can be useful in certain data-management programming situations.

Conformability

issymmetric(*A*), issymmetriconly(*A*):
A:	$r \times c$
result:	1×1

Diagnostics

issymmetric(*A*) returns 0 if A is not square. If A is 0×0, it is symmetric.

issymmetriconly(*A*) returns 0 if A is not square. If A is 0×0, it is symmetriconly.

Charles Hermite (1822–1901) was born in Dieuze in eastern France and early showed an aptitude for mathematics, publishing two papers before entering university. He started studying at the Ecole Polytechnique but left after 1 year because of difficulties from a defect in his right foot. However, Hermite was soon appointed to a post at the same institution and later at the Sorbonne. He made outstanding contributions to number theory, algebra, and especially analysis, and he published the first proof that e is transcendental. Hermite's name carries on in Hermite polynomials, Hermite's differential equation, Hermite's formula of interpolation, and Hermitian matrices.

Reference

James, I. 2002. *Remarkable Mathematicians: From Euler to von Neumann.* Cambridge: Cambridge University Press.

Also see

[M-5] **makesymmetric( )** — Make square matrix symmetric (Hermitian)

[M-5] **reldif( )** — Relative/absolute difference

[M-4] **utility** — Matrix utility functions

Title

[M-5] isview( ) — Whether matrix is view

Syntax

> *real scalar*   isview(*transmorphic matrix X*)

Description

isview(*X*) returns 1 if *X* is a view and 0 otherwise.

See [M-6] **Glossary** for a definition of a view.

Remarks

View matrices are created by st_view(); see [M-5] **st_view( )**.

Conformability

```
isview(X):
        X:      r × c
   result:      1 × 1
```

Diagnostics

None.

Also see

[M-5] **eltype( )** — Element type and organizational type of object

[M-4] **utility** — Matrix utility functions

Title

Syntax

transmorphic matrix J(*real scalar r*, *real scalar c*, *scalar val*)

transmorphic matrix J(*real scalar r*, *real scalar c*, *matrix mat*)

Description

J(*r*, *c*, *val*) returns an $r \times c$ matrix with each element equal to *val*.

J(*r*, *c*, *mat*) returns an $(r*\text{rows}(mat)) \times (c*\text{cols}(mat))$ matrix with elements equal to *mat*.

The first, J(*r*, *c*, *val*), is how J() is commonly used. The first is nothing more than a special case of the second, J(*r*, *c*, *mat*), when *mat* is 1×1.

Remarks

Remarks are presented under the following headings:

> First syntax: J(r, c, val), val a scalar
> Second syntax: J(r, c, mat), mat a matrix

First syntax: J(r, c, val), val a scalar

J(*r*, *c*, *val*) creates matrices of constants. For example, J(2, 3, 0) creates

	1	2	3
1	0	0	0
2	0	0	0

J() must be typed in uppercase.

J() can create any type of matrix:

function	returns
J(2, 3, 4)	2×3 real matrix, each element $= 4$
J(2, 3, 4+5i)	2×3 complex matrix, each element $= 4 + 5i$
J(2, 3, "hi")	2×3 string matrix, each element $=$ "hi"
J(2, 3, &x)	2×3 pointer matrix, each element $=$ address of x

Also, J() can create void matrices:

J(0, 0, .)	0×0 real
J(0, 1, .)	0×1 real
J(1, 0, .)	1×0 real
J(0, 0, 1i)	0×0 complex
J(0, 1, 1i)	0×1 complex
J(1, 0, 1i)	1×0 complex
J(0, 0, "")	0×0 string
J(0, 1, "")	0×1 string
J(1, 0, "")	1×0 string
J(0, 0, NULL)	0×0 pointer
J(0, 1, NULL)	0×1 pointer
J(1, 0, NULL)	1×0 pointer

When J(r, c, val) is used to create a void matrix, the particular value of the third argument does not matter. Its element type, however, determines the type of matrix produced. Thus, J(0, 0, .), J(0, 0, 1), and J(0, 0, 1/3) all create the same result: a 0×0 real matrix. Similarly, J(0, 0, ""), J(0, 0, "name"), and J(0, 0, "?") all create the same result: a 0×0 string matrix. See [M-2] **void** to learn how void matrices are used.

Second syntax: J(r, c, mat), mat a matrix

J(r, c, mat) is a generalization of J(r, c, val). When the third argument is a matrix, that matrix is replicated in the result. For instance, if X is (1,2\3,4), then J(2, 3, X) creates

	1	2	3	4	5	6
1	1	2	1	2	1	2
2	3	4	3	4	3	4
3	1	2	1	2	1	2
4	3	4	3	4	3	4

J(r, c, val) is a special case of J(r, c, mat); it just happens that mat is 1×1.

The matrix created has $r*\text{rows}(mat)$ rows and $c*\text{cols}(mat)$ columns.

Note that J(r, c, mat) creates a void matrix if any of r, c, rows(mat), or cols(mat) are zero.

Conformability

J(*r*, *c*, *val*):

r:	1×1
c:	1×1
val:	1×1
result:	$r \times c$

J(*r*, *c*, *mat*):

r:	1×1
c:	1×1
mat:	$m \times n$
result:	$r*m \times c*n$

Diagnostics

J(*r*, *c*, *val*) and J(*r*, *c*, *mat*) abort with error if $r < 0$ or $c < 0$, or if $r \geq$. or $c \geq$ .. Arguments *r* and *c* are interpreted as `trunc(`*r*`)` and `trunc(`*c*`)`.

Also see

[M-5] **missingof( )** — Appropriate missing value

[M-4] **standard** — Functions to create standard matrices

Title

[M-5] **Kmatrix( )** — Commutation matrix

Syntax

> *real matrix* Kmatrix(*real scalar m, real scalar n*)

Description

Kmatrix(*m*, *n*) returns the *mn* × *mn* commutation matrix K for which K*vec(*X*) = vec(*X'*), where *X* is an *mxn* matrix.

Remarks

Commutation matrices are frequently used in computing derivatives of functions of matrices. Section 9.2 of Lütkepohl (1996) lists many useful properties of commutation matrices.

Conformability

Kmatrix(*m*, *n*):
          m: 1 × 1
          n: 1 × 1
   result: *mn* × *mn*

Diagnostics

Kmatrix(*m*, *n*) aborts with error if either *m* or *n* is less than 0 or is missing. *m* and *n* are interpreted as trunc(*m*) and trunc(*n*).

Reference

Lütkepohl, H. 1996. *Handbook of Matrices.* New York: Wiley.

Also see

[M-5] **Dmatrix( )** — Duplication matrix

[M-5] **Lmatrix( )** — Elimination matrix

[M-5] **vec( )** — Stack matrix columns

[M-4] **standard** — Functions to create standard matrices

Title

Syntax

> *void* _flopin(*numeric matrix A*)
>
> *void* lapack_function(...)
>
> *void* _flopout(*numeric matrix A*)

where *lapack_function* may be

LA_DGEBAK()	LA_ZGEBAK()
LA_DGEBAL()	LA_ZGEBAL()
LA_DGEES()	LA_ZGEES()
LA_DGEEV()	LA_ZGEEV()
LA_DGEHRD()	LA_ZGEHRD()
LA_DGGBAK()	LA_ZGGBAK()
LA_DGGBAL()	LA_ZGGBAL()
LA_DGGHRD()	LA_ZGGHRD()
LA_DHGEQZ()	LA_ZHGEQZ()
LA_DHSEIN()	LA_ZHSEIN()
LA_DHSEQR()	LA_ZHSEQR()
LA_DLAMCH()	
LA_DORGHR()	
LA_DSYEVX()	
LA_DTGSEN()	LA_ZTGSEN()
LA_DTGEVC()	LA_ZTGEVC()
LA_DTREVC()	LA_ZTREVC()
LA_DTRSEN()	LA_ZTRSEN()
	LA_ZUNGHR()

Description

LA_DGEBAK(), LA_ZGEBAK(), LA_DGEBAL(), LA_ZGEBAL(), ... are LAPACK functions in original, as-is form; see [M-1] **LAPACK**. These functions form the basis for many of Mata's linear-algebra capabilities. Mata functions such as cholesky(), svd(), and eigensystem() are implemented using these functions; see [M-4] **matrix**. Those functions are easier to use. The LA_*() functions provide more capability.

_flopin() and _flopout() convert matrices to and from the form required by the LA_*() functions.

Remarks

LAPACK stands for Linear Algebra PACKage and is a freely available set of Fortran 90 routines for solving systems of simultaneous equations, eigenvalue problems, and singular-value problems. The original Fortran routines have six-letter names like DGEHRD, DORGHR, and so on. The Mata functions LA_DGEHRD( ), LA_DORGHR( ), etc., are a subset of the LAPACK double-precision real and complex routine. All LAPACK double-precision functions will eventually be made available.

Documentation for the LAPACK routines can be found at http://www.netlib.org/lapack/, although we recommend obtaining *LAPACK Users' Guide* by Anderson et al. (1999).

Remarks are presented under the following headings:

> *Mapping calling sequence from Fortran to Mata*
> *Flopping: Preparing matrices for LAPACK*
> *Warning on the use of rows( ) and cols( ) after _flopin( )*
> *Warning: It is your responsibility to check info*
> *Example*

Mapping calling sequence from Fortran to Mata

LAPACK functions are named with first letter S, D, C, or Z. S means single-precision real, D means double-precision real, C means single-precision complex, and Z means double-precision complex. Mata provides the D* and Z* functions. The LAPACK documentation is in terms of S* and C*. Thus, to find the documentation for LA_DGEHRD, you must look up SGEHRD in the original documentation.

The documentation (Anderson et al. 1999, 227) reads, in part,

> SUBROUTINE SGEHRD(N, ILO, IHI, A, LDA, TAU, WORK, LWORK, INFO)
> INTEGER   IHI, ILO, INFO, LDA, LWORK, N
> REAL   A(LDA, *), TAU(*), WORK(LWORK)

and the documentation states that SGEHDR reduces a real, general matrix, $\mathbf{A}$, to upper Hessenberg form, $\mathbf{H}$, by an orthogonal similarity transformation: $\mathbf{Q}' \times \mathbf{A} \times \mathbf{Q} = \mathbf{H}$.

The corresponding Mata function, LA_DGEHRD( ), has the same arguments. In Mata, arguments ihi, ilo, info, lda, lwork, and n are *real scalars*. Argument A is a *real matrix*, and arguments tau and work are *real vectors*.

You can read the rest of the original documentation to find out what is to be placed (or returned) in each argument. It turns out that $\mathbf{A}$ is assumed to be dimensioned LDA × *something* and that the routine works on $\mathbf{A}(1, 1)$ (using Fortran notation) through $\mathbf{A}(N, N)$. The routine also needs work space, which you are to supply in vector WORK. In the standard LAPACK way, LAPACK offers you a choice: you can preallocate WORK, in which case you have to choose a fairly large dimension for it, or you can do a query to find out how large the dimension needs to be for this particular problem. If you preallocate, the documentation reveals that the WORK must be of size N, and you set LWORK equal to N. If you wish to query, then you make WORK of size 1 and set LWORK equal to -1. The LAPACK routine will then return in the first element of WORK the optimal size. Then you call the function again with WORK allocated to be the optimal size and LWORK set to equal the optimal size.

Concerning Mata, the above works. You can follow the LAPACK documentation to the letter. Use J( ) to allocate matrices or vectors. Alternatively, you can specify all sizes as missing value (.), and Mata will fill in the appropriate value based on the assumption that you are using the entire matrix.

Thus, in LA_DGEHRD(), you could specify lda as missing, and the function would run as if you had specified lda equal to cols(A). You could specify n as missing, and the function would run as if you had specified n as rows(A).

Work areas, however, are treated differently. You can follow the standard LAPACK convention outlined above; or you can specify the sizes of work areas (lwork) and specify the work areas themselves (work) as missing values, and Mata will allocate the work areas for you. The allocation will be as you specified.

One feature provided by some LAPACK functions is not supported by the Mata implementation. If a function allows a function pointer, you may not avail yourself of that option.

Flopping: Preparing matrices for LAPACK

The LAPACK functions provided in Mata are the original LAPACK functions. Mata, which is C based, stores matrices rowwise. LAPACK, which is Fortran based, stores matrices columnwise. Mata and Fortran also disagree on how complex matrices are to be organized.

Functions _flopin() and _flopout() handle these issues. Coding _flopin(A) changes matrix A from the Mata convention to the LAPACK convention. Coding _flopout(A) changes A from the LAPACK convention to the Mata convention.

The LA_*() functions do not do this for you because LAPACK often takes two or three LAPACK functions run in sequence to achieve the desired result, and it would be a waste of computer time to switch conventions between calls.

Warning on the use of rows( ) and cols( ) after _flopin( )

Be careful using the rows() and cols() functions. rows() of a flopped matrix returns the logical number of columns and cols() of a flopped matrix returns the logical number of rows!

The danger of confusion is especially great when using J() to allocate work areas. If a LAPACK function requires a work area of $r \times c$, you code,

     _LA_*function*(..., J(c, r, .), ...)

Warning: It is your responsibility to check info

The LAPACK functions do not abort with error on failure. They instead store 0 in info (usually the last argument) if successful and store an error code if not successful. The error code is usually negative and indicates the argument that is a problem.

Example

The following example uses the LAPACK function DGEHRD to obtain the Hessenberg form of matrix **A**. We will begin with

```
        1    2    3    4

  1     1    2    3    4
  2     4    5    6    7
  3     7    8    9   10
  4     8    9   10   11
```

The first step is to use _flopin() to put **A** in LAPACK order:

```
: _flopin(A)
```

Next we make a work-space query to get the optimal size of the work area.

```
: LA_DGEHRD(., 1, 4, A, ., tau=., work=., lwork=-1, info=0)
: lwork = work[1,1]
: lwork
  128
```

After putting the work-space size in lwork, we can call LA_DGEHRD() again to perform the Hessenberg decomposition:

```
: LA_DGEHRD(., 1, 4, A, ., tau=., work=., lwork, info=0)
```

LAPACK function DGEHRD saves the result in the upper triangle and the first subdiagonal of **A**. We must use _flopout() to change that back to Mata order, and finally, we extract the result:

```
: _flopout(A)
: A = A-sublowertriangle(A, 2)
: A
                1            2            3            4

  1             1    -5.370750529    .0345341258    .3922322703
  2   -11.35781669    25.18604651    -4.40577178   -.6561483899
  3             0    -1.660145888   -.1860465116    .1760901813
  4             0              0    -8.32667e-16   -5.27356e-16
```

Reference

Anderson, E., Z. Bai, C. Bischof, S. Blackford, J. Demmel, J. J. Dongarra, J. Du Croz, A. Greenbaum, S. Hammarling, A. McKenney, and D. Sorensen. 1999. *LAPACK Users' Guide*. 3rd ed. Philadelphia: Society for Industrial and Applied Mathematics.

Also see

[M-1] **LAPACK** — The LAPACK linear-algebra routines

[R] **copyright lapack** — LAPACK copyright notification

[M-4] **matrix** — Matrix functions

Title

[M-5] liststruct( ) — List structure's contents

Syntax

void liststruct(*struct whatever matrix x*)

Description

liststruct() lists *x*'s contents, where *x* is an instance of structure *whatever*.

Remarks

liststruct() is often useful in debugging.

The dimension and type of all elements are listed, and the values of scalars are also shown.

Conformability

liststruct(*x*):
$$x: \quad r \times c$$
$$result: \quad void$$

Diagnostics

None.

Also see

[M-2] **struct** — Structures

[M-4] **io** — I/O functions

Title

Syntax

real matrix Lmatrix(*real scalar n*)

Description

Lmatrix(*n*) returns the $n(n + 1)/2 \times n^2$ elimination matrix L for which L*vec(X) = vech(X), where X is an $n \times n$ symmetric matrix.

Remarks

Elimination matrices are frequently used in computing derivatives of functions of symmetric matrices. Section 9.6 of Lütkepohl (1996) lists many useful properties of elimination matrices.

Conformability

Lmatrix(*n*):
$$n: \quad 1 \times 1$$
$$result: \quad n(n + 1)/2 \times n^2$$

Diagnostics

Lmatrix(*n*) aborts with error if *n* is less than 0 or is missing. *n* is interpreted as trunc(*n*).

Reference

Lütkepohl, H. 1996. *Handbook of Matrices*. New York: Wiley.

Also see

[M-5] **Dmatrix( )** — Duplication matrix

[M-5] **Kmatrix( )** — Commutation matrix

[M-5] **vec( )** — Stack matrix columns

[M-4] **standard** — Functions to create standard matrices

Title

Syntax

> *real matrix*   logit(*real matrix X*)
>
> *real matrix*   invlogit(*real matrix X*)
>
> *real matrix*   cloglog(*real matrix X*)
>
> *real matrix*   invcloglog(*real matrix X*)

Description

logit(X) returns the log of the odds ratio of the elements of X, $\ln\{x/(1-x)\}$.

invlogit(X) returns the inverse of the logit() of the elements of X, $\exp(x)/\{1+\exp(x)\}$.

cloglog(X) returns the complementary log-log of the elements of X, $\ln\{-\ln(1-x)\}$.

invcloglog(X) returns the elementwise inverse of cloglog() of the elements of X, $1-\exp\{-\exp(x)\}$.

Conformability

All functions return a matrix of the same dimension as input containing element-by-element calculated results.

Diagnostics

logit(X) and cloglog(X) return missing when $x \le 0$ or $x \ge 1$.

Also see

[M-4] **statistical** — Statistical functions

Title

Syntax

numeric matrix   lowertriangle(numeric matrix A [, numeric scalar d])

numeric matrix   uppertriangle(numeric matrix A [, numeric scalar d])

void   _lowertriangle(numeric matrix A [, numeric scalar d])

void   _uppertriangle(numeric matrix A [, numeric scalar d])

where argument d is optional.

Description

lowertriangle( ) returns the lower triangle of A.

uppertriangle( ) returns the upper triangle of A.

_lowertriangle( ) replaces A with its lower triangle.

_uppertriangle( ) replaces A with its upper triangle.

Remarks

Remarks are presented under the following headings:

> Optional argument d
> Nonsquare matrices

Optional argument d

Optional argument d specifies the treatment of the diagonal. Specifying $d>=.$, or not specifying d at all, means no special treatment; if

$$A = \begin{bmatrix} 1 & 2 & 3 \\ 4 & 5 & 6 \\ 7 & 8 & 9 \end{bmatrix}$$

then

$$\text{lowertriangle}(A, .) = \begin{bmatrix} 1 & 0 & 0 \\ 4 & 5 & 0 \\ 7 & 8 & 9 \end{bmatrix}$$

558

If a nonmissing value is specified for d, however, that value is substituted for each element of the diagonal, for example,

$$\text{lowertriangle}(A, 1) = \begin{bmatrix} 1 & 0 & 0 \\ 4 & 1 & 0 \\ 7 & 8 & 1 \end{bmatrix}$$

Nonsquare matrices

lowertriangle() and uppertriangle() may be used with nonsquare matrices. If

$$A = \begin{bmatrix} 1 & 2 & 3 & 4 \\ 5 & 6 & 7 & 8 \\ 9 & 10 & 11 & 12 \end{bmatrix}$$

then

$$\text{lowertriangle}(A) = \begin{bmatrix} 1 & 0 & 0 \\ 5 & 6 & 0 \\ 9 & 10 & 11 \end{bmatrix}$$

and

$$\text{uppertriangle}(A) = \begin{bmatrix} 1 & 2 & 3 & 4 \\ 0 & 6 & 7 & 8 \\ 0 & 0 & 11 & 12 \end{bmatrix}$$

_lowertriangle() and _uppertriangle(), however, may not be used with nonsquare matrices.

Conformability

lowertriangle(A, d):
         A: $r \times c$
         d: 1×1 (optional)
    result: $r \times \min(r, c)$

uppertriangle(A, d):
         A: $r \times c$
         d: 1×1 (optional)
    result: $\min(r, c) \times c$

_lowertriangle(A, d), _uppertriangle(A, d):

    input:
         A: $n \times n$
         d: 1×1 (optional)

    output:
         A: $n \times n$

Diagnostics

None.

Also see

[M-4] **manipulation** — Matrix manipulation

Title

Syntax

> *void* lud(*numeric matrix A*, *L*, *U*, *p*)
>
> *void* _lud(*numeric matrix L*, *U*, *p*)
>
> *void* _lud_la(*numeric matrix A*, *q*)

where

1. *A* may be real or complex and need not be square.

2. The types of *L*, *U*, *p*, and *q* are irrelevant; results are returned there.

Description

lud(*A*, *L*, *U*, *p*) returns the LU decomposition (with partial pivoting) of *A* in *L* and *U* along with a permutation vector *p*. The returned results are such that $A=L[p,.]*U$ up to roundoff error.

_lud(*L*, *U*, *p*) is similar to lud(), except that it conserves memory. The matrix to be decomposed is passed in *L*, and the same storage location is overwritten with the calculated *L* matrix.

_lud_la(*A*, *q*)·is the [M-1] **LAPACK** routine that the above functions use to calculate the LU decomposition. See *LAPACK routine* below.

Remarks

Remarks are presented under the following headings:

> *LU decomposition*
> *LAPACK routine*

LU decomposition

The LU decomposition of matrix *A* can be written

$$P'A = LU$$

where P' is a permutation matrix that permutes the rows of *A*. *L* is lower triangular and *U* is upper triangular. The decomposition can also be written

$$A = PLU$$

because, given that *P* is a permutation matrix, $P^{-1} = P'$.

Rather than returning P directly, returned is p corresponding to P. Lowercase p is a column vector that contains the subscripts of the rows in the desired order. That is,

$$PL = L[p,.]$$

The advantage of this is that p requires less memory than P and the reorganization, should it be desired, can be performed more quickly; see [M-1] **permutation**. In any case, the formula defining the LU decomposition can be written

$$A = L[p,.]*U$$

One can also write

$$B = LU, \text{ where } B[p,.] = A$$

LAPACK routine

_lud_la(A, q) is the interface into the [M-1] **LAPACK** routines that the above functions use to calculate the LU decomposition. You may use it directly if you wish.

Matrix A is decomposed, and the decomposition is placed back in A. U is stored in the upper triangle (including the diagonal) of A. L is stored in the lower triangle of A, and it is understood that L is supposed to have ones on its diagonal. q is a column vector recording the row swaps that account for the pivoting. This is the same information as stored in p, but in a different format.

q records the row swaps to be made. For instance, q = (1\2\2) means that (start at the end) the third row is to be swapped with the second row, then the second row is to stay where it is, and finally the first row is to stay where it is. q can be converted into p by the following logic:

```
p = 1::rows(q)
for (i=rows(q); i>=1; i--) {
        hold = p[i]
        p[i] = p[q[i]]
        p[q[i]] = hold
}
```

Conformability

lud(A, L, U, p):

> *input*:
>> A: $r \times c$
>
> *output*:
>> L: $r \times m$, $m = \min(r, c)$
>> U: $m \times c$
>> p: $r \times 1$

_lud(L, U, p):

    input:

             L: $r \times c$

    output:

             L: $r \times m$, $m = \min(r, c)$

             U: $m \times c$

             p: $r \times 1$

_lud_la(A, q):

    input:

             A: $r \times c$

    output:

             A: $r \times c$

             q: $r \times 1$

Diagnostics

lud(A, L, U, p) returns missing results if A contains missing values; L will have missing values below the diagonal, 1s on the diagonal, and 0s above the diagonal; U will have missing values on and above the diagonal and 0s below. Thus if there are missing values, $U[1,1]$ will contain missing.

_lud(L, U, p) sets L and U as described above if A contains missing values.

_lud_la(A, q) aborts with error if A is a view.

Also see

[M-5] **det( )** — Determinant of matrix

[M-5] **lusolve( )** — Solve AX=B for X using LU decomposition

[M-4] **matrix** — Matrix functions

Title

[M-5] **luinv( )** — Square matrix inversion

Syntax

> *numeric matrix*   `luinv`(*numeric matrix A*)
>
> *numeric matrix*   `luinv`(*numeric matrix A , real scalar tol*)
>
> *void*   `_luinv`(*numeric matrix A*)
>
> *void*   `_luinv`(*numeric matrix A , real scalar tol*)
>
> *real scalar*   ·   `_luinv_la`(*numeric matrix A , b*)

Description

`luinv`(*A*) and `luinv`(*A*, *tol*) return the inverse of real or complex, square matrix *A*.

`_luinv`(*A*) and `_luinv`(*A*, *tol*) do the same thing except that, rather than returning the inverse matrix, they overwrite the original matrix *A* with the inverse.

In all cases, optional argument *tol* specifies the tolerance for determining singularity; see *Remarks* below.

`_luinv_la`(*A*, *b*) is the interface to the [M-1] **LAPACK** routines that do the work. The output *b* is a real scalar, which is 1 if the LAPACK routine used a blocked algorithm and 0 otherwise.

Remarks

These routines calculate the inverse of *A*. The inverse matrix A^{-1} of *A* satisfies the conditions

$$AA^{-1} = I$$
$$A^{-1}A = I$$

A is required to be square and of full rank. See [M-5] **qrinv( )** and [M-5] **pinv( )** for generalized inverses of nonsquare or rank-deficient matrices. See [M-5] **invsym( )** for inversion of real, symmetric matrices.

`luinv`(*A*) is logically equivalent to `lusolve`(*A*, `I`(`rows`(*A*))); see [M-5] **lusolve( )** for details and for use of the optional *tol* argument.

Conformability

`luinv`(*A*, *tol*):

A:	$n \times n$	
tol:	1×1	(optional)
result:	$n \times n$	

_luinv(*A*, *tol*):

   input:

               A: $n \times n$

              tol: 1×1 (optional)

   output:

               A: $n \times n$

_luinv_la(*A*, *b*):

   input:

               A: $n \times n$

   output:

               A: $n \times n$

               b: 1×1

          result: 1×1

Diagnostics

The inverse returned by these functions is real if *A* is real and is complex if *A* is complex. If you use these functions with a singular matrix, returned will be a matrix of missing values. The determination of singularity is made relative to *tol*. See *Tolerance* under *Remarks* in [M-5] **lusolve( )** for details.

luinv(*A*) and _luinv(*A*) return a matrix containing missing if *A* contains missing values.

_luinv(*A*) aborts with error if *A* is a view.

_luinv_la(*A*, *b*) should not be used directly; use _luinv( ).

See [M-5] **lusolve( )** and [M-1] **tolerance** for information on the optional *tol* argument.

Also see

[M-5] **invsym( )** — Symmetric real matrix inversion

[M-5] **cholinv( )** — Symmetric, positive-definite matrix inversion

[M-5] **qrinv( )** — Generalized inverse of matrix via QR decomposition

[M-5] **pinv( )** — Moore–Penrose pseudoinverse

[M-5] **lusolve( )** — Solve AX=B for X using LU decomposition

[M-5] **lud( )** — LU decomposition

[M-4] **matrix** — Matrix functions

[M-4] **solvers** — Functions to solve AX=B and to obtain A inverse

Title

[M-5] lusolve( ) — Solve AX=B for X using LU decomposition

Syntax

numeric matrix	lusolve(*numeric matrix A*, *numeric matrix B*)
numeric matrix	lusolve(*numeric matrix A*, *numeric matrix B*, *real scalar tol*)
void	_lusolve(*numeric matrix A*, *numeric matrix B*)
void	_lusolve(*numeric matrix A*, *numeric matrix B*, *real scalar tol*)
real scalar	_lusolve_la(*numeric matrix A*, *numeric matrix B*)
real scalar	_lusolve_la(*numeric matrix A*, *numeric matrix B*, *real scalar tol*)

Description

lusolve(*A*, *B*) solves *AX=B* and returns *X*. lusolve() returns a matrix of missing values if *A* is singular.

lusolve(*A*, *B*, *tol*) does the same thing but allows you to specify the tolerance for declaring that *A* is singular; see *Tolerance* under *Remarks* below.

_lusolve(*A*, *B*) and _lusolve(*A*, *B*, *tol*) do the same thing except that, rather than returning the solution *X*, they overwrite *B* with the solution and, in the process of making the calculation, they destroy the contents of *A*.

_lusolve_la(*A*, *B*) and _lusolve_la(*A*, *B*, *tol*) are the interfaces to the [M-1] **LAPACK** routines that do the work. They solve *AX=B* for *X*, returning the solution in *B* and, in the process, using as workspace (overwriting) *A*. The routines return 1 if *A* was singular and 0 otherwise. If *A* was singular, *B* is overwritten with a matrix of missing values.

Remarks

The above functions solve *AX=B* via LU decomposition and are accurate. An alternative is qrsolve() (see [M-5] **qrsolve( )**), which uses QR decomposition. The difference between the two solutions is not, practically speaking, accuracy. When *A* is of full rank, both routines return equivalent results, and the LU approach is quicker, using approximately $O(2/3n^3)$ operations rather than $O(4/3n^3)$, where *A* is $n \times n$.

The difference arises when *A* is singular. Then the LU-based routines documented here return missing values. The QR-based routines documented in [M-5] **qrsolve( )** return a generalized (least squares) solution.

For more information on LU and QR decomposition, see [M-5] **lud( )** and see [M-5] **qrd( )**.

Remarks are presented under the following headings:

> *Derivation*
> *Relationship to inversion*
> *Tolerance*

Derivation

We wish to solve for X

$$AX = B \tag{1}$$

Perform LU decomposition on A so that we have $A = PLU$. Then (1) can be written

$$PLUX = B$$

or, premultiplying by P' and remembering that $P'P = I$,

$$LUX = P'B \tag{2}$$

Define

$$Z = UX \tag{3}$$

Then (2) can be rewritten

$$LZ = P'B \tag{4}$$

It is easy to solve (4) for Z because L is a lower-triangular matrix. Once Z is known, it is easy to solve (3) for X because U is upper triangular.

Relationship to inversion

Another way to solve

$$AX = B$$

is to obtain A^{-1} and then calculate

$$X = A^{-1}B$$

It is, however, better to solve $AX = B$ directly because fewer numerical operations are required, and the result is therefore more accurate and obtained in less computer time.

Indeed, rather than thinking about how solving a system of equations can be implemented via inversion, it is more productive to think about how inversion can be implemented via solving a system of equations. Obtaining A^{-1} amounts to solving

$$AX = I$$

Thus lusolve( ) (or any other solve routine) can be used to obtain inverses. The inverse of A can be obtained by coding

```
: Ainv = lusolve(A, I(rows(A)))
```

In fact, we provide luinv( ) (see [M-5] **luinv( )**) for obtaining inverses via LU decomposition, but luinv( ) amounts to making the above calculation, although a little memory is saved because the matrix I is never constructed.

Hence, everything said about lusolve( ) applies equally to luinv( ).

Tolerance

The default tolerance used is

$$eta = (1e\text{-}13)*\text{trace}(\text{abs}(U))/n$$

where U is the upper-triangular matrix of the LU decomposition of A: $n \times n$. A is declared to be singular if any diagonal element of U is less than or equal to eta.

If you specify $tol > 0$, the value you specify is used to multiply eta. You may instead specify $tol \leq 0$, and then the negative of the value you specify is used in place of eta; see [M-1] **tolerance**.

So why not specify $tol = 0$? You do not want to do that because, as matrices become close to being singular, results can become inaccurate. Here is an example:

```
: rseed(12345)

: A = lowertriangle(runiform(4,4))
: A[3,3] = 1e-15

: trux = runiform(4,1)

: b = A*trux

: /* the above created an Ax=b problem, and we have placed the true
>     value of x in trux.  We now obtain the solution via lusolve( )
>     and compare trux with the value obtained:
> */
: x = lusolve(A, b, 0)

: trux, x
```

```
      1      .7997150919     .7997150919      ← The discussed numerical
      2      .9102488109     .9102488109        instability can cause this
      3       .442547889     .3230683012        output to vary a little
      4       .756650276     .8673158447        across different computers
```

We would like to see the second column being nearly equal to the first—the estimated x being nearly equal to the true x—but there are substantial differences.

Even though the difference between x and xtrue is substantial, the difference between them is small in the prediction space:

```
: A*trux-b, A*x-b
                   1                2

      1           0        -2.77556e-17
      2           0                   0
      3           0        -1.11022e-16
      4           0                   0
```

What made this problem so difficult was the line A[3,3] = 1e-15. Remove that and you would find that the maximum difference between x and trux would be 2.22045e–16.

The degree to which the residuals A*x-b are a reliable measure of the accuracy of x depends on the condition number of the matrix, which can be obtained by [M-5] **cond( )**, which for A, is 3.2947e+15. If the matrix is well conditioned, small residuals imply an accurate solution for x. If the matrix is ill conditioned, small residuals are not a reliable indicator of accuracy.

Another way to check the accuracy of x is to set $tol = 0$ and to see how well x could be obtained were $x = $ x:

```
: x  = lusolve(A, b, 0)
: x2 = lusolve(A, A*x, 0)
```

If x and x2 are virtually the same, then you can safely assume that x is the result of a numerically accurate calculation. You might compare x and x2 with mreldif(x2,x); see [M-5] **reldif( )**. In our example, mreldif(x2,x) is .03, a large difference.

If *A* is ill conditioned, then small changes in *A* or *B* can lead to radical differences in the solution for *X*.

Conformability

lusolve(A, B, *tol*):

>*input*:

>>| | | |
>>|---|---|---|
>>| *A*: | $n \times n$ | |
>>| *B*: | $n \times k$ | |
>>| *tol*: | 1×1 | (optional) |

>*output*:

>>| | |
>>|---|---|
>>| *result*: | $n \times k$ |

_lusolve(A, B, *tol*):

>*input*:

>>| | | |
>>|---|---|---|
>>| *A*: | $n \times n$ | |
>>| *B*: | $n \times k$ | |
>>| *tol*: | 1×1 | (optional) |

>*output*:

>>| | |
>>|---|---|
>>| *A*: | 0×0 |
>>| *B*: | $n \times k$ |

_lusolve_la(A, B, *tol*):

>*input*:

>>| | | |
>>|---|---|---|
>>| *A*: | $n \times n$ | |
>>| *B*: | $n \times k$ | |
>>| *tol*: | 1×1 | (optional) |

>*output*:

>>| | |
>>|---|---|
>>| *A*: | 0×0 |
>>| *B*: | $n \times k$ |
>>| *result*: | 1×1 |

Diagnostics

lusolve(A, B, ...), _lusolve(A, B, ...), and _lusolve_la(A, B, ...) return a result containing missing if *A* or *B* contain missing values. The functions return a result containing all missing values if *A* is singular.

_lusolve(A, B, ...) and _lusolve_la(A, B, ...) abort with error if *A* or *B* is a view.

_lusolve_la(A, B, ...) should not be used directly; use _lusolve().

Also see

[M-5] **luinv( )** — Square matrix inversion

[M-5] **lud( )** — LU decomposition

[M-5] **solvelower( )** — Solve AX=B for X, A triangular

[M-5] **cholsolve( )** — Solve AX=B for X using Cholesky decomposition

[M-5] **qrsolve( )** — Solve AX=B for X using QR decomposition

[M-5] **svsolve( )** — Solve AX=B for X using singular value decomposition

[M-4] **matrix** — Matrix functions

[M-4] **solvers** — Functions to solve AX=B and to obtain A inverse

Title

[M-5] makesymmetric( ) — Make square matrix symmetric (Hermitian)

Syntax

numeric matrix   makesymmetric(*numeric matrix A*)

void   _makesymmetric(*numeric matrix A*)

Description

makesymmetric(*A*) returns *A* made into a symmetric (Hermitian) matrix by reflecting elements below the diagonal.

_makesymmetric(*A*) does the same thing but stores the result back in *A*.

Remarks

If *A* is real, elements below the diagonal are copied into their corresponding above-the-diagonal position.

If *A* is complex, the conjugate of the elements below the diagonal are copied into their corresponding above-the-diagonal positions, and the imaginary part of the diagonal is set to zero.

Whether *A* is real or complex, roundoff error can make matrix calculations that are supposed to produce symmetric matrices produce matrices that vary a little from symmetry, and makesymmetric() can be used to correct the situation.

Conformability

makesymmetric(*A*):

A:	$n \times n$
result:	$n \times n$

_makesymmetric(*A*):

A:	$n \times n$

Diagnostics

makesymmetric(*A*) and _makesymmetric(*A*) abort with error if *A* is not square. Also, _makesymmetric() aborts with error if *A* is a view.

Also see

[M-5] **issymmetric( )** — Whether matrix is symmetric (Hermitian)

[M-4] **manipulation** — Matrix manipulation

Title

Syntax

numeric matrix	matexpsym(*numeric matrix A*)
numeric matrix	matlogsym(*numeric matrix A*)
void	_matexpsym(*numeric matrix A*)
void	_matlogsym(*numeric matrix A*)

Description

matexpsym(*A*) returns the matrix exponential of the symmetric (Hermitian) matrix *A*.

matlogsym(*A*) returns the matrix natural logarithm of the symmetric (Hermitian) matrix *A*.

_matexpsym(*A*) and _matlogsym(*A*) do the same thing as matexpsym() and matlogsym(), but instead of returning the result, they store the result in *A*.

Remarks

Do not confuse matexpsym(*A*) with exp(*A*), nor matlogsym(*A*) with log(*A*).

matexpsym(2*matlogsym(*A*)) produces the same result as *A**A. exp() and log() return elementwise results.

Exponentiated results and logarithms are obtained by extracting the eigenvectors and eigenvalues of *A*, performing the operation on the eigenvalues, and then rebuilding the matrix. That is, first *X* and *L* are found such that

$$AX = X * \mathrm{diag}(L) \tag{1}$$

For symmetric (Hermitian) matrix *A*, *X* is orthogonal, meaning $X'X = XX' = I$. Thus

$$A = X * \mathrm{diag}(L) * X' \tag{2}$$

matexpsym(*A*) is then defined

$$A = X * \mathrm{diag}(\exp(L)) * X' \tag{3}$$

and matlogsym(*A*) is defined

$$A = X * \mathrm{diag}(\log(L)) * X' \tag{4}$$

(1) is obtained via symeigensystem(); see [M-5] **eigensystem( )**.

Conformability

matexpsym(A),   matlogsym(A):

A:	$n \times n$
result:	$n \times n$

_matexpsym(A), _matlogsym(A):

input:

A:	$n \times n$

output:

A:	$n \times n$

Diagnostics

matexpsym(A), matlogsym(A), _matexpsym(A), and _matlogsym(A) return missing results if A contains missing values.

Also:

1. These functions do not check that A is symmetric or Hermitian. If A is a real matrix, only the lower triangle, including the diagonal, is used. If A is a complex matrix, only the lower triangle and the real parts of the diagonal elements are used.

2. These functions return a matrix of the same storage type as A.

   For symatlog(A), this means that if A is real and the result cannot be expressed as a real, a matrix of missing values is returned. If you want the generalized solution, code matlogsym(C(A)). This is the same rule as with scalars: log(-1) evaluates to missing, but log(C(-1)) is 3.14159265i.

3. These functions are guaranteed to return a matrix that is numerically symmetric, Hermitian, or symmetriconly if theory states that the matrix should be symmetric, Hermitian, or symmetriconly. See [M-5] **matpowersym( )** for a discussion of this issue.

   For the functions here, real function exp(x) is defined for all real values of x (ignoring overflow), and thus the matrix returned by matexpsym() will be symmetric (Hermitian).

   The same is not true for matlogsym(). log(x) is not defined for $x = 0$, so if any of the eigenvalues of A are 0 or very small, a matrix of missing values will result. Also, log(x) is complex for $x < 0$, and thus if any of the eigenvalues are negative, the resulting matrix will be (1) missing if A is real stored as real, (2) symmetriconly if A contains reals stored as complex, and (3) general if A is complex.

Also see

[M-5] **matpowersym( )** — Powers of a symmetric matrix

[M-5] **eigensystem( )** — Eigenvectors and eigenvalues

[M-4] **matrix** — Matrix functions

Title

[M-5] matpowersym( ) — Powers of a symmetric matrix

Syntax

> *numeric matrix*   `matpowersym`(*numeric matrix A, real scalar p*)
>
> *void*   `_matpowersym`(*numeric matrix A, real scalar p*)

Description

`matpowersym`(*A, p*) returns A^p for symmetric matrix or Hermitian matrix A. The matrix returned is real if A is real and complex is A is complex.

`_matpowersym`(*A, p*) does the same thing, but instead of returning the result, it stores the result in A.

Remarks

Do not confuse `matpowersym`(*A, p*) and *A*:^*p*. If *p*==2, the first returns *A*A* and the second returns A with each element squared.

Powers can be positive, negative, integer, or noninteger. Thus `matpowersym`(*A*, .5) is a way to find the square-root matrix R such that *R*R*==*A*, and `matpowersym`(*A*, -1) is a way to find the inverse. For inversion, you could obtain the result more quickly using other routines.

Powers are obtained by extracting the eigenvectors and eigenvalues of A, raising the eigenvalues to the specified power, and then rebuilding the matrix. That is, first X and L are found such that

$$AX = X * \mathrm{diag}(L) \tag{1}$$

For symmetric (Hermitian) matrix A, X is orthogonal, meaning $X'X = XX' = I$. Thus

$$A = X * \mathrm{diag}(L)X' \tag{2}$$

A^p is then defined

$$A = X * \mathrm{diag}(L{:}^{\wedge}p) * X' \tag{3}$$

(1) is obtained via `symeigensystem`( ); see [M-5] **eigensystem( )**.

Conformability

`matpowersym`(*A, p*):

A:	$n \times n$
p:	1×1
result:	$n \times n$

574

_matpowersym(A, p):

  input:

        A: $n \times n$

        p: 1×1

  output:

        A: $n \times n$

Diagnostics

matpowersym(A, p) and _matpowersym(A, p) return missing results if A contains missing values.

Also:

1. These functions do not check that A is symmetric or Hermitian. If A is a real matrix, only the lower triangle, including the diagonal, is used. If A is a complex matrix, only the lower triangle and the real parts of the diagonal elements are used.

2. These functions return a matrix of the same storage type as A. That means that if A is real and A^p cannot be expressed as a real, a matrix of missing values is returned. If you want the generalized solution, code matpowersym(C(A), p). This is the same rule as with scalars: (-1)^.5 is missing, but C(-1)^.5 is 1i.

3. These functions are guaranteed to return a matrix that is numerically symmetric, Hermitian, or symmetriconly if theory states that the matrix should be symmetric, Hermitian, or symmetriconly.

   Concerning theory, the returned result is not necessarily symmetric (Hermitian). The eigenvalues L of a symmetric (Hermitian) matrix are real. If L:^p are real, then the returned matrix will be symmetric (Hermitian), but otherwise, it will not. Think of a negative eigenvalue and $p = .5$: this results in a complex eigenvalue for A^p. Then if the original matrix was real (the eigenvectors were real), the resulting matrix will be symmetriconly. If the original matrix was complex (the eigenvectors were complex), the resulting matrix will have no special structure.

Also see

[M-5] **matexpsym( )** — Exponentiation and logarithms of symmetric matrices

[M-5] **eigensystem( )** — Eigenvectors and eigenvalues

[M-4] **matrix** — Matrix functions

Title

[M-5] **mean( )** — Means, variances, and correlations

Syntax

real rowvector   mean $(X[\, , \, w])$

real matrix   variance $(X[\, , \, w])$

real matrix   quadvariance $(X[\, , \, w])$

real matrix   meanvariance $(X[\, , \, w])$

real matrix   quadmeanvariance $(X[\, , \, w])$

real matrix   correlation $(X[\, , \, w])$

real matrix   quadcorrelation $(X[\, , \, w])$

where

   X: *real matrix* X (rows are observations, columns are variables)

   w: *real colvector* w and is optional

Description

mean (X, w) returns the weighted-or-unweighted column means of data matrix X. mean() uses quad precision in forming sums and so is very accurate.

variance (X, w) returns the weighted-or-unweighted variance matrix of X. In the calculation, means are removed and those means are calculated in quad precision, but quad precision is not otherwise used.

quadvariance (X, w) returns the weighted-or-unweighted variance matrix of X. Calculation is highly accurate; quad precision is used in forming all sums.

meanvariance (X, w) returns mean (X,w)\variance (X,w).

quadmeanvariance (X, w) returns mean (X,w)\quadvariance (X,w).

correlation (X, w) returns the weighted-or-unweighted correlation matrix of X. correlation() obtains the variance matrix from variance().

quadcorrelation (X, w) returns the weighted-or-unweighted correlation matrix of X. quadcorrelation() obtains the variance matrix from quadvariance().

In all cases, w specifies the weight. Omit w, or specify w as 1 to obtain unweighted means.

In all cases, rows of X or w that contain missing values are omitted from the calculation, which amounts to casewise deletion.

Remarks

1. There is no quadmean( ) function because mean( ), in fact, is quadmean( ). The fact that mean( ) defaults to the quad-precision calculation reflects our judgment that the extra computational cost in computing means in quad precision is typically justified.

2. The fact that variance( ) and correlation( ) do not default to using quad precision for their calculations reflects our judgment that the extra computational cost is typically not justified. The emphasis on this last sentence is on the word *typically*.

   It is easier to justify means in part because the extra computational cost is less: there are only k means but $k(k+1)/2$ variances and covariances.

3. If you need just the mean or just the variance matrix, call mean( ) or variance( ) (or quadvariance( )). If you need both, there is a CPU-time savings to be had by calling meanvariance( ) instead of the two functions separately (or quadmeanvariance( ) instead of calling mean( ) and quadvariance( )).

   The savings is not great—one mean( ) calculation is saved—but the greater rows(X), the greater the savings.

   Upon getting back the combined result, it can be efficiently broken into its components via

   ```
   : var   = meanvariance(X)
   : means = var[1,.]
   : var   = var[|2,1  .,.|]
   ```

Conformability

mean(X, w):

X:	$n \times k$			
w:	$n \times 1$	or	1×1	(optional, $w = 1$ assumed)
result:	$1 \times k$			

variance(X, w), quadvariance(X, w), correlation(X, w), quadcorrelation(X, w):

X:	$n \times k$			
w:	$n \times 1$	or	1×1	(optional, $w = 1$ assumed)
result:	$k \times k$			

meanvariance(X, w), quadmeanvariance(X, w):

X:	$n \times k$			
w:	$n \times 1$	or	1×1	(optional, $w = 1$ assumed)
result:	$(k+1) \times k$			

Diagnostics

All functions omit from the calculation rows that contain missing values unless all rows contain missing values. Then the returned result contains all missing values.

Also see

[M-4] **statistical** — Statistical functions

Title

Syntax

real scalar   mindouble()

real scalar   maxdouble()

real scalar   smallestdouble()

Description

mindouble() returns the largest negative, nonmissing value.

maxdouble() returns the largest positive, nonmissing value.

smallestdouble() returns the smallest full-precision value of e, $e > 0$. The largest full-precision value of e, $e < 0$ is $-$smallestdouble().

Remarks

All nonmissing values x fulfill mindouble() $\leq x \leq$ maxdouble().

All missing values m fulfill $m >$ maxdouble()

Missing values also fulfill $m \geq$.

On all computers on which Stata and Mata are currently implemented, which are computers following IEEE standards:

function	exact hexadecimal value	approximate decimal value
mindouble()	-1.ffffffffffffffX+3ff	$-1.7977e+308$
smallestdouble()	+1.0000000000000X-3fe	$2.2251e{-}308$
epsilon(1)	+1.0000000000000X-034	$2.2205e{-}016$
maxdouble()	+1.ffffffffffffffX+3fe	$8.9885e+307$

The smallest missing value (. $<$ .a $< \cdots <$ .z) is +1.0000000000000X+3ff.

Do not confuse smallestdouble() with the more interesting value epsilon(1). smallestdouble() is the smallest full-precision value of e, $e > 0$. epsilon(1) is the smallest value of e, $e+1 > 1$; see [M-5] **epsilon( )**.

Conformability

mindouble( ), maxdouble( ), smallestdouble( ):
        result: 1×1

Diagnostics

None.

Reference

Linhart, J. M. 2008. Mata Matters: Overflow, underflow and the IEEE floating-point format. *Stata Journal* 8: 255–268.

Also see

[M-5] **epsilon( )** — Unit roundoff error (machine precision)

[M-4] **utility** — Matrix utility functions

Title

Syntax

> *void* `minindex`(*real vector v*, *real scalar k*, *i*, *w*)

> *void* `maxindex`(*real vector v*, *real scalar k*, *i*, *w*)

Results are returned in *i* and *w*.

> *i* will be a *real colvector*.

> *w* will be a $K \times 2$ *real matrix*, $K \leq |k|$.

Description

`minindex`(*v, k, i, w*) returns in *i* and *w* the indices of the *k* minimums of *v*.

`maxindex`(*v, k, i, w*) does the same, except that it returns the indices of the *k* maximums.

`minindex`() may be called with $k < 0$; it is then equivalent to `maxindex`().

`maxindex`() may be called with $k < 0$; it is then equivalent to `minindex`().

Remarks

Remarks are presented under the following headings:

> *Use of functions when v has all unique values*
> *Use of functions when v has repeated (tied) values*
> *Summary*

Remarks are cast in terms of `minindex`() but apply equally to `maxindex`().

Use of functions when v has all unique values

Consider $v = (3, 1, 5, 7, 6)$.

1. `minindex`(v, 1, i, w) returns i = 2, which means that v[2] is the minimum value in v.

2. `minindex`(v, 2, i, w) returns i = $(2, 1)'$, which means that v[2] is the minimum value of v and that v[1] is the second minimum.

. . .

5. `minindex`(v, 5, i, w) returns i = $(2, 1, 3, 5, 4)'$, which means that the ordered values in v are v[2], v[1], v[3], v[5], and v[4].

6. `minindex`(v, 6, i, w), `minindex`(v, 7, i, w), and so on, return the same as (5), because there are only five minimums in a five-element vector.

When v has unique values, the values returned in w are irrelevant.

- In (1), w will be $(1, 1)$.

- In (2), w will be $(1, 1 \backslash 2, 1)$.

- ...

- In (5), w will be $(1, 1 \backslash 2, 1 \backslash 3, 1 \backslash 4, 1 \backslash 5, 1)$.

The second column of w records the number of tied values. Since the values in v are unique, the second column of w will be ones. If you have a problem where you are uncertain whether the values in v are unique, code

```
if (!allof(w[,2], 1)) {
        /* uniqueness assumption false */
}
```

Use of functions when v has repeated (tied) values

Consider $v = (3, 2, 3, 2, 3, 3)$.

1. minindex(v, 1, i, w) returns i = (2, 4)', which means that there is one minimum value and that it is repeated in two elements of v, namely, v[2] and v[4].

   Here, w will be $(1, 2)$, but you can ignore that. There are two values in i corresponding to the same minimum.

   When k==1, rows(i) equals the number of observations in v corresponding to the minimum, as does w[1,2].

2. minindex(v, 2, i, w) returns $i = (2, 4, 1, 3, 5, 6)'$ and $w = (1, 2 \backslash 3, 4)$.

   Begin with w. The first row of w is $(1, 2)$, which states that the indices of the first minimums of v start at i[1] and consist of two elements. Thus the indices of the first minimums are i[1] and i[2] (the minimums are v[i[1]] and v[i[2]], which of course are equal).

   The second row of w is $(3, 4)$, which states that the indices of the second minimums of v start at i[3] and consist of four elements: i[3], i[4], i[5], and i[6] (which are 1, 3, 5, and 6).

   In summary, rows(w) records the number of minimums returned. w[m,1] records where in i the mth minimum begins (it begins at i[w[m,1]]). w[m,2] records the total number of tied values. Thus one could step across the minimums and the tied values by coding

```
minindex(v, k, i, w)
for (m=1; m<=rows(w); m++) {
        for (j=w[m,1]; j<w[m,1]+w[m,2]; j++) {
                /* i[j] is the index in v of an mth minimum */
        }
}
```

3. minindex(v, 3, i, w), minindex(v, 4, i, w), and so on, return the same as (2) because, with v = (3, 2, 3, 2, 3, 3), there are only two minimums.

Summary

Consider minindex(v, k, i, w). Returned will be

$$w = \begin{bmatrix} i1 & n1 \\ i2 & n2 \\ . & . \\ . & . \end{bmatrix} \quad w : K \times 2, \quad K \leq |k|$$

$$i = \begin{bmatrix} j1 \\ j2 \\ j3 \\ j4 \\ . \\ . \\ . \end{bmatrix} \begin{array}{l} \leftarrow i[i1] \text{ is start of first minimums} \\ \\ \\ \leftarrow i[i2] \text{ is start of second minimums} \\ \\ \text{etc.} \\ i : 1 \times m, \quad m = n1 + n2 + \dots \end{array}$$

has $n1$ values

has $n2$ values

$j1, j2, \dots$, are indices into v.

Conformability

minindex(v, k, i, w), maxindex(v, k, i, w):

   input:

        v: $n \times 1$ or $1 \times n$

        k: 1×1

   output:

        i: $L \times 1, \quad L \geq K$

        w: $K \times 2, \quad K \leq |k|$

Diagnostics

minindex(v, k, i, w) and maxindex(v, k, i, w) abort with error if i or w is a view.

In minindex(v, k, i, w) and maxindex(v, k, i, w), missing values in v are ignored in obtaining minimums and maximums.

In the examples above, we have shown input vector v as a row vector. It can also be a column vector; it makes no difference.

In minindex(v, k, i, w), input argument k specifies the number of minimums to be obtained. k may be zero. If k is negative, $-k$ maximums are obtained.

Similarly, in maxindex(v, k, i, w), input argument k specifies the number of maximums to be obtained. k may be zero. If k is negative, $-k$ minimums are obtained.

minindex() and maxindex() are designed for use when k is small relative to length(v); otherwise, see order() in [M-5] **sort()**.

Also see

[M-5] **minmax( )** — Minimums and maximums

[M-4] **utility** — Matrix utility functions

Title

Syntax

real colvector   rowmin(*real matrix X*)

real rowvector colmin(*real matrix X*)

real scalar   min(*real matrix X*)

real colvector   rowmax(*real matrix X*)

real rowvector colmax(*real matrix X*)

real scalar   max(*real matrix X*)

real matrix   rowminmax(*real matrix X*)

real matrix   colminmax(*real matrix X*)

real rowvector minmax(*real matrix X*)

real matrix   rowminmax(*real matrix X*, *real scalar usemiss*)

real matrix   colminmax(*real matrix X*, *real scalar usemiss*)

real rowvector minmax(*real matrix X*, *real scalar usemiss*)

real colvector   rowmaxabs(*numeric matrix A*)

real rowvector colmaxabs(*numeric matrix A*)

Description

These functions return the indicated minimums and maximums of X.

rowmin(X) returns the minimum of each row of X, colmin(X) returns the minimum of each column, and min(X) returns the overall minimum. Elements of X that contain missing are ignored.

rowmax(X) returns the maximum of each row of X, colmax(X) returns the maximum of each column, and max(X) returns the overall maximum. Elements of X that contain missing are ignored.

rowminmax(X) returns the minimum and maximum of each row of X in an $r \times 2$ matrix; colminmax(X) returns the minimum and maximum of each column in a $2 \times c$ matrix, and minmax(X) returns the overall minimum and maximum. Elements of X that contain missing are ignored.

The two-argument versions of rowminmax(), colminmax(), and minmax() allow you to specify how missing values are to be treated. Specifying a second argument with value 0 is the same as using the single-argument versions of the functions. In the two-argument versions, if the second argument is not zero, missing values are treated like all other values in determining the minimums and maximums: *nonmissing* $<$. $<$ .a $<$ .b $< \cdots <$ .z.

rowmaxabs(A) and colmaxabs(A) return the same result as rowmax(abs(A)) and colmax(abs(A)). The advantage is that matrix abs(A) is never formed or stored, and so these functions use less memory.

Conformability

rowmin(X), rowmax(X):

$$X: \quad r \times c$$
$$result: \quad r \times 1$$

colmin(X), colmax(X):

$$X: \quad r \times c$$
$$result: \quad 1 \times c$$

min(X), max(X):

$$X: \quad r \times c$$
$$result: \quad 1 \times 1$$

rowminmax(X, *usemiss*):

$$X: \quad r \times c$$
$$usemiss: \quad 1 \times 1$$
$$result: \quad r \times 2$$

colminmax(X, *usemiss*)

$$X: \quad r \times c$$
$$usemiss: \quad 1 \times 1$$
$$result: \quad 2 \times c$$

minmax(X, *usemiss*)

$$X: \quad r \times c$$
$$usemiss: \quad 1 \times 1$$
$$result: \quad 1 \times 2$$

rowmaxabs(A):

$$A: \quad r \times c$$
$$result: \quad r \times 1$$

colmaxabs(A):

$$A: \quad r \times c$$
$$result: \quad 1 \times c$$

Diagnostics

row*() functions return missing value for the corresponding minimum or maximum when the entire row contains missing.

col*() functions return missing value for the corresponding minimum or maximum when the entire column contains missing.

min() and max() return missing value when the entire matrix contains missing.

Also see

[M-5] **minindex( )** — Indices of minimums and maximums

[M-4] **mathematical** — Important mathematical functions

[M-4] **utility** — Matrix utility functions

Title

Syntax

real rowvector	colmissing(*numeric matrix X*)
real colvector	rowmissing(*numeric matrix X*)
real scalar	missing(*numeric matrix X*)
real rowvector	colnonmissing(*numeric matrix X*)
real colvector	rownonmissing(*numeric matrix X*)
real scalar	nonmissing(*numeric matrix X*)
real scalar	hasmissing(*numeric matrix X*)

Description

These functions return the indicated count of missing or nonmissing values.

colmissing(X) returns the count of missing values of each column of X, rowmissing(X) returns the count of missing values of each row, and missing(X) returns the overall count.

colnonmissing(X) returns the count of nonmissing values of each column of X, rownonmissing(X) returns the count of nonmissing values of each row, and nonmissing(X) returns the overall count.

hasmissing(X) returns 1 if X has a missing value or 0 if X does not have a missing value.

Remarks

$$\text{colnonmissing}(X) = \text{rows}(X) :- \text{colmissing}(X)$$
$$\text{rownonmissing}(X) = \text{cols}(X) :- \text{rowmissing}(X)$$
$$\text{nonmissing}(X) = \text{rows}(X)*\text{cols}(X) - \text{missing}(X)$$

Conformability

colmissing(X), colnonmissing(X):
          X: $r \times c$
     result: $1 \times c$

rowmissing(X), rownonmissing(X):
          X: $r \times c$
     result: $r \times 1$

missing(X), nonmissing(X), hasmissing(X):
          X: $r \times c$
     result: 1×1

Diagnostics

None.

Also see

[M-5] **editmissing( )** — Edit matrix for missing values

[M-4] **utility** — Matrix utility functions

Title

[M-5] **missingof( )** — Appropriate missing value

Syntax

$transmorphic\ scalar$ missingof($transmorphic\ matrix\ A$)

Description

missingof(A) returns a missing of the same element type as A:

- if A is real, a real missing is returned;
- if A is complex, a complex missing is returned;
- if A is pointer, NULL is returned;
- if A is string, "" is returned.

Remarks

missingof() is useful when creating empty matrices of the same type as another matrix; e.g.,

newmat = J(rows(x), cols(x), missingof(x))

Conformability

missingof(A)
        A: $r \times c$
    $result$: 1×1

Diagnostics

None.

Also see

[M-4] **utility** — Matrix utility functions

Title

[M-5] mod( ) — Modulus

Syntax

> *real matrix*   mod(*real matrix x*, *real matrix y*)

Description

mod(x, y) returns the elementwise modulus of x with respect to y. mod() is defined

$$\text{mod}(x, y) = x - y * \text{trunc}(x/y)$$

Conformability

mod(x, y):

x:	$r_1 \times c_1$	
y:	$r_2 \times c_2$,	x and y r-conformable
result:	$\max(r_1, r_2) \times \max(c_1, c_2)$	(elementwise calculation)

Diagnostics

mod(x, y) returns missing when either argument is missing or when $y = 0$.

Also see

[M-4] **scalar** — Scalar mathematical functions

Title

[M-5] **moptimize( )** — Model optimization

Syntax

If you are reading this entry for the first time, skip down to Description and to Remarks, and more especially, to Mathematical statement of the moptimize( ) problem under Remarks.

Syntax is presented under the following headings:

Step 1: Initialization

M = moptimize_init()

Step 2: Definition of maximization or minimization problem

In each of the functions, the last argument is optional. If specified, the function sets the value and returns void. If not specified, no change is made, and instead what is currently set is returned.

(varies) `moptimize_init_which(M, { "max" | "min" })`

(varies) `moptimize_init_evaluator(M, &functionname())`

(varies) `moptimize_init_evaluator(M, "programname")`

(varies) `moptimize_init_evaluatortype(M, evaluatortype)`

(varies) `moptimize_init_negH(M, { "off" | "on" })`

(varies) `moptimize_init_touse(M, "tousevarname")`

(varies) `moptimize_init_view(M, { "on" | "off" })`

(varies) `moptimize_init_ndepvars(M, D)`

(varies) `moptimize_init_depvar(M, j, y)`

(varies) `moptimize_init_eq_n(M, m)`

(varies) `moptimize_init_eq_indepvars(M, i, X)`

(varies) `moptimize_init_eq_cons(M, i, { "on" | "off" })`

(varies) `moptimize_init_eq_offset(M, i, o)`

(varies) `moptimize_init_eq_exposure(M, i, t)`

(varies) `moptimize_init_eq_name(M, i, name)`

(varies) `moptimize_init_eq_colnames(M, i, names)`

(varies) `moptimize_init_eq_coefs(M, i, b0)`

(varies) `moptimize_init_constraints(M, Cc)`

(varies) `moptimize_init_search(M, { "on" | "off" })`

(varies) `moptimize_init_search_random(M, { "off" | "on" })`

(varies) `moptimize_init_search_repeat(M, nr)`

(varies) `moptimize_init_search_bounds(M, i, minmax)`

(varies) `moptimize_init_search_rescale(M, { "on" | "off" })`

(varies) moptimize_init_weight(*M*, *w*)

(varies) moptimize_init_weighttype(*M*, *weighttype*)

(varies) moptimize_init_cluster(*M*, *c*)

(varies) moptimize_init_svy(*M*, { "off" | "on" })

(varies) moptimize_init_by(*M*, *by*)

(varies) moptimize_init_nuserinfo(*M*, *n_user*)

(varies) moptimize_init_userinfo(*M*, *l*, *Z*)

(varies) moptimize_init_technique(*M*, *technique*)

(varies) moptimize_init_vcetype(*M*, *vcetype*)

(varies) moptimize_init_nmsimplexdeltas(*M*, *delta*)

(varies) moptimize_init_gnweightmatrix(*M*, *W*)

(varies) moptimize_init_singularHmethod(*M*, *singularHmethod*)

(varies) moptimize_init_conv_maxiter(*M*, *maxiter*)

(varies) moptimize_init_conv_warning(*M*, { "on" | "off" })

(varies) moptimize_init_conv_ptol(*M*, *ptol*)

(varies) moptimize_init_conv_vtol(*M*, *vtol*)

(varies) moptimize_init_conv_nrtol(*M*, *nrtol*)

(varies) moptimize_init_conv_ignorenrtol(*M*, { "off" | "on" })

(varies) moptimize_init_iterid(*M*, *id*)

(varies) moptimize_init_valueid(*M*, *id*)

(varies) moptimize_init_tracelevel(*M*, *tracelevel*)

(varies) moptimize_init_trace_ado(*M*, { "off" | "on" })

(varies) moptimize_init_trace_dots(*M*, { "off" | "on" })

(varies) moptimize_init_trace_value(*M*, { "on" | "off" })

(varies) moptimize_init_trace_tol(*M*, { "off" | "on" })

(varies) moptimize_init_trace_coefs(*M*, { "off" | "on" })

(varies) moptimize_init_trace_step(*M*, { "off" | "on" })

(varies) moptimize_init_trace_gradient(*M*, { "off" | "on" })

(varies) moptimize_init_trace_Hessian(*M*, { "off" | "on" })

(varies) moptimize_init_evaluations(M, { "off" | "on" })

(varies) moptimize_init_verbose(M, { "on" | "off" })

Step 3: Perform optimization or perform a single function evaluation

void   moptimize(M)

real scalar _moptimize(M)

void   moptimize_evaluate(M)

real scalar _moptimize_evaluate(M)

Step 4: Post, display, or obtain results

void   moptimize_result_post(M [, vcetype])

void   moptimize_result_display([M [, vcetype]])

real scalar   moptimize_result_value(M)

real scalar   moptimize_result_value0(M)

real rowvector moptimize_result_coefs(M [, i])

string matrix   moptimize_result_colstripe(M [, i])

real matrix   moptimize_result_scores(M)

real rowvector moptimize_result_gradient(M [, i])

real matrix   moptimize_result_Hessian(M [, i])

real matrix   moptimize_result_V(M [, i])

string scalar   moptimize_result_Vtype(M)

real matrix   moptimize_result_V_oim(M [, i])

real matrix   moptimize_result_V_opg(M [, i])

real matrix   moptimize_result_V_robust(M [, i])

real scalar	`moptimize_result_iterations(`*M*`)`
real scalar	`moptimize_result_converged(`*M*`)`
real colvector	`moptimize_result_iterationlog(`*M*`)`
real rowvector	`moptimize_result_evaluations(`*M*`)`
real scalar	`moptimize_result_errorcode(`*M*`)`
string scalar	`moptimize_result_errortext(`*M*`)`
real scalar	`moptimize_result_returncode(`*M*`)`
void	`moptimize_ado_cleanup(`*M*`)`

Utility functions for use in all steps

void   `moptimize_query(`*M*`)`

real matrix   `moptimize_util_eq_indices(`*M*`, `*i* $\left[\,,\ i2\right]$`)`

(varies)   `moptimize_util_depvar(`*M*`, `*j*`)`
        returns y set by `moptimize_init_depvar(`*M*`, `*j*`, `*y*`)`, *which is usually a real colvector*

real colvector   `moptimize_util_xb(`*M*`, `*b*`, `*i*`)`

real scalar   `moptimize_util_sum(`*M*`, `*real colvector v*`)`

real rowvector   `moptimize_util_vecsum(`*M*`, `*i*`, `*real colvector s*`, `*real scalar value*`)`

real matrix   `moptimize_util_matsum(`*M*`, `*i*`, `*i2*`, `*real colvector s*`,
        real scalar value`)`

real matrix   `moptimize_util_matbysum(`*M*`, `*i*`, `*real colvector a*`, `*real colvector b*`,
        real scalar value`)`

real matrix   `moptimize_util_matbysum(`*M*`, `*i*`, `*i2*`, `*real colvector a*`,
        real colvector b`, `*real colvector c*`, `*real scalar value*`)`

Definition of M

M, if it is declared, should be declared transmorphic. *M* is obtained from `moptimize_init()` and then passed as an argument to the other `moptimize()` functions.

    `moptimize_init()` returns *M*, called an `moptimize()` problem handle. The function takes no arguments. *M* holds the information about the optimization problem.

Setting the sample

Various `moptimize_init_*()` functions set values for dependent variables, independent variables, etc. When you set those values, you do that either by specifying Stata variable names or by specifying Mata matrices containing the data themselves. Function `moptimize_init_touse()` specifies the sample to be used when you specify Stata variable names.

> `moptimize_init_touse(M, "`*tousevarname*`")` specifies the name of the variable in the Stata dataset that marks the observations to be included. Observations for which the Stata variable is nonzero are included. The default is `""`, meaning all observations are to be used.

> You need to specify *tousevarname* only if you specify Stata variable names in the other `moptimize_init_*()` functions, and even then it is not required. Setting `tousevar` when you specify the data themselves via Mata matrices, whether views or not, has no effect.

> `moptimize_init_view(M, {"on" | "off"})` specifies whether, when you specify Stata variable names in the other `moptimize_init_*()` functions, that data should be accessed as a view or as a copy. The default is `"on"`, meaning data are accessed via views. Using views makes `moptimize()` use less memory but run slightly slower.

Specifying dependent variables

D and j index dependent variables:

index	description
D	number of dependent variables, $D \geq 0$
j	dependent variable index, $1 \leq j \leq D$

D and j are real scalars.

You set the dependent variables one at a time. In a particular optimization problem, you may have no dependent variables or have more dependent variables than equations.

> `moptimize_init_depvar(M, j, y)` sets the jth dependent variable to be y. y may be a string scalar containing a Stata variable name that in turn contains the values of the jth dependent variable, or y may be a real colvector directly containing the values.

> `moptimize_init_ndepvars(M, D)` sets the total number of dependent variables. You can set D before defining dependent variables, and that speeds execution slightly, but it is not necessary because D is automatically set to the maximum j.

Specifying independent variables

Independent variables are defined within parameters or, equivalently, equations. The words parameter and equation mean the same thing. m, i, and $i2$ index parameters:

index	description
m	number of parameters (equations), $m \geq 1$
i	equation index, $1 \leq i \leq m$
$i2$	equation index, $1 \leq i2 \leq m$

m, i, and $i2$ are real scalars.

The function to be optimized is $f(p_1, p_2, \ldots, p_m)$. The ith parameter (equation) is defined as

$$pi = Xi \times bi' + oi + \ln(ti) :+ b0i$$

where

pi: $Ni \times 1$	(ith parameter)
Xi: $Ni \times ki$	(Ni observations on ki independent variables)
bi: $1 \times ki$	(coefficients to be fit)
oi: $Ni \times 1$	(exposure/offset in offset form, optional)
ti: $Ni \times 1$	(exposure/offset in exposure form, optional)
$b0i$: 1×1	(constant or intercept, optional)

Any of the terms may be omitted. The most common forms for a parameter are $pi = Xi \times bi' + b0i$ (standard model), $pi = Xi \times bi'$ (no-constant model), and $pi = b0i$ (constant-only model).

In addition, define b: $1 \times K$ as the entire coefficient vector, which is to say,

$$b = (b1, [b01,] \quad b2, [b02,] \quad \ldots)$$

That is, because bi is $1 \times ki$ for $i = 1, 2, \ldots, m$, then b is $1 \times K$, where $K = \sum_i ki + ci$, where ci is 1 if equation i contains an intercept and is 0 otherwise. Note that bi does not contain the constant or intercept, if there is one, but b contains all the coefficients, including the intercepts. b is called *the full set of coefficients*.

Parameters are defined one at a time by using the following functions:

moptimize_init_eq_n(*M*, *m*) sets the number of parameters. Use of this function is optional; *m* will be automatically determined from the other moptimize_init_eq_*() functions you issue.

moptimize_init_eq_indepvars(*M*, *i*, *X*) sets *X* to be the data (independent variables) for the ith parameter. *X* may be a $1 \times ki$ string rowvector containing Stata variable names, or *X* may be a string scalar containing the same names in space-separated format, or *X* may be an $Ni \times ki$ real matrix containing the data for the independent variables. Specify *X* as "" to omit term $Xi \times bi'$, for instance, as when fitting a constant-only model. The default is "".

moptimize_init_eq_cons(*M*, *i*, { "on" | "off" }) specifies whether the equation for the ith parameter includes $b0i$, a constant or intercept. Specify "on" to include $b0i$, "off" to exclude it. The default is "on".

moptimize_init_eq_offset(*M*, *i*, *o*) specifies oi in the equation for the ith parameter. *o* may be a string scalar containing a Stata variable name, or *o* may be an $Ni \times 1$ real colvector containing the offsets. The default is "", meaning term oi is omitted. Parameters may not have both oi and $\ln(ti)$ terms.

moptimize_init_eq_exposure(*M*, *i*, *t*) specifies ti in term $\ln(ti)$ of the equation for the ith parameter. *t* may be a string scalar containing a Stata variable name, or *t* may be an $Ni \times 1$ real colvector containing the exposure values. The default is "", meaning term $\ln(ti)$ is omitted.

moptimize_init_eq_name(*M*, *i*, *name*) specifies a string scalar, *name*, to be used in the output to label the ith parameter. The default is to use an automatically generated name.

moptimize_init_eq_colnames(M, i, *names*) specifies a $1 \times ki$ string rowvector, *names*, to be used in the output to label the coefficients for the *i*th parameter. The default is to use automatically generated names.

Specifying constraints

Linear constraints may be placed on the coefficients, b, which may be either within equation or between equations.

moptimize_init_constraints(M, Cc) specifies an $R \times K + 1$ real matrix, Cc, that places R linear restrictions on the $1 \times K$ full set of coefficients, b. Think of Cc as being (C,c), C: $R \times K$ and c: $R \times 1$. Optimization will be performed subject to the constraint $Cb' = c$. The default is no constraints.

Specifying weights or survey data

You may specify weights, and once you do, everything is automatic, assuming you implement your evaluator by using the provided utility functions.

moptimize_init_weight(M, w) specifies the weighting variable or data. w may be a string scalar containing a Stata variable name, or w may be a real colvector directly containing the weight values. The default is "", meaning no weights.

moptimize_init_weighttype(M, *weighttype*) specifies how w is to be treated. *weighttype* may be "fweight", "aweight", "pweight", or "iweight". You may set w first and then *weighttype*, or the reverse. If you set w without setting *weighttype*, then "fweight" is assumed. If you set *weighttype* without setting w, then *weighttype* is ignored. The default *weighttype* is "fweight".

Alternatively, you may inherit the full set of survey settings from Stata by using moptimize_init_svy(). If you do this, do not use moptimize_init_weight(), moptimize_init_weighttype(), or moptimize_init_cluster(). When you use the survey settings, everything is nearly automatic, assuming you use the provided utility functions to implement your evaluator. The proviso is that your evaluator must be of evaluatortype lf, e, v, or q.

moptimize_init_svy(M, { "off" | "on" }) specifies whether Stata's survey settings should be used. The default is "off". Using the survey settings changes the default *vcetype* to "svy", which is equivalent to "robust".

Specifying clusters and panels

Clustering refers to possible nonindependence of the observations within groups called clusters. A cluster variable takes on the same value within a cluster and different values across clusters. After setting the cluster variable, there is nothing special you have to do, but be aware that clustering is allowed only if you use a type lf, e, v, or q evaluator. moptimize_init_cluster() allows you to set a cluster variable.

Panels refer to likelihood functions or other objective functions that can only be calculated at the panel level, for which there is no observation-by-observation decomposition. Unlike clusters, these panel likelihood functions are difficult to calculate and require the use of type d or v evaluators. A panel variable takes on the same value within a panel and different values across panels. moptimize_init_by() allows you to set a panel variable.

You may set both a cluster variable and a panel variable, but be careful because, for most likelihood functions, panels are mathematically required to be nested within cluster.

moptimize_init_cluster(M, c) specifies a cluster variable. c may be a string scalar containing a Stata variable name, or c may be a real colvector directly containing the cluster values. The default is "", meaning no clustering. If clustering is specified, the default *vcetype* becomes "robust".

moptimize_init_by(M, *by*) specifies a panel variable and specifies that only panel-level calculations are meaningful. *by* may be a string scalar containing a Stata variable name, or *by* may be a real colvector directly containing the panel ID values. The default is "", meaning no panels. If panels are specified, the default *vcetype* remains unchanged, but if the opg variance estimator is used, the opg calculation is modified so that it is clustered at the panel level.

Specifying optimization technique

Technique refers to the numerical methods used to solve the optimization problem. The default is Newton–Raphson maximization.

moptimize_init_which(M, { "max" | "min" }) sets whether the maximum or minimum of the objective function is to be found. The default is "max".

moptimize_init_technique(M, *technique*) specifies the technique to be used to find the coefficient vector b that maximizes or minimizes the objective function. Allowed values are

technique	description
"nr"	modified Newton–Raphson
"dfp"	Davidon–Fletcher–Powell
"bfgs"	Broyden–Fletcher–Goldfarb–Shanno
"bhhh"	Berndt–Hall–Hall–Hausman
"nm"	Nelder–Mead
"gn"	Gauss–Newton (quadratic optimization)

The default is "nr".

You can switch between "nr", "dfp", "bfgs", and "bhhh" by specifying two or more of them in a space-separated list. By default, moptimize() will use an algorithm for five iterations before switching to the next algorithm. To specify a different number of iterations, include the number after the technique. For example, specifying moptimize_init_technique(M, "bhhh 10 nr 1000") requests that moptimize() perform 10 iterations using the Berndt–Hall–Hall–Hausman algorithm, followed by 1,000 iterations using the modified Newton-Raphson algorithm, and then switch back to Berndt–Hall–Hall–Hausman for 10 iterations, and so on. The process continues until convergence or until *maxiter* is exceeded.

moptimize_init_singularHmethod(M, *singularHmethod*) specifies the action to be taken during optimization if the Hessian is found to be singular and the *technique* requires the Hessian be of full rank. Allowed values are

singularHmethod	description
`"m-marquardt"`	modified Marquardt algorithm
`"hybrid"`	mixture of steepest descent and Newton

The default is `"m-marquardt"`.

`"hybrid"` is equivalent to `ml`'s `difficult` option; see [R] **ml**.

`moptimize_init_nmsimplexdeltas(M, delta)` is for use with Nelder–Mead, also known as technique `nm`. This function sets the values of *delta* to be used, along with the initial parameters, to build the simplex required by Nelder–Mead. Use of this function is required only in the Nelder–Mead case. The values in *delta* must be at least 10 times larger than *ptol*. The initial simplex will be $\{p, p + (d_1, 0, \ldots 0), p + (0, d_2, 0, \ldots, 0), \ldots, p + (0, 0, \ldots, 0, d_K)\}$.

Specifying initial values

Initial values are values you optionally specify that via a search procedure result in starting values that are then used for the first iteration of the optimization technique. That is,

$$\text{initial values} \xrightarrow[\text{(searching)}]{} \text{starting values} \xrightarrow[\text{technique)}]{\text{(optimization}} \text{final results}$$

Initial values are specified parameter by parameter.

`moptimize_init_eq_coefs(M, i, b0)` sets the initial values of the coefficients for the *i*th parameter to be *b0*: $1 \times ki$. The default is $(0, 0, \ldots, 0)$.

The following functions control whether searching is used to improve on the initial values to produce better starting values. In addition to searching a predetermined set of hardcoded starting values, there are two other methods that can be used to improve on the initial values: random and rescaling. By default, random is off and rescaling is on. You can use one, the other, or both.

`moptimize_init_search(M, {"on" | "off"})` determines whether any attempts are to be made to improve on the initial values via a search technique. The default is `"on"`. If you specify `"off"`, the initial values become the starting values.

`moptimize_init_search_random(M, {"off" | "on"})` determines whether the random method of improving initial values is to be attempted. The default is `"off"`. Use of the random method is recommended when the initial values are or might be infeasible. Infeasible means that the function cannot be evaluated, which mechanically corresponds to the user-written evaluator returning a missing value. The random method is seldom able to improve on feasible initial values. It works well when the initial values are or might be infeasible.

`moptimize_init_search_repeat(M, nr)` controls how many times random values are tried if the random method is turned on. The default is 10.

`moptimize_init_search_bounds(M, i, minmax)` specifies the bounds for the random search. *minmax* is a 1×2 real rowvector containing the minimum and maximum values for the *i*th parameter (equation). The default is $(., .)$, meaning no lower and no upper bounds.

`moptimize_init_search_rescale(M, {"on" | "off"})` determines whether rescaling is attempted. The default is `"on"`. Rescaling is a deterministic (not random) method. It also usually improves initial values, and usually reduces the number of subsequent iterations required by the optimization technique.

Performing one evaluation of the objective function

`moptimize_evaluate(`*M*`)` and `_moptimize_evaluate(`*M*`)` perform one evaluation of the function evaluated at the initial values. Results can be accessed by using `moptimize_result()`, including first- and second-derivative-based results.

`moptimize_evaluate()` and `_moptimize_evaluate()` do the same thing, differing only in that `moptimize_evaluate()` aborts with a nonzero return code if things go badly, whereas `_moptimize_evaluate()` returns the real scalar error code. An infeasible initial value is an error.

The evaluation is performed at the initial values, not the starting values, and this is true even if search is turned on. If you want to perform an evaluation at the starting values, then perform optimization with *maxiter* set to 0.

Performing optimization of the objective function

`moptimize(`*M*`)` and `_moptimize(`*M*`)` perform optimization. Both routines do the same thing; they differ only in their behavior when things go badly. `moptimize()` returns nothing and aborts with error. `_moptimize()` returns a real scalar error code. `moptimize()` is best for interactive use and often adequate for use in programs that do not want to consider the possibility that optimization might fail.

The optimization process is as follows:

1. The initial values are used to create starting values. The value of the function at the starting values is calculated. If that results in a missing value, the starting values are declared infeasible. `moptimize()` aborts with return code 430; `_moptimize()` returns a nonzero error code, which maps to 430 via `moptimize_result_returncode()`. This step is called iteration 0.

2. The starting values are passed to the technique to produce better values. Usually this involves the technique calculating first and second derivatives, numerically or analytically, and then stepping multiple times in the appropriate direction, but techniques can vary on this. In general, the technique performs what it calls one iteration, the result of which is to produce better values. Those new values then become the starting values and the process repeats.

   An iteration is said to fail if the new coefficient vector is infeasible (results in a missing value). Then attempts are made to recover and, if those attempts are successful, optimization continues. If they fail, `moptimize()` aborts with error and `_moptimize()` returns a nonzero error code.

   Other problems may arise, such as singular Hessians or the inability to find better values. Various fix-ups are made and optimization continues. These are not failures.

   This step is called iterations 1, 2, and so on.

3. Step 2 continues either until the process converges or until the maximum number of iterations (*maxiter*) is exceeded. Stopping due to *maxiter* is not considered an error. Upon completion, programmers should check `moptimize_result_converged()`.

If optimization succeeds, which is to say, if `moptimize()` does not abort or `_moptimize()` returns 0, you can use the `moptimize_result()` functions to access results.

Tracing optimization

moptimize( ) and _moptimize( ) will produce output like

$$\text{Iteration 0:} \quad \text{f(p)} = \ldots\ldots\ldots$$
$$\text{Iteration 1:} \quad \text{f(p)} = \ldots\ldots\ldots$$

You can change the $f(p)$ to be "log likelihood" or whatever else you want. You can also change "Iteration".

moptimize_init_iterid(M, *id*) sets the string to be used to label the iterations in the iteration log. *id* is a string scalar. The default is "Iteration".

moptimize_init_valueid(M, *id*) sets the string to be used to label the objective function value in the iteration log. *id* is a string scalar. The default is "f(p)".

Additional results can be displayed during optimization, which can be useful when you are debugging your evaluator. This is called tracing the execution.

moptimize_init_tracelevel(M, *tracelevel*) specifies the output to be displayed during the optimization process. Allowed values are

tracelevel	to be displayed each iteration
"none"	nothing
"value"	function value
"tolerance"	previous + convergence values
"coefs"	previous + parameter values
"params"	same as "coefs"
"step"	previous + stepping information
"gradient"	previous + gradient vector
"hessian"	previous + Hessian matrix

The default is "value".

Setting *tracelevel* is a shortcut. The other trace functions allow you to turn on and off individual features. In what follows, the documented defaults are the defaults when *tracelevel* is "value".

moptimize_init_trace_ado(M, { "off" | "on" }) traces the execution of evaluators written as ado-files. This topic is not discussed in this manual entry. The default is "off".

moptimize_init_trace_dots(M, { "off" | "on" }) displays a dot each time your evaluator is called. The default is "off".

moptimize_init_trace_value(M, { "on" | "off" }) displays the function value at the start of each iteration. The default is "on".

moptimize_init_trace_tol(M, { "off" | "on" }) displays the value of the calculated result that is compared with the effective convergence criterion at the end of each iteration. The default is "off".

moptimize_init_trace_coefs(M, { "off" | "on" }) displays the coefficients at the start of each iteration. The default is "off".

`moptimize_init_trace_step(`*M*`, {` `"off"` `|` `"on"` `})` displays the steps within iteration. Listed are the value of objective function along with the word forward or backward. The default is `"off"`.

`moptimize_init_trace_gradient(`*M*`, {` `"off"` `|` `"on"` `})` displays the gradient vector at the start of each iteration. The default is `"off"`.

`moptimize_init_trace_Hessian(`*M*`, {` `"off"` `|` `"on"` `})` displays the Hessian matrix at the start of each iteration. The default is `"off"`.

Specifying convergence criteria

Convergence is based on several rules controlled by four parameters: *maxiter*, *ptol*, *vtol*, and *nrtol*. The first rule is not a convergence rule, but a stopping rule, and it is controlled by *maxiter*.

`moptimize_init_conv_maxiter(`*M*`, `*maxiter*`)` specifies the maximum number of iterations. If this number is exceeded, optimization stops and results are posted where they are accessible by using the `moptimize_result_*()` functions, just as if convergence had been achieved. `moptimize_result_converged()`, however, is set to 0 rather than 1. The default *maxiter* is Stata's `c(maxiter)`, which is usually 16,000.

`moptimize_init_conv_warning(`*M*`, {` `"on"` `|` `"off"` `})` specifies whether the warning message "convergence not achieved" is to be displayed when this stopping rule is invoked. The default is `"on"`.

Usually, convergence occurs before the stopping rule comes into effect. The convergence criterion is a function of three real scalar values: *ptol*, *vtol*, and *nrtol*. Let

$$
\begin{aligned}
b &= \text{full set of coefficients} \\
b\_prior &= \text{value of } b \text{ from prior iteration} \\
v &= \text{value of objective function} \\
v\_prior &= \text{value of } v \text{ from prior iteration} \\
g &= \text{gradient vector from this iteration} \\
H &= \text{Hessian matrix from this iteration}
\end{aligned}
$$

Define, for maximization,

$$
\begin{aligned}
&C\_ptol: & \texttt{mreldif}(b,\ b\_prior) &\leq ptol \\
&C\_vtol: & \texttt{reldif}(v,\ v\_prior) &\leq vtol \\
&C\_nrtol: & g \times \texttt{invsym}(-H) \times g' &< nrtol \\
&C\_concave: & -H \text{ is positive semidefinite}
\end{aligned}
$$

For minimization, think in terms of maximization of $-f(p)$. Convergence is declared when

$$(C\_ptol \mid C\_vtol)\ \&\ C\_nrtol\ \&\ C\_concave$$

The above applies in cases of derivative-based optimization, which currently is all techniques except `"nm"` (Nelder–Mead). In the Nelder–Mead case, the criterion is

$$
\begin{aligned}
&C\_ptol: & \texttt{mreldif}(\text{vertices of } R) &\leq ptol \\
&C\_vtol: & \texttt{reldif}(R) &\leq vtol
\end{aligned}
$$

where R is the minimum and maximum values on the simplex. Convergence is declared when $C\_ptol \mid C\_vtol$.

The values of *ptol*, *vtol*, and *nrtol* are set by the following functions:

moptimize_init_conv_ptol(*M, ptol*) sets *ptol*. The default is 1e–6.

moptimize_init_conv_vtol(*M, vtol*) sets *vtol*. The default is 1e–7.

moptimize_init_conv_nrtol(*M, nrtol*) sets *nrtol*. The default is 1e–5.

moptimize_init_conv_ignorenrtol(*M,* { "off" | "on" }) sets whether *C_nrtol* should always be treated as true, which in effect removes the *nrtol* criterion from the convergence rule. The default is "off".

Accessing results

Once you have successfully performed optimization, or you have successfully performed a single function evaluation, you may display results, post results to Stata, or access individual results.

To display results, use moptimize_result_display().

moptimize_result_display(*M*) displays estimation results. Standard errors are shown using the default *vcetype*.

moptimize_result_display(*M, vcetype*) displays estimation results. Standard errors are shown using the specified *vcetype*.

Also there is a third syntax for use after results have been posted to Stata, which we will discuss below.

moptimize_result_display() without arguments (not even *M*) displays the estimation results currently posted in Stata.

vcetype specifies how the variance–covariance matrix of the estimators (VCE) is to be calculated. Allowed values are

vcetype	description
""	use default for technique
"oim"	observed information matrix
"opg"	outer product of gradients
"robust"	Huber/White/sandwich estimator
"svy"	survey estimator; equivalent to robust

The default *vcetype* is oim except for technique bhhh, where it is opg. If survey, pweights, or clusters are used, the default becomes robust or svy.

As an aside, if you set moptimize_init_vcetype() during initialization, that changes the default.

moptimize_init_vcetype(*M, vcetype*), *vcetype* being a string scalar, resets the default *vcetype*.

To post results to Stata, use moptimize_result_post().

moptimize_result_post(*M*) posts estimation results to Stata where they can be displayed with Mata function moptimize_result_post() (without arguments) or with Stata command ereturn display (see [P] **ereturn**). The posted VCE will be of the default *vcetype*.

moptimize_result_post(M, $vcetype$) does the same thing, except the VCE will be of the specified $vcetype$.

The remaining moptimize_result_*() functions simply return the requested result. It does not matter whether results have been posted or previously displayed.

moptimize_result_value(M) returns the real scalar value of the objective function.

moptimize_result_value0(M) returns the real scalar value of the objective function at the starting values.

moptimize_result_coefs(M $[$, i $]$) returns the $1 \times (ki + ci)$ coefficient rowvector for the ith equation. If $i \geq$. or argument i is omitted, the $1 \times K$ full set of coefficients is returned.

moptimize_result_colstripe(M $[$, i $]$) returns a $(ki + ci) \times 2$ string matrix containing, for the ith equation, the equation names in the first column and the coefficient names in the second. If $i \geq$. or argument i is omitted, the result is $K \times 2$.

moptimize_result_scores(M) returns an $N \times m$ (evaluator types lf and e), or an $N \times K$ (evaluator type v), or an $L \times K$ (evaluator type q) real matrix containing the observation-by-observation scores. For all other evaluator types, J(0,0,.) is returned. For evaluator types lf and e, scores are defined as the derivative of the objective function with respect to the parameters. For evaluator type v, scores are defined as the derivative of the objective function with respect to the coefficients. For evaluator type q, scores are defined as the derivatives of the L independent elements with respect to the coefficients.

moptimize_result_gradient(M $[$, i $]$) returns the $1 \times (ki + ci)$ gradient rowvector for the ith equation. If $i \geq$. or argument i is omitted, the $1 \times K$ gradient corresponding to the full set of coefficients is returned. Gradient is defined as the derivative of the objective function with respect to the coefficients.

moptimize_result_Hessian(M $[$, i $]$) returns the $(ki + ci) \times (ki + ci)$ Hessian matrix for the ith equation. If $i \geq$. or argument i is omitted, the $K \times K$ Hessian corresponding to the full set of coefficients is returned. The Hessian is defined as the second derivative of the objective function with respect to the coefficients.

moptimize_result_V(M $[$, i $]$) returns the appropriate $(ki + ci) \times (ki + ci)$ submatrix of the full variance matrix calculated according to the default $vcetype$. If $i \geq$. or argument i is omitted, the full $K \times K$ variance matrix corresponding to the full set of coefficients is returned.

moptimize_result_Vtype(M) returns a string scalar containing the default $vcetype$.

moptimize_result_V_oim(M $[$, i $]$) returns the appropriate $(ki + ci) \times (ki + ci)$ submatrix of the full variance matrix calculated as the inverse of the negative Hessian matrix (the observed information matrix). If $i \geq$. or argument i is omitted, the full $K \times K$ variance matrix corresponding to the full set of coefficients is returned.

moptimize_result_V_opg(M $[$, i $]$) returns the appropriate $(ki + ci) \times (ki + ci)$ submatrix of the full variance matrix calculated as the inverse of the outer product of the gradients. If $i \geq$. or argument i is omitted, the full $K \times K$ variance matrix corresponding to the full set of coefficients is returned. If moptimize_result_V_opg() is used with evaluator types other than lf, e, v, or q, an appropriately dimensioned matrix of zeros is returned.

moptimize_result_V_robust(M $[$, i $]$) returns the appropriate $(ki + ci) \times (ki + ci)$ submatrix of the full variance matrix calculated via the sandwich estimator. If $i \geq$. or argument i is omitted, the full $K \times K$ variance matrix corresponding to the full set of coefficients is returned.

If `moptimize_result_V_robust()` is used with evaluator types other than lf, e, v, or q, an appropriately dimensioned matrix of zeros is returned.

`moptimize_result_iterations(M)` returns a real scalar containing the number of iterations performed.

`moptimize_result_converged(M)` returns a real scalar containing 1 if convergence was achieved and 0 otherwise.

`moptimize_result_iterationlog(M)` returns a real colvector containing the values of the objective function at the end of each iteration. Up to the last 20 iterations are returned, one to a row.

`moptimize_result_errorcode(M)` returns the real scalar containing the error code from the most recently run optimization or function evaluation. The error code is 0 if there are no errors. This function is useful only after `_moptimize()` or `_moptimize_evaluate()` because the nonunderscore versions aborted with error if there were problems.

`moptimize_result_errortext(M)` returns a string scalar containing the error text corresponding to `moptimize_result_errorcode()`.

`moptimize_result_returncode(M)` returns a real scalar containing the Stata return code corresponding to `moptimize_result_errorcode()`.

The following error codes and their corresponding Stata return codes are for `moptimize()` only. To see other error codes and their corresponding Stata return codes, see [M-5] **optimize( )**.

Error code	Return code	Error text
400	1400	could not find feasible values
401	491	Stata program evaluator returned an error
402	198	views are required when the evaluator is a Stata program
403	198	Stata program evaluators require a touse variable

Stata evaluators

The following function is useful only when your evaluator is a Stata program instead of a Mata function.

`moptimize_ado_cleanup(M)` removes all the global macros with the ML_ prefix. A temporary weight variable is also dropped if weights were specified.

Advanced functions

These functions are not really advanced, they are just seldomly used.

`moptimize_init_verbose(M, { "on" | "off" })` specifies whether error messages are to be displayed. The default is "on".

`moptimize_init_evaluations(M, { "off" | "on" })` specifies whether the system is to count the number of times the evaluator is called. The default is "off".

`moptimize_result_evaluations`(*M*) returns a 1×3 real rowvector containing the number of times the evaluator was called, assuming `moptimize_init_evaluations()` was set on. Contents are the number of times called for the purposes of 1) calculating the objective function, 2) calculating the objective function and its first derivative, and 3) calculating the objective function and its first and second derivatives. If `moptimize_init_evaluations()` was set off, returned is (0,0,0).

Syntax of evaluators

An *evaluator* is a program you write that calculates the value of the function being optimized and optionally calculates the function's first and second derivatives. The evaluator you write is called by the `moptimize()` functions.

There are five styles in which the evaluator can be written, known as types `lf`, `d`, `e`, `v`, and `q`. *evaluatortype*, optionally specified in `moptimize_init_evaluatortype()`, specifies the style in which the evaluator is written. Allowed values are

evaluatortype	description
`"lf"`	*function*() returns $N \times 1$ colvector value
`"d0"`	*function*() returns scalar value
`"d1"`	same as `"d0"` and returns gradient rowvector
`"d2"`	same as `"d1"` and returns Hessian matrix
`"d1debug"`	same as `"d1"` but checks gradient
`"d2debug"`	same as `"d2"` but checks gradient and Hessian
`"e1"`	*function*() returns scalar value and returns equation-level score matrix
`"e2"`	same as `"e1"` and returns Hessian matrix
`"e1debug"`	same as `"e1"` but checks gradient
`"e2debug"`	same as `"e2"` but checks gradient and Hessian
`"v0"`	*function*() returns $N \times 1$ colvector value
`"v1"`	same as `"v0"` and returns score matrix
`"v2"`	same as `"v1"` and returns Hessian matrix
`"v1debug"`	same as `"v1"` but checks gradient
`"v2debug"`	same as `"v2"` but checks gradient and Hessian
`"q0"`	*function*() returns colvector value
`"q1"`	same as `"q0"` and returns score matrix
`"q1debug"`	same as `"q1"` but checks gradient

The default is `"lf"` if not set.
`"q"` evaluators are used with technique `"gn"`.
Returned gradients are $1 \times K$ rowvectors.
Returned Hessians are $K \times K$ matrices.

Examples of each of the evaluator types are outlined below.

You must tell `moptimize()` the identity and type of your evaluator, which you do by using the `moptimize_init_evaluator()` and `moptimize_init_evaluatortype()` functions.

`moptimize_init_evaluator(`*M*, &*functionname*`())` sets the identity of the evaluator function that you write in Mata.

`moptimize_init_evaluator(`*M*, `"`*programname*`")` sets the identity of the evaluator program that you write in Stata.

`moptimize_init_evaluatortype(`*M*, *evaluatortype*`)` informs `moptimize()` of the style of evaluator you have written. *evaluatortype* is a string scalar from the table above. The default is `"lf"`.

`moptimize_init_negH(`*M*, { `"off"` | `"on"` }`)` sets whether the evaluator you have written returns H or $-H$, the Hessian or the negative of the Hessian, if it returns a Hessian at all. This is for backward compatibility with prior versions of Stata's `ml` command (see [R] **ml**). Modern evaluators return H. The default is `"off"`.

Syntax of type lf evaluators

`lfeval(`*M*, *b*, $\overline{fv}$`)`:

    inputs:
          M: problem definition
          b: coefficient vector
    outputs:
          fv: $N \times 1$, $N = $ # of observations

Notes:

1. The objective function is $f() = $ `colsum(`*fv*`)`.

2. In the case where $f()$ is a log-likelihood function, the values of the log likelihood must be summable over the observations.

3. For use with any technique except `gn`.

4. May be used with robust, clustering, and survey.

5. Returns *fv* containing missing (*fv* = .) if evaluation is not possible.

Syntax of type d evaluators

`deval(`*M*, `todo`, *b*, $\overline{fv}$, $\overline{g}$, $\overline{H}$`)`:

    inputs:
          M: problem definition
         todo: real scalar containing 0, 1, or 2
          b: coefficient vector
    outputs:
          fv: real scalar
          g: $1 \times K$, gradients, $K = $ # of coefficients
          H: $K \times K$, Hessian

Notes:

1. The objective function is $f() = fv$.

2. For use with any log-likelihood function, or any function.

3. For use with any technique except gn and bhhh.

4. Cannot be used with robust, clustering, and survey.

5. *deval*( ) must always fill in *fv*, and fill in *g* if *todo* $\geq$ 1, and fill in *H* if *todo* = 2. For type d0, *todo* will always be 0. For type d1 and d1debug, *todo* will be 0 or 1. For type d2 and d2debug, *todo* will be 0, 1, or 2.

6. Returns *fv* = . if evaluation is not possible.

Syntax of type e evaluators

`eeval(`*M*, *todo*, *b*, $\overline{fv}$, $\overline{S}$, $\overline{H}$`):`

> *inputs*:

	M:	problem definition
	todo:	real scalar containing 0, 1, or 2
	b:	coefficient vector

> *outputs*:

	fv:	real scalar
	S:	$N \times m$, scores, m = # of equations (parameters) N = # of observations
	H:	$K \times K$, Hessian, K = # of coefficients

Notes:

1. The objective function is $f() = fv$.

2. Type e is a variation of type d that optionally allows robust, clustering, and survey. Although e could be used with an arbitrary function, it is intended for use when $f()$ is a log-likelihood function and the log-likelihood values are summable over the observations.

3. For use with any technique except gn.

4. May be used with robust, clustering, and survey.

5. Always returns *fv*, returns *S* if *todo* $\geq$ 1, and returns *H* if *todo* = 2. For type e1 and e1debug, *todo* will be 0 or 1. For type e2 and e2debug, *todo* will be 0, 1, or 2. There is no e0.

6. Returns *fv* = . if evaluation is not possible.

Syntax of type v evaluators

`veval(`*M*, *todo*, *b*, $\overline{fv}$, $\overline{S}$, $\overline{H}$`):`

> *inputs*:

	M:	problem definition
	todo:	real scalar containing 0, 1, or 2
	b:	coefficient vector

> *outputs*:

	fv:	$L \times 1$, values, L = # of independent elements
	S:	$L \times K$, scores, K = # of coefficients
	H:	$K \times K$, Hessian

Notes:

1. The objective function is $f() = \texttt{colsum}(fv)$.

2. Type v is a variation on type e that relaxes the requirement that the log-likelihood function be summable over the observations. v is especially useful for fitting panel-data models with technique bhhh. Then L is the number of panels.

3. For use with any technique except gn.

4. May be used with robust, clustering, and survey.

5. Always returns fv, returns S if $todo \geq 1$, and returns H if $todo = 2$. For type v0, $todo$ will always be 0. For type v1 and v1debug, $todo$ will be 0 or 1. For type v2 and v2debug, $todo$ will be 0, 1, or 2.

6. Returns $fv = .$ if evaluation is not possible.

Syntax of type q evaluators

> qeval(M, $todo$, b, $\bar{r}$, $\bar{S}$)

inputs:

M:	problem definition
$todo$:	real scalar containing 0 or 1
b:	coefficient vector

outputs:

r:	$L \times 1$ of independent elements
S:	$L \times m$, scores, $m = $ # of parameters

Notes:

1. Type q is for quadratic optimization. The objective function is $f() = r'Wr$, where r is returned by *qeval*() and W has been previously set by using moptimize_init_gnweightmatrix(), described below.

2. For use only with techniques gn and nm.

3. Always returns r and returns S if $todo = 1$. For type q0, $todo$ will always be 0. For type q1 and q1debug, $todo$ will be 0 or 1. There is no type q2.

4. Returns r containing missing, or $r = .$ if evaluation is not possible.

Use moptimize_init_gnweightmatrix() during initialization to set matrix W.

> moptimize_init_gnweightmatrix(M, W) sets real matrix W: $L \times L$, which is used only by type q evaluators. The objective function is $r'Wr$. If W is not set and if observation weights w are set by using moptimize_init_weight(), then $W = \text{diag}(w)$. If w is not set, then W is the identity matrix.

> moptimize() does not produce a robust VCE when you set W with moptimize_init_gnweight().

Passing extra information to evaluators

In addition to the arguments the evaluator receives, you may arrange that extra information be sent to the evaluator. Specify the extra information to be sent by using moptimize_init_userinfo().

moptimize_init_userinfo(M, l, Z) specifies that the lth piece of extra information is Z. l is a real scalar. The first piece of extra information should be 1; the second piece, 2; and so on. Z can be anything. No copy of Z is made.

moptimize_init_nuserinfo(M, n_user) specifies the total number of extra pieces of information to be sent. Setting n_user is optional; it will be automatically determined from the moptimize_init_userinfo() calls you issue.

Inside your evaluator, you access the information by using moptimize_util_userinfo().

moptimize_util_userinfo(M, l) returns the Z set by moptimize_init_userinfo().

Utility functions

There are various utility functions that are helpful in writing evaluators and in processing results returned by the moptimize_result_*() functions.

The first set of utility functions are useful in writing evaluators, and the first set return results that all evaluators need.

moptimize_util_depvar(M, j) returns an $Nj \times 1$ colvector containing the values of the jth dependent variable, the values set by moptimize_init_depvar(M, j, ...).

moptimize_util_xb(M, b, i) returns the $Ni \times 1$ colvector containing the value of the ith parameter, which is usually $Xi \times bi'$:+ $b0i$, but might be as complicated as $Xi \times bi' + oi +$ $\ln(ti)$:+ $b0i$.

Once the inputs of an evaluator have been processed, the following functions assist in making the calculations required of evaluators.

moptimize_util_sum(M, v) returns the "sum" of colvector v. This function is for use in evaluators that require you to return an overall objective function value rather than observation-by-observation results. Usually, moptimize_util_sum() returns sum(v), but in cases where you have specified a weight by using moptimize_init_weight() or there is an implied weight due to use of moptimize_init_svy(), the appropriately weighted sum is returned. Use moptimize_util_sum() to sum log-likelihood values.

moptimize_util_vecsum(M, i, s, *value*) is like moptimize_util_sum(), but for use with gradients. The gradient is defined as the vector of partial derivatives of $f()$ with respect to the coefficients bi. Some evaluator types require that your evaluator be able to return this vector. Nonetheless, it is usually easier to write your evaluator in terms of parameters rather than coefficients, and this function handles the mapping of parameter gradients to the required coefficient gradients.

Input s is an $Ni \times 1$ colvector containing df/dpi for each observation. df/dpi is the partial derivative of the objective function, but with respect to the ith parameter rather than the ith set of coefficients. moptimize_util_vecsum() takes s and returns the $1 \times (ki + ci)$ summed gradient. Also weights, if any, are factored into the calculation.

If you have more than one equation, you will need to call `moptimize_util_vecsum()` m times, once for each equation, and then concatenate the individual results into one vector.

value plays no role in `moptimize_util_vecsum()`'s calculations. *value*, however, should be specified as the result obtained from `moptimize_util_sum()`. If that is inconvenient, make *value* any nonmissing value. If the calculation from parameter space to vector space cannot be performed, or if your original parameter space derivatives have any missing values, *value* will be changed to missing. Remember, when a calculation cannot be made, the evaluator is to return a missing value for the objective function. Thus storing the value of the objective function in *value* ensures that your evaluator will return missing if it is supposed to.

`moptimize_util_matsum(M, i, i2, s, value)` is similar to `moptimize_util_vecsum()`, but for Hessians (matrix of second derivatives).

Input s is an $Ni \times 1$ colvector containing $d^2f/dpidpi2$ for each observation. `moptimize_util_matsum()` returns the $(ki + ci) \times (ki2 + ci2)$ summed Hessian. Also weights, if any, are factored into the calculation.

If you have $m > 1$ equations, you will need to call `moptimize_util_matsum()` $m \times (m+1)/2$ times and then join the results into one symmetric matrix.

value plays no role in the calculation and works the same way it does in `moptimize_util_vecsum()`.

`moptimize_util_matbysum()` is an added helper for making `moptimize_util_matsum()` calculations in cases where you have panel data and the log-likelihood function's values exists only at the panel level. `moptimize_util_matbysum(M, i, a, b, value)` is for making diagonal calculations and `moptimize_util_matbysum(M, i, i2, a, b, c, value)` is for making off-diagonal calculations.

This is an advanced topic; see Gould, Pitblado, and Sribney (2006, 117–119) for a full description of it. In applying the chain rule to translate results from parameter space to coefficient space, `moptimize_util_matsum()` can be used to make some of the calculations, and `moptimize_util_matbysum()` can be used to make the rest. *value* plays no role and works just as it did in the other helper functions. `moptimize_util_matbysum()` is for use sometimes when *by* has been set, which is done via `moptimize_init_by(M, by)`. `moptimize_util_matbysum()` is never required unless *by* has been set.

The other utility functions are useful inside or outside of evaluators. One of the more useful is `moptimize_util_eq_indices()`, which allows two or three arguments.

`moptimize_util_eq_indices(M, i)` returns a 1×2 vector that can be used with range subscripts to extract the portion relevant for the ith equation from any $1 \times K$ vector, that is, from any vector conformable with the full coefficient vector.

`moptimize_util_eq_indices(M, i, i2)` returns a 2×2 matrix that can be used with range subscripts to exact the portion relevant for the ith and $i2$th equations from any $K \times K$ matrix, that is, from any matrix with rows and columns conformable with the full variance matrix.

For instance, let b be the $1 \times K$ full coefficient vector, perhaps obtained by being passed into an evaluator, or perhaps obtained from b = `moptimize_result_coefs(M)`. Then b[|`moptimize_util_eq_indices(M, i)`|] is the $1 \times (ki + ci)$ vector of coefficients for the ith equation.

Let V be the $K \times K$ full variance matrix obtained by $V = \texttt{moptimize\_result\_V}(M)$. Then $\texttt{V[|moptimize\_util\_eq\_indices}(M, i, i)|]$ is the $(ki + ci) \times (ki + ci)$ variance matrix for the ith equation. $\texttt{V[|moptimize\_util\_eq\_indices}(M, i, j)|]$ is the $(ki + ci) \times (kj + cj)$ covariance matrix between the ith and jth equations.

Finally, there is one more utility function that may help when you become confused: $\texttt{moptimize\_query}()$.

> $\texttt{moptimize\_query}(M)$ displays in readable form everything you have set via the $\texttt{moptimize\_init\_*}()$ functions, along with the status of the system.

Description

The $\texttt{moptimize}()$ functions find coefficients $(\mathbf{b}_1, \mathbf{b}_2, \ldots, \mathbf{b}_m)$ that maximize or minimize $f(\mathbf{p}_1, \mathbf{p}_2, \ldots, \mathbf{p}_m)$, where $\mathbf{p}i = \mathbf{X}i \times \mathbf{b}i'$, a linear combination of $\mathbf{b}i$ and the data. The user of $\texttt{moptimize}()$ writes a Mata function or Stata program to evaluate $f(\mathbf{p}_1, \mathbf{p}_2, \ldots, \mathbf{p}_m)$. The data can be in Mata matrices or in the Stata dataset currently residing in memory.

$\texttt{moptimize}()$ is especially useful for obtaining solutions for maximum likelihood models, minimum chi-squared models, minimum squared-residual models, and the like.

Remarks

Remarks are presented under the following headings:

> *Relationship of moptimize( ) to Stata's ml and to Mata's optimize( )*
> *Mathematical statement of the moptimize( ) problem*
> *Filling in moptimize( ) from the mathematical statement*
> *The type lf evaluator*
> *The type d, e, v, and q evaluators*
> *Example using type d*
> *Example using type e*

Relationship of moptimize( ) to Stata's ml and to Mata's optimize( )

$\texttt{moptimize}()$ is Mata's and Stata's premier optimization routine. This is the routine used by most of the official optimization-based estimators implemented in Stata.

That said, Stata's $\texttt{ml}$ command—see [R] **ml**—provides most of the capabilities of Mata's $\texttt{moptimize}()$, and $\texttt{ml}$ is easier to use. In fact, $\texttt{ml}$ uses $\texttt{moptimize}()$ to perform the optimization, and $\texttt{ml}$ amounts to little more than a shell providing a friendlier interface. If you have a maximum-likelihood model you wish to fit, we recommend you use $\texttt{ml}$ instead of $\texttt{moptimize}()$. Use $\texttt{moptimize}()$ when you need or want to work in the Mata environment, or when you wish to implement a specialized system for fitting a class of models.

Also make note of Mata's $\texttt{optimize}()$ function; see [M-5] **optimize( )**. $\texttt{moptimize}()$ finds coefficients $(\mathbf{b}_1, \mathbf{b}_2, \ldots, \mathbf{b}_m)$ that maximize or minimize $f(\mathbf{p}_1, \mathbf{p}_2, \ldots, \mathbf{p}_m)$, where $\mathbf{p}i = \mathbf{X}i \times \mathbf{b}i$. $\texttt{optimize}()$ handles a simplified form of the problem, namely, finding constant $(p_1, p_2, \ldots, pm)$ that maximizes or minimizes $f()$. $\texttt{moptimize}()$ is the appropriate routine for fitting a Weibull model, but if all you need to estimate are the fixed parameters of the Weibull distribution for some population, $\texttt{moptimize}()$ is overkill and $\texttt{optimize}()$ will prove easier to use.

These three routines are all related. Stata's ml uses moptimize() to do the numerical work. moptimize(), in turn, uses optimize() to perform certain calculations, including the search for parameters. There is nothing inferior about optimize() except that it cannot efficiently deal with models in which parameters are given by linear combinations of coefficients and data.

Mathematical statement of the moptimize( ) problem

We mathematically describe the problem moptimize() solves not merely to fix notation and ease communication, but also because there is a one-to-one correspondence between the mathematical notation and the moptimize*() functions. Simply writing your problem in the following notation makes obvious the moptimize*() functions you need and what their arguments should be.

In what follows, we are going to simplify the mathematics a little. For instance, we are about to claim $pi = Xi \times bi$:+ ci, when in the syntax section, you will see that $pi = Xi \times bi + oi + \ln(ti)$:+ ci. Here we omit oi and $\ln(ti)$ because they are seldom used. We will omit some other details, too. The statement of the problem under *Syntax*, above, is the full and accurate statement. We will also use typefaces a little differently. In the syntax section, we use italics following programming convention. In what follows, we will use boldface for matrices and vectors, and italics for scalars so that you can follow the math more easily. So in this section, we will write $\mathbf{b}i$, whereas under syntax we would write *bi*; regardless of typeface, they mean the same thing.

Function moptimize() finds coefficients

$$\mathbf{b} = ((\mathbf{b}_1, c_1), (\mathbf{b}_2, c_2), \dots, (\mathbf{b}_m, cm))$$

where

$$\mathbf{b}_1: 1 \times k_1, \qquad \mathbf{b}_2: 1 \times k_2, \qquad \dots, \qquad \mathbf{b}_m: 1 \times k_m$$
$$c_1: 1 \times 1, \qquad c_2: 1 \times 1, \qquad \dots, \qquad c_m: 1 \times 1$$

that maximize or minimize function

$$f(\mathbf{p}_1, \mathbf{p}_2, \dots, \mathbf{p}_m; \qquad \mathbf{y}_1, \mathbf{y}_2, \dots, \mathbf{y}_D)$$

where

$$\mathbf{p}_1 = \mathbf{X}_1 \times \mathbf{b}_1' \text{ :+ } c_1, \qquad \mathbf{X}_1 : N_1 \times k_1$$
$$\mathbf{p}_2 = \mathbf{X}_2 \times \mathbf{b}_2' \text{ :+ } c_2, \qquad \mathbf{X}_2 : N_2 \times k_2$$
$$\cdot$$
$$\cdot$$
$$\mathbf{p}_m = \mathbf{X}_m \times \mathbf{b}_m' \text{ :+ } c_m, \qquad \mathbf{X}_m : N_m \times k_m$$

and where $\mathbf{y}_1, \mathbf{y}_2, \dots, \mathbf{y}_D$ are of arbitrary dimension.

Usually, $N_1 = N_2 = \dots = N_m$, and the model is said to be fit on data of N observations. Similarly, column vectors $\mathbf{y}_1, \mathbf{y}_2, \dots, \mathbf{y}_D$ are usually called dependent variables, and each is also of N observations.

As an example, let's write the maximum-likelihood estimator for linear regression in the above notation. We begin by stating the problem in the usual way, but in Mata-ish notation:

Given data $\mathbf{y}$: $N \times 1$ and $\mathbf{X}$: $N \times k$, obtain $((\mathbf{b}, c), s^2)$ to fit

$$\mathbf{y} = \mathbf{X} \times \mathbf{b}' :+ c + \mathbf{u}$$

where the elements of $\mathbf{u}$ are distributed $N(0, s^2)$. The log-likelihood function is

$$\ln L = \sum_j \ln(\texttt{normalden}(\mathbf{y}_j - (\mathbf{X}_j \times \mathbf{b}' :+ c),\ 0,\ \texttt{sqrt}(s^2)))$$

where $\texttt{normalden}(x, \textit{mean}, \textit{sd})$ returns the density at x of the Gaussian normal with the specified mean and standard deviation; see [M-5] **normal( )**.

The above is a two-parameter or, equivalently, two-equation model in $\texttt{moptimize()}$ jargon. There may be many coefficients, but the likelihood function can be written in terms of two parameters, namely $\mathbf{p}_1 = \mathbf{X} \times \mathbf{b}' :+ c$ and $\mathbf{p}_2 = s^2$. Here is the problem stated in the $\texttt{moptimize()}$ notation:

Find coefficients

$$\mathbf{b} = ((\mathbf{b}_1, c_1), (c_2))$$

where

$$\mathbf{b}_1: 1 \times k$$
$$c_1: 1 \times 1, \qquad c_2: 1 \times 1$$

that maximize

$$f(\mathbf{p}_1, \mathbf{p}_2;\ \mathbf{y}) = \sum \ln(\texttt{normalden}(\mathbf{y} - \mathbf{p}_1, 0, \texttt{sqrt}(\mathbf{p}_2)))$$

where

$$\mathbf{p}_1 = \mathbf{X} \times \mathbf{b}_1' :+ c_1, \qquad \mathbf{X} : N \times k$$
$$\mathbf{p}_2 = c_2$$

and where y is $N \times 1$.

Notice that, in this notation, the regression coefficients $(\mathbf{b}_1, c_1)$ play a secondary role, namely, to determine $\mathbf{p}_1$. That is, the function, $f()$, to be optimized—a log-likelihood function here—is written in terms of $\mathbf{p}_1$ and $\mathbf{p}_2$. The program you will write to evaluate $f()$ will be written in terms of $\mathbf{p}_1$ and $\mathbf{p}_2$, thus abstracting from the particular regression model being fit. Whether the regression is mpg on weight or log income on age, education, and experience, your program to calculate $f()$ will remain unchanged. All that will change are the definitions of $\mathbf{y}$ and $\mathbf{X}$, which you will communicate to $\texttt{moptimize()}$ separately.

There is another advantage to this arrangement. We can trivially generalize linear regression without writing new code. Note that the variance s^2 is given by $\mathbf{p}_2$, and currently, we have $\mathbf{p}_2 = c_2$, that is, a constant. $\texttt{moptimize()}$ allows parameters to be constant, but it equally allows them to be given by a linear combination. Thus rather than defining $\mathbf{p}_2 = c_2$, we could define $\mathbf{p}_2 = \mathbf{X}_2 \times \mathbf{b}_2' :+ c_2$. If we did that, we would have a second linear equation that allowed the variance to vary observation by observation. As far as $\texttt{moptimize()}$ is concerned, that problem is the same as the original problem.

Filling in moptimize( ) from the mathematical statement

The mathematical statement of our sample problem is the following:

Find coefficients

$$\mathbf{b} = ((\mathbf{b}_1, c_1), (c_2))$$

$$\mathbf{b}_1: 1 \times k$$
$$c_1: 1 \times 1, \qquad c_2 : 1 \times 1$$

that maximize

$$f(\mathbf{p}_1, \mathbf{p}_2; \mathbf{y}) = \sum \ln(\texttt{normalden}(\mathbf{y} - \mathbf{p}_1, 0, \texttt{sqrt}(\mathbf{p}_2)))$$

where

$$\mathbf{p}_1 = \mathbf{X} \times \mathbf{b}_1' \ :+ c_1, \qquad \mathbf{X} : N \times k$$
$$\mathbf{p}_2 = c_2$$

and where y is $N \times 1$.

The corresponding code to perform the optimization is

```
. sysuse auto
. mata:
: function linregeval(transmorphic M, real rowvector b,
                      real colvector lnf)
{
        real colvector  p1, p2
        real colvector  y1

        p1 = moptimize_util_xb(M, b, 1)
        p2 = moptimize_util_xb(M, b, 2)
        y1 = moptimize_util_depvar(M, 1)

        lnf = ln(normalden(y1:-p1, 0, sqrt(p2)))
}
: M = moptimize_init()
: moptimize_init_evaluator(M, &linregeval())
: moptimize_init_depvar(M, 1, "mpg")
: moptimize_init_eq_indepvars(M, 1, "weight foreign")
: moptimize_init_eq_indepvars(M, 2, "")
: moptimize(M)
: moptimize_result_display(M)
```

Here is the result of running the above code:

```
. sysuse auto
(1978 Automobile Data)

. mata:
                                                    ─── mata (type end to exit) ───
: function linregeval(transmorphic M, real rowvector b, real colvector lnf)
> {
>           real colvector  p1, p2
>           real colvector  y1
>
>           p1 = moptimize_util_xb(M, b, 1)
>           p2 = moptimize_util_xb(M, b, 2)
>           y1 = moptimize_util_depvar(M, 1)
>             .
>           lnf = ln(normalden(y1:-p1, 0, sqrt(p2)))
> }
: M = moptimize_init()

: moptimize_init_evaluator(M, &linregeval())

: moptimize_init_depvar(M, 1, "mpg")

: moptimize_init_eq_indepvars(M, 1, "weight foreign")

: moptimize_init_eq_indepvars(M, 2, "")

: moptimize(M)
initial:        f(p) =      -<inf>  (could not be evaluated)
feasible:       f(p) = -12949.708
rescale:        f(p) = -243.04355
rescale eq:     f(p) = -236.58999
Iteration 0:    f(p) = -236.58999  (not concave)
Iteration 1:    f(p) = -227.46735
Iteration 2:    f(p) = -205.73496  (backed up)
Iteration 3:    f(p) = -195.72762
Iteration 4:    f(p) = -194.20885
Iteration 5:    f(p) = -194.18313
Iteration 6:    f(p) = -194.18306
Iteration 7:    f(p) = -194.18306

: moptimize_result_display(M)
```

	Number of obs	=	74

mpg	Coef.	Std. Err.	z	P>|z|	[95% Conf. Interval]
eq1					
weight	-.0065879	.0006241	-10.56	0.000	-.007811 -.0053647
foreign	-1.650029	1.053958	-1.57	0.117	-3.715749 .4156903
_cons	41.6797	2.121197	19.65	0.000	37.52223 45.83717
eq2					
_cons	11.13746	1.830987	6.08	0.000	7.54879 14.72613

(Continued on next page)

The type lf evaluator

Let's now interpret the code we wrote, which was

```
: function linregeval(transmorphic M, real rowvector b,
                      real colvector lnf)
{
        real colvector  p1, p2
        real colvector  y1

        p1 = moptimize_util_xb(M, b, 1)
        p2 = moptimize_util_xb(M, b, 2)
        y1 = moptimize_util_depvar(M, 1)

        lnf = ln(normalden(y1:-p1, 0, sqrt(p2)))
}
: M = moptimize_init()
: moptimize_init_evaluator(M, &linregeval())
: moptimize_init_depvar(M, 1, "mpg")
: moptimize_init_eq_indepvars(M, 1, "weight foreign")
: moptimize_init_eq_indepvars(M, 2, "")
: moptimize(M)
: moptimize_result_display(M)
```

We first defined the function to evaluate our likelihood function—we named the function lin-regeval(). The name was of our choosing. After that, we began an optimization problem by typing M = moptimize_init(), described the problem with moptimize_init_*() functions, performed the optimization by typing moptimize(), and displayed results by using mopti-mize_result_display().

Function linregeval() is an example of a type lf evaluator. There are several different evaluator types, including d0, d1, d2, d3, through q1. Of all of them, type lf is the easiest to use and is the one moptimize() uses unless we tell it differently. What makes lf easy is that we need only calculate the likelihood function; we are not required to calculate its derivatives. A description of lf appears under the heading *Syntax of type lf evaluators* under *Syntax* above.

In the syntax diagrams, you will see that type lf evaluators receive three arguments, *M*, *b*, and *fv*, although in linregeval(), we decided to call them M, b, and lnf. The first two arguments are inputs, and your evaluator is expected to fill in the third argument with observation-by-observation values of the log-likelihood function.

The input arguments are M and b. M is the problem handle, which we have not explained yet. Basically, all evaluators receive M as the first argument and are expected to pass M to any moptimize*() subroutines that they call. M in fact contains all the details of the optimization problem. The second argument, b, is the entire coefficient vector, which in the linregeval() case will be all the coefficients of our regression, the constant (intercept), and the variance of the residual. Those details are unimportant. Instead, your evaluator will pass *M* and *b* to moptimize() utility programs that will give you back what you need.

Using one of those utilities is the first action our linregeval() evaluator performs:

```
        p1 = moptimize_util_xb(M, b, 1)
```

That returns observation-by-observation values of the first parameter, namely, $X \times b_1$:+ c_1. mopti-mize_util_xb(x, b, 1) returns the first parameter because the last argument specified is 1. We obtained the second parameter similarly:

```
p2 = moptimize_util_xb(M, b, 2)
```

To evaluate the likelihood function, we also need the dependent variable. Another `moptimize*()` utility returns that to us:

```
y1 = moptimize_util_depvar(M, 1)
```

Having p1, p2, and y1, we are ready to fill in the log-likelihood values:

```
lnf = ln(normalden(y1:-p1, 0, sqrt(p2)))
```

For a type `lf` evaluator, you are to return observation-by-observation values of the log-likelihood function; `moptimize()` itself will sum them to obtain the overall log likelihood. That is exactly what the line `lnf = ln(normalden(y1:-p1, 0, sqrt(p2)))` did. Note that y1 is $N \times 1$, p1 is $N \times 1$, and p2 is $N \times 1$, so the `lnf` result we calculate is also $N \times 1$. Most of the other evaluator types are expected to return a scalar equal to the overall value of the function. That, however, is the only peculiarity of type `lf`. Everything else we have told you and are about to tell you always applies.

With the evaluator defined, we can estimate a linear regression by typing

```
: M = moptimize_init()
: moptimize_init_evaluator(M, &linregeval())
: moptimize_init_depvar(M, 1, "mpg")
: moptimize_init_eq_indepvars(M, 1, "weight foreign")
: moptimize_init_eq_indepvars(M, 2, "")
: moptimize(M)
: moptimize_result_display(M)
```

All estimation problems begin with

```
M = moptimize_init()
```

The returned value M is called a problem handle, and from that point on, you will pass M to every other `moptimize()` function you call. M contains the details of your problem. If you were to list *M*, you would see something like

```
: M
  0x15369a
```

$0\times15369a$ is in fact the address where all those details are stored. Exactly how M works does not matter, but it is important that you understand what M is. M is your problem. In a more complicated problem, you might need to perform nested optimizations. You might have one optimization problem, and right in the middle of it, even right in the middle of evaluating its log-likelihood function, you might need to set up and solve another optimization problem. You can do that. The first problem you would set up as M1 = `moptimize_init()`. The second problem you would set up as M2 = `moptimize_init()`. `moptimize()` would not confuse the two problems because it would know to which problem you were referring by whether you used M1 or M2 as the argument of the `moptimize()` functions. As another example, you might have one optimization problem, M = `moptimize_init()`, and halfway through it, decide you want to try something wild. You could code M2 = M, thus making a copy of the problem, use the `moptimize*()` functions with M2, and all the while your original problem would remain undisturbed.

Having obtained a problem handle, that is, having coded M = `moptimize_init()`, you now need to fill in the details of your problem. You do that with the `moptimize_init_*()` functions. The order in which you do this does not matter. We first informed `moptimize()` of the identity of the evaluator function:

```
: moptimize_init_evaluator(M, &linregeval())
```

We must also inform `moptimize()` as to the type of evaluator function `linregeval()` is, which we could do by coding

```
: moptimize_init_evaluatortype(M, "lf")
```

We did not bother, however, because type `lf` is the default.

After that, we need to inform `moptimize()` as to the identity of the dependent variables:

```
: moptimize_init_depvar(M, 1, "mpg")
```

Dependent variables play no special role in `moptimize()`; they are merely something that are remembered so that they can be passed to the evaluator function that we write. One problem might have no dependent variables and another might have lots of them. `moptimize_init_depvar(M, i, y)`'s second argument specifies which dependent variable is being set. There is no requirement that the number of dependent variables match the number of equations. In the linear regression case, we have one dependent variable and two equations.

Next we set the independent variables, or equivalently, the mapping of coefficients, into parameters. When we code

```
: moptimize_init_eq_indepvars(M, 1, "weight foreign")
```

we are stating that there is a parameter, $\mathbf{p}_1 = \mathbf{X}_1 \times \mathbf{b}_1$:+ c_1, and that $\mathbf{X}_1 = (\texttt{weight}, \texttt{foreign})$. Thus $\mathbf{b}_1$ contains two coefficients, that is, $\mathbf{p}_1 = (\texttt{weight}, \texttt{foreign}) \times (b_{11}, b_{12})'$:+ c_1. Actually, we have not yet specified whether there is a constant, c_1, on the end, but if we do not specify otherwise, the constant will be included. If we want to suppress the constant, after coding `moptimize_init_eq_indepvars(M, 1, "weight foreign")`, we would code `moptimize_init_eq_cons(M, 1, "off")`. The 1 says first equation, and the "off" says to turn the constant off.

As an aside, we coded `moptimize_init_eq_indepvars(M, 1, "weight foreign")` and so specified that the independent variables were the Stata variables `weight` and `foreign`, but the independent variables do not have to be in Stata. If we had a 74×2 matrix named data in Mata that we wanted to use, we would have coded `moptimize_init_eq_indepvars(M, 1, data)`.

To define the second parameter, we code

```
: moptimize_init_eq_indepvars(M, 2, "")
```

Thus we are stating that there is a parameter, $\mathbf{p}_2 = \mathbf{X}_2 \times \mathbf{b}_2$:+ c_2, and that $\mathbf{X}_2$ does not exist, leaving $\mathbf{p}_2 = c_2$, meaning that the second parameter is a constant.

Our problem defined, we code

```
: moptimize(M)
```

to obtain the solution, and we code

```
: moptimize_result_display(M)
```

to see the results. There are many `moptimize_result_*()` functions for use after the solution is obtained.

The type d, e, v, and q evaluators

Above we wrote our evaluator function in the style of type lf. moptimize() provides four other evaluator types—called types d, e, v, and q—and each have their uses.

Using type lf above, we were required to calculate the observation-by-observation log likelihoods and that was all. Using another type of evaluator, say, type d, we are required to calculate the overall log likelihood, and optionally, its first derivatives, and optionally, its second derivatives. The corresponding evaluator types are called d0, d1, and d2. Type d is better than type lf because if we do calculate the derivatives, then moptimize() can execute more quickly and it can produce a slightly more accurate result (more accurate because numerical derivatives are not involved). These speed and accuracy gains justify type d1 and d2, but what about type d0? For many optimization problems, type d0 is redundant and amounts to nothing more than a slight variation on type lf. In these cases, type d0's justification is that if we want to write a type d1 or type d2 evaluator, then it is usually easiest to start by writing a type d0 evaluator. Make that work, and then add to the code to convert our type d0 evaluator into a type d1 evaluator; make that work, and then, if we are going all the way to type d2, add the code to convert our type d1 evaluator into a type d2 evaluator.

For other optimization problems, however, there is a substantive reason for type d0's existence. Type lf requires observation-by-observation values of the log-likelihood function, and for some likelihood functions, those simply do not exist. Think of a panel-data model. There may be observations within each of the panels, but there is no corresponding log-likelihood value for each of them. The log-likelihood function is defined only across the entire panel. Type lf cannot handle problems like that. Type d0 can.

That makes type d0 seem to be a strict improvement on type lf. Type d0 can handle any problem that type lf can handle, and it can handle other problems to boot. Where both can handle the problem, the only extra work to use type d0 is that we must sum the individual values we produce, and that is not difficult. Type lf, however, has other advantages. If you write a type lf evaluator, then without writing another line of code, you can obtain the robust estimates of variance, adjust for clustering, account for survey design effects, and more. moptimize() can do all that because it has the results of an observation-by-observation calculation. moptimize() can break into the assembly of those observation-by-observation results and modify how that is done. moptimize() cannot do that for d0.

So there are advantages both ways.

Another provided evaluator type is type e. Type e is a variation on type d. It also comes in the subflavors e0, e1, and e2, except that there is no e0 because it would be identical to d0. Type e allows you to make observation-level derivative calculations, which means that results can be obtained more quickly and more accurately. Type e is designed to always work where lf is appropriate, which means panel-data estimators are excluded. In return, it provides all the ancillary features provided by type lf, meaning that robust standard errors, clustering, and survey-data adjustments are available. You write the evaluator function in a slightly different style when you use type e rather than type d.

Type v is a variation on type e that relaxes the requirement that the log-likelihood function be summable over the observations. Thus type v can work with panel-data models and resurrect the features of robust standard errors, clustering, and survey-data adjustments. Type v evaluators, however, are more difficult to write than type e evaluators.

Type q is for the special case of quadratic optimization. You either need it, and then only type q will do, or you do not.

Example using type d

Let's return to our linear regression maximum-likelihood estimator. To remind you, this is a two-parameter model, and the log-likelihood function is

$$f(\mathbf{p}_1, \mathbf{p}_2; \mathbf{y}) = \sum \ln(\texttt{normalden}(\mathbf{y} - \mathbf{p}_1, 0, \texttt{sqrt}(\mathbf{p}_2)))$$

This time, however, we are going to parameterize the variance parameter $\mathbf{p}_2$ as the log of the standard deviation, so we will write

$$f(\mathbf{p}_1, \mathbf{p}_2; \mathbf{y}) = \sum \ln(\texttt{normalden}(\mathbf{y} - \mathbf{p}_1, 0, \texttt{exp}(\mathbf{p}_2)))$$

It does not make any conceptual difference which parameterization we use, but the log parameterization converges a little more quickly, and the derivatives are easier, too. We are going to implement a type d2 evaluator for this function. To save you from pulling out pencil and paper, let's tell you the derivatives:

$$df/d\mathbf{p}_1 \;=\; \mathbf{z}\texttt{:/s}$$
$$df/d\mathbf{p}_2 \;=\; \mathbf{z}\texttt{:\^2:-1}$$

$$d^2f/d\mathbf{p}_1^2 \;=\; -1\texttt{:/s:\^2}$$
$$d^2f/d\mathbf{p}2^2 \;=\; -2 \times \mathbf{z}\texttt{:\^2}$$
$$d^2f/d\mathbf{p}d\mathbf{p}_2 \;=\; -2 \times \mathbf{z}\texttt{:/s}$$

where

$$\mathbf{z} = (\mathbf{y} : -\mathbf{p}_1)\texttt{:/s}$$

$$\mathbf{s} = \exp(\mathbf{p}_2)$$

The d2 evaluator function for this problem is

```
function linregevald2(transmorphic M, real scalar todo,
                      real rowvector b, fv, g, H)
{
        y1  = moptimize_calc_depvar(M, 1)
        p1  = moptimize_calc_xb(M, b, 1)
        p2  = moptimize_calc_xb(M, b, 2)
        s   = exp(p2)
        z   = (y1:-p1):/s
        fv  = moptimize_util_sum(M, ln(normalden(y1:-p1, 0, s)))
        if (todo>=1) {
                s1  = z:/s
                s2  = z:^2 :- 1
                g1  = moptimize_util_vecsum(M, 1, s1, fv)
                g2  = moptimize_util_vecsum(M, 2, s2, fv)
                g   = (g1, g2)
                if (todo==2) }
                        h11 = -1:/s:^2
                        h22 = -2*z:^2
                        h12 = -2*z:/s
                        H11 = moptimize_util_matsum(M, 1,1, h11, fv)
                        H22 = moptimize_util_matsum(M, 2,2, h22, fv)
                        H12 = moptimize_util_matsum(M, 1,2, h12, fv)
                        H   = (H11, H12 \ H12', H22)
                        }
        }
}
```

The code to fit a model of mpg on weight and foreign reads

```
: M = optimize_init()
: moptimize_init_evaluator(M, &linregevald2())
: moptimize_init_evaluatortype(M, "d2")
: moptimize_init_depvar(M, 1, "mpg")
: moptimize_init_eq_indepvars(M, 1, "weight foreign")
: moptimize_init_eq_indepvars(M, 2, "")
: moptimize()
: moptimize_result_display(M)
```

By the way, function linregevald2() will work not only with type d2, but also with types d1 and d0. Our function has the code to calculate first and second derivatives, but if we use type d1, todo will never be 2 and the second derivative part of our code will be ignored. If we use type d0, todo will never be 1 or 2 and so the first and second derivative parts of our code will be ignored. You could delete the unnecessary blocks if you wanted.

It is also worth trying the above code with types d1debug and d2debug. Type d1debug is like d1; the second derivative code will not be used. Also type d1debug will almost ignore the first derivative code. Our program will be asked to make the calculation, but moptimize() will not use the results except to report a comparison of the derivatives we calculate with numerically calculated derivatives. That way, we can check that our program is right. Once we have done that, we move to type d2debug, which will check our second-derivative calculation.

Example using type e

The e2 evaluator function for the linear-regression problem is almost identical to the type d2 evaluator. It differs in that rather than return the gradient vector, we return the observation-level scores that when summed produce the gradient. That the conversion was from d2 to e2 possibly had to do with the observation-by-observation nature of the linear-regression problem; if the evaluator was not going to be implemented as lf, it always should have been implemented as e1 or e2 instead of d1 or d2. In the d2 evaluator above, we went to extra work—summing the scores—the result of which was to eliminate moptimize() features such as being able to automatically adjust for clusters and survey data. In a more appropriate type d problem—a problem for which a type e evaluator could not have been implemented—those scores never would have been available in the first place.

The e2 evaluator is

```
        function linregevale2(transmorphic M, real scalar todo,
                        real rowvector b, fv, S, H)
        {
                y1   = moptimize_calc_depvar(M, 1)
                p1   = moptimize_calc_xb(M, b, 1)
                p2   = moptimize_calc_xb(M, b, 2)

                s    = exp(p2)
                z    = (y1:-p1):/s

                fv   = moptimize_util_sum(M, ln(normalden(y1:-p1, 0, s)))

                if (todo>=1) c -(
                        s1   = z:/s
                        s2   = z:^2 :- 1
                        S    = (s1, s2)
                        if (todo==2) {
                                h11 = -1:/s:^2
                                h22 = -2*z:^2
                                h12 = -2*z:/s
                                H11 = moptimize_util_matsum(M, 1,1, h11, fv)
                                H22 = moptimize_util_matsum(M, 2,2, h22, fv)
                                H12 = moptimize_util_matsum(M, 1,2, h12, fv)
                                H   = (H11, H12 \ H12', H22)
                        }
                }
        }
```

The code to fit a model of mpg on weight and foreign reads nearly identically to the code we used in the type d2 case. We must specify the name of our type e2 evaluator and specify that it is type e2:

```
        : M = optimize_init()
        : moptimize_init_evaluator(M, &linregevale2())
        : moptimize_init_evaluatortype(M, "e2")
        : moptimize_init_depvar(M, 1, "mpg")
        : moptimize_init_eq_indepvars(M, 1, "weight foreign")
        : moptimize_init_eq_indepvars(M, 2, "")
        : moptimize()
        : moptimize_result_display(M)
```

Conformability

See *Syntax* above.

Diagnostics

All functions abort with error when used incorrectly.

moptimize( ) aborts with error if it runs into numerical difficulties. _moptimize( ) does not; it instead returns a nonzero error code.

The moptimize_result*( ) functions abort with error if they run into numerical difficulties when called after moptimize( ) or moptimize_evaluate( ). They do not abort when run after _moptimize( ) or _moptimize_evaluate( ). They instead return a properly dimensioned missing result and set moptimize_result_errorcode( ) and moptimize_result_errortext( ).

Reference

Gould, W. W., J. S. Pitblado, and W. M. Sribney. 2006. *Maximum Likelihood Estimation with Stata*. 3rd ed. College Station, TX: Stata Press.

Also see

[M-5] **optimize( )** — Function optimization

[M-4] **mathematical** — Important mathematical functions

[M-4] **statistical** — Statistical functions

Title

[M-5] more( ) — Create –more– condition

Syntax

> *void*   more()
>
> *real scalar* setmore()
>
> *void*   setmore(*real scalar onoff*)
>
> *void*   setmoreonexit(*real scalar onoff*)

Description

more() displays --more-- and waits for a key to be pressed. That is, more() does that if more is on, which it usually is, and it does nothing otherwise. more can be turned on and off by Stata's set more command or by the functions below.

setmore() returns whether more is on or off, encoded as 1 and 0.

setmore(*onoff*) sets more on if *onoff* $\neq$ 0 and sets more off otherwise.

setmoreonexit(*onoff*) sets more on or off when the current execution ends. It has no effect on the current setting. The specified setting will take effect when control is passed back to the Mata prompt or to the calling ado-file or to Stata itself, and it will take effect regardless of whether execution ended because of a return, exit(), error, or abort. Only the first call to setmoreonexit() has that effect. Later calls have no effect whatsoever.

Remarks

setmoreonexit() is used to ensure that the more setting is restored if a program wants to temporarily reset it:

        setmoreonexit(setmore())
        setmore(0)

Only the first invocation of setmoreonexit() has any effect. This way, a subroutine that is used in various contexts might also contain

        setmoreonexit(setmore())
        setmore(0)

and that will not cause the wrong more setting to be restored if an earlier routine had already done that and yet still cause the right setting to be restored if the subroutine is the first to issue setmoreonexit().

Conformability

more() takes no arguments and returns *void*.

setmore():
     result: 1×1

setmore(*onoff*), setmoreonexit(*onoff*):
     onoff: 1×1
     result: *void*

Diagnostics

None.

Also see

[P] **more** — Pause until key is pressed

[M-4] **io** — I/O functions

Title

[M-5] **_negate( )** — Negate real matrix

Syntax

void _negate(*real matrix X*)

Description

_negate(*X*) speedily replaces $X = -X$.

Remarks

Coding _negate(X) executes more quickly than coding X = -X.

However, coding

        B = A
        _negate(B)

does not execute more quickly than coding

        B = -A

Conformability

_negate(*X*):
            X: $r \times c$
      result: *void*

Diagnostics

None. *X* may be a view.

Also see

[M-4] **utility** — Matrix utility functions

628

Title

Syntax

> *real scalar* norm(*numeric matrix A*)
>
> *real scalar* norm(*numeric matrix A*, *real scalar p*)

Description

norm(*A*) returns norm(*A*, 2).

norm(*A*, *p*) returns the value of the norm of *A* for the specified *p*. The possible values and the meaning of *p* depend on whether *A* is a vector or a matrix.

When *A* is a vector, norm(*A*, *p*) returns

> sum(abs(*A*):^*p*) ^ (1/*p*)   if $1 \leq p < .$
>
> max(abs(*A*))   if $p \geq .$

When *A* is a matrix, returned is

p	norm(*A*, *p*)
0	sqrt(trace(conj(*A*)'*A*))
1	max(colsum(abs(*A*)))
2	max(svdsv(*A*))
.	max(rowsum(abs(*A*)))

Remarks

norm(*A*) and norm(*A*, *p*) calculate vector norms and matrix norms. *A* may be real or complex and need not be square when it is a matrix.

The formulas presented above are not the actual ones used in calculation. In the vector-norm case when $1 \leq p < .$, the formula is applied to *A*:/max(abs(*A*)) and the result then multiplied by max(abs(*A*)). This prevents numerical overflow. A similar technique is used in calculating the matrix norm for $p = 0$, and that technique also avoids storage of conj(*A*)'*A*.

Conformability

norm(*A*):
>| | |
>|---|---|
>| *A*: | $r \times c$ |
>| *result*: | 1×1 |

norm(*A*, *p*):
>| | |
>|---|---|
>| *A*: | $r \times c$ |
>| *p*: | 1×1 |
>| *result*: | 1×1 |

Diagnostics

The norm( ) is defined to return 0 if A is void and missing if any element of A is missing.

norm(A, p) aborts with error if p is out of range. When A is a vector, p must be greater than or equal to 1. When A is a matrix, p must be 0, 1, 2, or . (missing).

norm(A) and norm(A, p) return missing if the 2-norm is requested and the singular value decomposition does not converge, an event not expected to occur; see [M-5] **svd( )**.

Also see

[M-4] **matrix** — Matrix functions

Title

Syntax

Gaussian normal

$$f = \texttt{normalden}(z)$$
$$f = \texttt{normalden}(x,\ sd)$$
$$f = \texttt{normalden}(x,\ mean,\ sd)$$
$$p = \texttt{normal}(z)$$
$$z = \texttt{invnormal}(p)$$
$$ln(f) = \texttt{lnnormalden}(z)$$
$$ln(f) = \texttt{lnnormalden}(x,\ sd)$$
$$ln(f) = \texttt{lnnormalden}(x,\ mean,\ sd)$$
$$ln(p) = \texttt{lnnormal}(z)$$

Binormal

$$p = \texttt{binormal}(z_1,\ z_2,\ rho)$$

Beta

$$f = \texttt{betaden}(a,\ b,\ x)$$
$$p = \texttt{ibeta}(a,\ b,\ x)$$
$$q = \texttt{ibetatail}(a,\ b,\ x)$$
$$x = \texttt{invibeta}(a,\ b,\ p)$$
$$x = \texttt{invibetatail}(a,\ b,\ q)$$

Binomial

$$pk = \texttt{binomialp}(n,\ k,\ pi)$$
$$p = \texttt{binomial}(n,\ k,\ pi)$$
$$q = \texttt{binomialtail}(n,\ k,\ pi)$$
$$pi = \texttt{invbinomial}(n,\ k,\ p)$$
$$pi = \texttt{invbinomialtail}(n,\ k,\ q)$$

Chi-squared

$$p = \texttt{chi2}(n,\ x)$$
$$q = \texttt{chi2tail}(n,\ x)$$
$$x = \texttt{invchi2}(n,\ p)$$
$$x = \texttt{invchi2tail}(n,\ q)$$

F

$$f = \texttt{Fden}(n_1,\ n_2,\ Fstat)$$
$$p = \texttt{F}(n_1,\ n_2,\ Fstat)$$
$$q = \texttt{Ftail}(n_1,\ n_2,\ Fstat)$$
$$Fstat = \texttt{invF}(n_1,\ n_2,\ p)$$
$$Fstat = \texttt{invFtail}(n_1,\ n_2,\ q)$$

Gamma

$$f = \texttt{gammaden}(a,\ b,\ g,\ x)$$
$$p = \texttt{gammap}(a,\ x)$$
$$q = \texttt{gammaptail}(a,\ x)$$
$$x = \texttt{invgammap}(a,\ p)$$
$$x = \texttt{invgammaptail}(a,\ q)$$
$$dg/da = \texttt{dgammapda}(a,\ x)$$
$$dg/dx = \texttt{dgammapdx}(a,\ x)$$
$$d2g/da2 = \texttt{dgammapdada}(a,\ x)$$
$$d2g/dadx = \texttt{dgammapdadx}(a,\ x)$$
$$d2g/dx2 = \texttt{dgammapdxdx}(a,\ x)$$

Hypergeometric

$$pk = \texttt{hypergeometricp}(N,\ K,\ n,\ k)$$
$$p = \texttt{hypergeometric}(N,\ K,\ n,\ k)$$

Negative binomial

$$pk = \texttt{nbinomialp}(n,\ k,\ pi)$$
$$p = \texttt{nbinomial}(n,\ k,\ pi)$$
$$q = \texttt{nbinomialtail}(n,\ k,\ pi)$$
$$pi = \texttt{invnbinomial}(n,\ k,\ p)$$
$$pi = \texttt{invnbinomialtail}(n,\ k,\ q)$$

Noncentral Beta

$$f = \texttt{nbetaden}(a,\ b,\ L,\ x)$$
$$p = \texttt{nibeta}(a,\ b,\ L,\ x)$$
$$x = \texttt{invnibeta}(a,\ b,\ L,\ p)$$

Noncentral chi-squared

$$p = \texttt{nchi2}(n,\ L,\ x)$$
$$x = \texttt{invnchi2}(n,\ L,\ p)$$
$$L = \texttt{npnchi2}(n,\ x,\ p)$$

Noncentral F

$$f = \texttt{nFden}(n_1, n_2, L, F)$$
$$q = \texttt{nFtail}(n_1, n_2, L, F)$$
$$F = \texttt{invnFtail}(n_1, n_2, L, q)$$

Poisson

$$pk = \texttt{poissonp}(mean, k)$$
$$p = \texttt{poisson}(mean, k)$$
$$q = \texttt{poissontail}(mean, k)$$
$$m = \texttt{invpoisson}(k, p)$$
$$m = \texttt{invpoissontail}(k, q)$$

Student's t

$$f = \texttt{tden}(n, t)$$
$$q = \texttt{ttail}(n, t)$$
$$t = \texttt{invttail}(n, q)$$

where

1. All functions return real and all arguments are real.

2. The left-hand-side notation is used to assist in interpreting the meaning of the returned value:

   f = density value
   pk= probability of discrete outcome $K = \Pr(K = k)$
   p = left cumulative
   $\quad = \Pr(-\infty < \textit{statistic} \le x)$ (continuous)
   $\quad = \Pr(0 \le K \le k)$ (discrete)
   q = right cumulative
   $\quad = 1 - p$ (continuous)
   $\quad = \Pr(K \ge k) = 1 - p + pk$ (discrete)

3. Hypergeometric distribution:

   N = number of objects in the population
   K = number of objects in the population with the characteristic of interest,
   $\quad\quad K < N$
   n = sample size, $n < N$
   k = number of objects in the sample with the characteristic of interest,
   $\quad\quad \max(0, n - N + K) \le k \le \min(K, n)$

4. Negative binomial distribution: $n > 0$ and may be nonintegral.

Description

The above functions return density values, cumulatives, reverse cumulatives, and in one case, derivatives of the indicated probability density function. These functions mirror the Stata functions of the same name and in fact are the Stata functions.

See [D] **functions** for function details. In the syntax diagram above, some arguments have been renamed in hope of aiding understanding, but the function arguments match one to one with the underlying Stata functions.

Remarks

Remarks are presented under the following headings:

> *R-conformability*
> *A note concerning invbinomial( ) and invbinomialtail( )*
> *A note concerning ibeta( )*
> *A note concerning gammap( )*

R-conformability

The above functions are usually used with scalar arguments and then return a scalar result:

```
: x = chi2(10, 12)
: x
  .7149434997
```

The arguments may, however, be vectors or matrices. For instance,

```
: x = chi2((10,11,12), 12)
: x
```

	1	2	3
1	.7149434997	.6363567795	.5543203586

```
: x = chi2(10, (12,12.5,13))
: x
```

	1	2	3
1	.7149434997	.7470146767	.7763281832

```
: x = chi2((10,11,12), (12,12.5,13))
: x
```

	1	2	3
1	.7149434997	.6727441644	.6309593164

In the last example, the numbers correspond to chi2(10,12), chi2(11,12.5), and chi2(12,13).

Arguments are required to be r-conformable (see [M-6] **Glossary**), and thus,

```
: x = chi2((10\11\12), (12,12.5,13))
: x
```

	1	2	3
1	.7149434997	.7470146767	.7763281832
2	.6363567795	.6727441644	.7066745906
3	.5543203586	.593595966	.6309593164

which corresponds to

	1	2	3
1	chi2(10,12)	chi2(10,12.5)	chi2(10,13)
2	chi2(11,12)	chi2(11,12.5)	chi2(11,13)
3	chi2(12,12)	chi2(12,12.5)	chi2(12,13)

A note concerning invbinomial( ) and invbinomialtail( )

invbinomial(n, k, pi) invbinomialtail(n, k, q) are useful for calculating confidence intervals for pi, the probability of a success. invbinomial( ) returns the probability pi such that the probability of observing k or fewer successes in n trials is p. invbinomialtail( ) returns the probability pi such that the probability of observing k or more successes in n trials is q.

A note concerning ibeta( )

ibeta(a, b, x) is known as the cumulative beta distribution, and it is known as the incomplete beta function $I_x(a, b)$.

A note concerning gammap( )

gammap(a, x) is known as the cumulative gamma distribution, and it is known as the incomplete gamma function $P(a, x)$.

Conformability

All functions require that arguments be r-conformable; see *R-conformability* above. Returned is matrix of max(argument rows) rows and max(argument columns) columns containing element-by-element calculated results.

Diagnostics

All functions return missing when arguments are out of range.

Also see

[M-4] **statistical** — Statistical functions

Title

[M-5] **optimize( )** — Function optimization

Syntax

S = optimize_init()

(varies) optimize_init_which(S [, { "max" | "min" }])

(varies) optimize_init_evaluator(S [, &*function*()])

(varies) optimize_init_evaluatortype(S [, *evaluatortype*])

(varies) optimize_init_negH(S, { "off" | "on" })

(varies) optimize_init_params(S [, *real rowvector initialvalues*])

(varies) optimize_init_nmsimplexdeltas(S [, *real rowvector delta*])

(varies) optimize_init_argument(S, *real scalar k* [, *X*])

(varies) optimize_init_narguments(S [, *real scalar K*])

(varies) optimize_init_cluster(S, S)

(varies) optimize_init_colstripe(S [, *stripe*])

(varies) optimize_init_technique(S [, *technique*])

(varies) optimize_init_gnweightmatrix(S, *W*)

(varies) optimize_init_singularHmethod(S [, *singularHmethod*])

(varies) optimize_init_conv_maxiter(S [, *real scalar max*])

(varies) optimize_init_conv_warning(S, { "on" | "off" })

(varies) optimize_init_conv_ptol(S [, *real scalar ptol*])

(varies) optimize_init_conv_vtol(S [, *real scalar vtol*])

(varies) optimize_init_conv_nrtol(S [, *real scalar nrtol*])

(varies) optimize_init_conv_ignorenrtol(S, { "off" | "on" })

(varies) optimize_init_iterid(S [, *string scalar id*])

(varies) optimize_init_valueid(S [, *string scalar id*])

(varies) optimize_init_tracelevel(S [, *tracelevel*])

(varies)	optimize_init_trace_dots(*S*, { "off" | "on" })
(varies)	optimize_init_trace_value(*S*, { "on" | "off" })
(varies)	optimize_init_trace_tol(*S*, { "off" | "on" })
(varies)	optimize_init_trace_params(*S*, { "off" | "on" })
(varies)	optimize_init_trace_step(*S*, { "off" | "on" })
(varies)	optimize_init_trace_gradient(*S*, { "off" | "on" })
(varies)	optimize_init_trace_Hessian(*S*, { "off" | "on" })
(varies)	optimize_init_evaluations(*S*, { "off" | "on" })
(varies)	optimize_init_constraints(*S* [, *real matrix Cc*])
(varies)	optimize_init_verbose(*S* [, *real scalar verbose*])

real rowvector	optimize(*S*)
real scalar	_optimize(*S*)
void	optimize_evaluate(*S*)
real scalar	_optimize_evaluate(*S*)

real rowvector	optimize_result_params(*S*)
real scalar	optimize_result_value(*S*)
real scalar	optimize_result_value0(*S*)
real rowvector	optimize_result_gradient(*S*)
real matrix	optimize_result_scores(*S*)
real matrix	optimize_result_Hessian(*S*)
real matrix	optimize_result_V(*S*)
string scalar	optimize_result_Vtype(*S*)
real matrix	optimize_result_V_oim(*S*)
real matrix	optimize_result_V_opg(*S*)
real matrix	optimize_result_V_robust(*S*)
real scalar	optimize_result_iterations(*S*)
real scalar	optimize_result_converged(*S*)
real colvector	optimize_result_iterationlog(*S*)
real rowvector	optimize_result_evaluations(*S*)
real scalar	optimize_result_errorcode(*S*)
string scalar	optimize_result_errortext(*S*)
real scalar	optimize_result_returncode(*S*)

void optimize_query(*S*)

where *S*, if it is declared, should be declared

transmorphic *S*

and where *evaluatortype* optionally specified in optimize_init_evaluatortype() is

evaluatortype	description
"d0"	*function*() returns *scalar* value
"d1"	same as "d0" and returns gradient *rowvector*
"d2"	same as "d1" and returns Hessian *matrix*
"d1debug"	same as "d1" but checks gradient
"d2debug"	same as "d2" but checks gradient and Hessian
"v0"	*function*() returns *colvector* value
"v1"	same as "v0" and returns score *matrix*
"v2"	same as "v1" and returns Hessian *matrix*
"v1debug"	same as "v1" but checks gradient
"v2debug"	same as "v2" but checks gradient and Hessian

The default is "d0" if not set.

and where *technique* optionally specified in optimize_init_technique() is

technique	description
"nr"	modified Newton–Raphson
"dfp"	Davidon–Fletcher–Powell
"bfgs"	Broyden–Fletcher–Goldfarb–Shanno
"bhhh"	Berndt–Hall–Hall–Hausman
"nm"	Nelder–Mead
"gn"	Gauss–Newton (quaratic optimization)

The default is "nr".

and where *singularHmethod* optionally specified in optimize_init_singularHmethod() is

singularHmethod	description
"m-marquardt"	modified Marquardt algorithm
"hybrid"	mixture of steepest descent and Newton

The default is "m-marquardt" if not set;
"hybrid" is equivalent to ml's difficult option; see [R] **ml**.

and where *tracelevel* optionally specified in optimize_init_tracelevel() is

tracelevel	To be displayed each iteration
"none"	nothing
"value"	function value
"tolerance"	previous + convergence values
"params"	previous + parameter values
"step"	previous + stepping information
"gradient"	previous + gradient vector
"hessian"	previous + Hessian matrix

The default is "value" if not set.

Description

These functions find parameter vector or scalar p such that function $f(p)$ is a maximum or a minimum.

optimize_init() begins the definition of a problem and returns S, a problem-description handle set to contain default values.

The optimize_init_*(S, ...) functions then allow you to modify those defaults. You use these functions to describe your particular problem: to set whether you wish maximization or minimization, to set the identity of function $f()$, to set initial values, and the like.

optimize(S) then performs the optimization. optimize() returns *real rowvector p* containing the values of the parameters that produce a maximum or minimum.

The optimize_result_*(S) functions can then be used to access other values associated with the solution.

Usually you would stop there. In other cases, you could restart optimization by using the resulting parameter vector as new initial values, change the optimization technique, and restart the optimization:

```
optimize_init_param(S, optimize_result_param(S))
optimize_init_technique(S, "dfp")
optimize(S)
```

Aside: The optimize_init_*(S, ...) functions have two modes of operation. Each has an optional argument that you specify to set the value and that you omit to query the value. For instance, the full syntax of optimize_init_params() is

> *void* optimize_init_params(S, *real rowvector initialvalues*)
>
> *real rowvector* optimize_init_params(S)

The first syntax sets the initial values and returns nothing. The second syntax returns the previously set (or default, if not set) initial values.

All the optimize_init_*(S, ...) functions work the same way.

Remarks

Remarks are presented under the following headings:

First example
Notation
Type d evaluators
Example of d0, d1, and d2
d1debug and d2debug
Type v evaluators
Example of v0, v1, and v2
Functions
   optimize_init( )
   optimize_init_which( )
   optimize_init_evaluator( ) and optimize_init_evaluatortype( )
   optimize_init_negH( )
   optimize_init_params( )
   optimize_init_nmsimplexdeltas( )
   optimize_init_argument( ) and optimize_init_narguments( )
   optimize_init_cluster( )
   optimize_init_colstripe( )
   optimize_init_technique( )
   optimize_init_gnweightmatrix( )
   optimize_init_singularHmethod( )
   optimize_init_conv_maxiter( )
   optimize_init_conv_warning( )
   optimize_init_conv_ptol( ), . . . _vtol( ), . . . _nrtol( )
   optimize_init_conv_ignorenrtol( )
   optimize_init_iterid( )
   optimize_init_valueid( )
   optimize_init_tracelevel( )
   optimize_init_trace_dots( ), . . . _value( ), . . . _tol( ), . . . _step( ), . . . _gradient( ), . . . _Hessian( )
   optimize_init_evaluations( )
   optimize_init_constraints( )
   optimize_init_verbose( )

   optimize( )
   _optimize( )
   optimize_evaluate( )
   _optimize_evaluate( )

   optimize_result_params( )
   optimize_result_value( ) and optimize_result_value0( )
   optimize_result_gradient( )
   optimize_result_scores( )
   optimize_result_Hessian( )
   optimize_result_V( ) and optimize_result_Vtype( )
   optimize_result_V_oim( ), . . . _opg( ), . . . _robust( )
   optimize_result_iterations( )
   optimize_result_converged( )
   optimize_result_iterationlog( )
   optimize_result_evaluations( )
   optimize_result_errorcode( ), . . . _errortext( ), and . . . _returncode( )

   optimize_query( )

First example

The optimization functions may be used interactively.

Below we use the functions to find the value of x that maximizes $y = \exp(-x^2 + x - 3)$:

```
: void myeval(todo, x,  y, g, H)
> {
>           y = exp(-x^2 + x - 3)
> }
note: argument todo unused
note: argument g unused
note: argument H unused

: S = optimize_init()

: optimize_init_evaluator(S, &myeval())

: optimize_init_params(S, 0)

: x = optimize(S)
Iteration 0:  f(p) = .04978707
Iteration 1:  f(p) = .04978708
Iteration 2:  f(p) = .06381186
Iteration 3:  f(p) = .06392786
Iteration 4:  f(p) = .06392786

: x
.5
```

Notation

We wrote the above in the way that mathematicians think, i.e., optimizing $y = f(x)$. Statisticians, on the other hand, think of optimizing $s = f(b)$. To avoid favoritism, we will write $v = f(p)$ and write the general problem with the following notation:

Maximize or minimize $v = f(p)$,

   v: a scalar

   p: $1 \times np$

subject to the constraint $Cp' = c$,

   C: $nc \times np$ ($nc = 0$ if no constraints)
   c: $nc \times 1$

where g, the gradient vector, is $g = f'(p) = df/dp$,

   g: $1 \times np$

and H, the Hessian matrix, is $H = f''(p) = d^2f/dpdp'$

   H: $np \times np$

Type d evaluators

You must write an evaluator function to calculate $f()$ before you can use the optimization functions. The example we showed above was of what is called a type d evaluator. Let's stay with that.

The evaluator function we wrote was

```
void myeval(todo, x,  y, g, H)
{
        y = exp(-x^2 + x - 3)
}
```

All type d evaluators open the same way,

  void evaluator(todo, x, y, g, H)

although what you name the arguments is up to you. We named the arguments the way that mathematicians think, although we could just as well have named them the way that statisticians think:

  void evaluator(todo, b, s, g, H)

To avoid favoritism, we will write them as

  void evaluator(todo, p, v, g, H)

i.e., we will think in terms of optimizing $v = f(p)$.

Here is the full definition of a type d evaluator:

---

void evaluator(real scalar todo, real rowvector p, v, g, H)

where v, g, and H are values to be returned:

  v: *real scalar*
  g: *real rowvector*
  H: *real matrix*

evaluator( ) is to fill in v given the values in p and optionally to fill in g and H, depending on the value of *todo*:

todo	Required action by *evaluator( )*
0	calculate $v = f(p)$ and store in v
1	calculate $v = f(p)$ and $g = f'(p)$ and store in v and g
2	calculate $v = f(p)$, $g = f'(p)$, and $H = f''(p)$ and store in v, g, and H

evaluator( ) may return $v=.$ if $f()$ cannot be evaluated at p. Then g and H need not be filled in even if requested.

An evaluator does not have to be able to do all of this. In the first example, `myeval()` could handle only *todo* $= 0$. There are three types of type d evaluators:

d type	Capabilities expected of *evaluator( )*
d0	can calculate $v = f(p)$
d1	can calculate $v = f(p)$ and $g = f'(p)$
d2	can calculate $v = f(p)$ and $g = f'(p)$ and $H = f''(p)$

`myeval()` is a type d0 evaluator. Type d0 evaluators are never asked to calculate g or H. Type d0 is the default type but, if we were worried that it was not, we could set the evaluator type before invoking `optimize()` by coding

```
optimize_init_evaluatortype(S, "d0")
```

Here are code outlines of the three types of evaluators:

```
void d0_evaluator(todo, p, v, g, H)
{
        v = ...
}
```

```
void d1_evaluator(todo, p, v, g, H)
{
        v = ...
        if (todo>=1) {
                g = ...
        }
}
```

```
void d2_evaluator(todo, p, v, g, H)
{
        v = ...
        if (todo>=1) {
                g = ...
                if (todo==2) {
                        H = ...
                }
        }
}
```

Example of d0, d1, and d2

We wish to find the p_1 and p_2 corresponding to the maximum of

$$v = \exp(-p_1^2 - p_2^2 - p_1 p_2 + p_1 - p_2 - 3)$$

A d0 solution to the problem would be

```
: void eval0(todo, p, v, g, H)
> {
>         v = exp(-p[1]^2 - p[2]^2 - p[1]*p[2] + p[1] - p[2] - 3)
> }
note: argument todo unused
note: argument g unused
note: argument h unused
: S = optimize_init()
: optimize_init_evaluator(S, &eval0())
```

```
: optimize_init_params(S, (0,0))

: p = optimize(S)
Iteration 0:  f(p) = .04978707  (not concave)
Iteration 1:  f(p) = .12513024
Iteration 2:  f(p) = .13495886
Iteration 3:  f(p) = .13533527
Iteration 4:  f(p) = .13533528

: p
        1    2

1       1   -1
```

A d1 solution to the problem would be

```
: void eval1(todo, p, v, g, H)
> {
>         v = exp(-p[1]^2 - p[2]^2 - p[1]*p[2] + p[1] - p[2] - 3)
>         if (todo==1) {
>                 g[1] = (-2*p[1] - p[2] + 1)*v
>                 g[2] = (-2*p[2] - p[1] - 1)*v
>         }
> }
note: argument H unused
: S = optimize_init()

: optimize_init_evaluator(S, &eval1())

: optimize_init_evaluatortype(S, "d1")          ←  important

: optimize_init_params(S, (0,0))

: p = optimize(S)
Iteration 0:  f(p) = .04978707  (not concave)
Iteration 1:  f(p) = .12513026
Iteration 2:  f(p) = .13496887
Iteration 3:  f(p) = .13533527
Iteration 4:  f(p) = .13533528

: p
        1    2

1       1   -1
```

The d1 solution is better than the d0 solution because it runs faster and usually is more accurate. Type d1 evaluators require more code, however, and deriving analytic derivatives is not always possible.

A d2 solution to the problem would be

```
: void eval2(todo, p, v, g, H)
> {
>         v = exp(-p[1]^2 - p[2]^2 - p[1]*p[2] + p[1] - p[2] - 3)
>         if (todo>=1) {
>                 g[1] = (-2*p[1] - p[2] + 1)*v
>                 g[2] = (-2*p[2] - p[1] - 1)*v
>                 if (todo==2) {
>                         H[1,1] = -2*v + (-2*p[1]-p[2]+1)*g[1]
>                         H[2,1] = -1*v + (-2*p[2]-p[1]-1)*g[1]
>                         H[2,2] = -2*v + (-2*p[2]-p[1]-1)*g[2]
>                         _makesymmetric(H)
>                 }
>         }
> }
: S = optimize_init()
```

```
: optimize_init_evaluator(S, &eval2())

: optimize_init_evaluatortype(S, "d2")            ← important

: optimize_init_params(S, (0,0))

: p = optimize(S)
Iteration 0:  f(p) = .04978707  (not concave)
Iteration 1:  f(p) = .12513026
Iteration 2:  f(p) = .13496887
Iteration 3:  f(p) = .13533527
Iteration 4:  f(p) = .13533528

: p
         1    2

   1 ┌              ┐
     │   1    -1    │
     └              ┘
```

A d2 solution is best because it runs fastest and usually is the most accurate. Type d2 evaluators require the most code, and deriving analytic derivatives is not always possible.

In the d2 evaluator eval2(), note our use of _makesymmetric(). Type d2 evaluators are required to return H as a symmetric matrix; filling in just the lower or upper triangle is not sufficient. The easiest way to do that is to fill in the lower triangle and then use _makesymmetric() to reflect the lower off-diagonal elements; see [M-5] **makesymmetric( )**.

d1debug and d2debug

In addition to evaluator types "d0", "d1", and "d2" that are specified in optimize_init_evaluatortype(*S*, *evaluatortype*), there are two more: "d1debug" and "d2debug". They assist in coding d1 and d2 evaluators.

In *Example of d0, d1, and d2* above, we admit that we did not correctly code the functions eval1() and eval2() at the outset, before you saw them. In both cases, that was because we had taken the derivatives incorrectly. The problem was not with our code but with our math. d1debug and d2debug helped us find the problem.

d1debug is an alternative to d1. When you code optimize_init_evaluatortype(*S*, "d1debug"), the derivatives you calculate are not taken seriously. Instead, optimize() calculates its own numerical derivatives and uses those. Each time optimize() does that, however, it compares your derivatives to the ones it calculated and gives you a report on how they differ. If you have coded correctly, they should not differ by much.

d2debug does the same thing, but for d2 evaluators. When you code optimize_init_evaluatortype(*S*, "d2debug"), optimize() uses numerical derivatives but, each time, optimize() gives you a report on how much your results for the gradient and for the Hessian differ from the numerical calculations.

For each comparison, optimize() reports just one number: the mreldif() (see [M-5] **reldif( )**) between your results and the numerical ones. When you have done things right, gradient vectors will differ by approximately 1e–12 or less and Hessians will differ by 1e–7 or less.

When differences are large, you will want to see not only the summary comparison but also the full vectors and matrices so that you can compare your results element by element with those calculated numerically. Sometimes the error is in one element and not the others. To do this, set the trace level with optimize_init_tracelevel(*S*, *tracelevel*) before issuing optimize(). Code optimize_init_tracelevel(*S*, "gradient") to get a full report on the gradient comparison, or set `optimize_init_tracelevel(S, "hessian")` to get a full report on the gradient comparison and the Hessian comparison.

Type v evaluators

In some statistical applications, you will find v0, v1, and v2 more convenient to code than d0, d1, and d2. The *v* stands for vector.

In statistical applications, one tends to think of a dataset of values arranged in matrix X, the rows of which are observations. A function $h(p, X[i, .])$ can be calculated for each row separately, and it is the sum of those resulting values that forms the function $f(p)$ that is to be maximized or minimized.

The v0, v1, and v2 methods are for such cases.

In a type d0 evaluator, you return scalar $v = f(p)$.

In a type v0 evaluator, you return a column vector v such that $\texttt{colsum}(v) = f(p)$.

In a type d1 evaluator, you return $v = f(p)$ and you return a row vector $g = f'(p)$.

In a type v1 evaluator, you return v such that $\texttt{colsum}(v) = f(p)$ and you return matrix g such that $\texttt{colsum}(g) = f'(p)$.

In a type d2 evaluator, you return $v = f(p)$, $g = f'(p)$, and you return $H = f''(p)$.

In a type v2 evaluator, you return v such that $\texttt{colsum}(v) = f(p)$, g such that $\texttt{colsum}(g) = f'(p)$, and you return $H = f''(p)$. This is the same H returned for d2.

The code outline for type v evaluators is the same as those for d evaluators. For instance, the outline for a v2 evaluator is

```
void v2_evaluator(todo, p, v, g, H)
{
        v = ...
        if (todo>=1) {
                g = ...
                if (todo==2) {
                        H = ...
                }
        }
}
```

The above is the same as the outline for d2 evaluators. All that differs is that v and g, which were *real scalar* and *real rowvector* in the d2 case, are now *real colvector* and *real matrix* in the v2 case. The same applies to v1 and v0.

The type v evaluators arise in statistical applications and, in such applications, there are data; i.e., just knowing p is not sufficient to calculate v, g, and H. Actually, that same problem can arise when coding type d evaluators as well.

You can pass extra arguments to evaluators, whether they be d0, d1, or d2 or v0, v1, or v2. The first line of all evaluators, regardless of style, is

$$void\ evaluator(todo, p, v, g, H)$$

If you code

> optimize_init_argument(*S*, 1, *X*)

the first line becomes

> *void evaluator(todo, p, X, v, g, H)*

If you code

> optimize_init_argument(*S*, 1, *X*)
> optimize_init_argument(*S*, 2, *Y*)

the first line becomes

> *void evaluator(todo, p, X, Y, v, g, H)*

and so on, up to nine extra arguments. That is, you can specify extra arguments to be passed to your function.

Example of v0, v1, and v2

You have the following data:

```
: x
          1

  1   .35
  2   .29
  3    .3
  4    .3
  5   .65
  6   .56
  7   .37
  8   .16
  9   .26
 10   .19
```

You believe that the data are the result of a beta distribution process with fixed parameters alpha and beta and you wish to obtain the maximum likelihood estimates of alpha and beta (*a* and *b* in what follows). The formula for the density of the beta distribution is

$$\text{density}(x) = \frac{\Gamma(a+b)}{\Gamma(a)\Gamma(b)} \, x^{a-1} \, (1-x)^{b-1}$$

The v0 solution to this problem is

```
: void lnbetaden0(todo, p,  x,  lnf, S, H)
> {
>         a   = p[1]
>         b   = p[2]
>         lnf = lngamma(a+b) :- lngamma(a) :- lngamma(b) :+
>               (a-1)*log(x) :+ (b-1)*log(1:-x)
> }
note: argument todo unused
note: argument S unused
note: argument H unused
```

```
: S = optimize_init()

: optimize_init_evaluator(S, &lnbetaden0())

: optimize_init_evaluatortype(S, "v0")

: optimize_init_params(S, (1,1))

: optimize_init_argument(S, 1, x)                    ← important

: p = optimize(S)
Iteration 0:  f(p) =          0
Iteration 1:  f(p) = 5.7294728
Iteration 2:  f(p) = 5.7646641
Iteration 3:  f(p) = 5.7647122
Iteration 4:  f(p) = 5.7647122

: p
                    1            2

  1 |  3.714209592   7.014926315  |
```

Note the following:

1. Rather than calling the returned value v, we called it lnf. You can name the arguments as you please.

2. We arranged for an extra argument to be passed by coding optimize_init_argument(S, 1, x). The extra argument is the vector x, which we listed previously for you. In our function, we received the argument as x, but we could have used a different name, just as we used lnf rather than v.

3. We set the evaluator type to "v0".

This being a statistical problem, we should be interested not only in the estimates p but also in their variance. We can get this from the inverse of the negative Hessian, which is the observed information matrix:

```
: optimize_result_V_oim(S)
[symmetric]
                    1            2

  1 |  2.556301184                |
  2 |  4.498194785   9.716647065  |
```

The v1 solution to this problem is

```
: void lnbetaden1(todo, p,  x,  lnf, S, H)
> {
>         a    = p[1]
>         b    = p[2]
>         lnf = lngamma(a+b) :- lngamma(a) :- lngamma(b) :+
>               (a-1)*log(x) :+ (b-1)*log(1:-x)
>         if (todo >= 1) }
>               S        = J(rows(x), 2, .)
>               S[.,1]   = log(x) :+ digamma(a+b) :- digamma(a)
>               S[.,2]   = log(1:-x) :+ digamma(a+b) :- digamma(b)
>         }
> }
note: argument H unused
: S = optimize_init()

: optimize_init_evaluator(S, &lnbetaden1())
```

```
: optimize_init_evaluatortype(S, "v1")

: optimize_init_params(S, (1,1))

: optimize_init_argument(S, 1, x)

: p = optimize(S)
Iteration 0:  f(p) =          0
Iteration 1:  f(p) = 5.7297061
Iteration 2:  f(p) = 5.7641349
Iteration 3:  f(p) = 5.7647121
Iteration 4:  f(p) = 5.7647122

: p
                      1              2

   1    3.714209343    7.014925751


: optimize_result_V_oim(S)
[symmetric]
                      1              2

   1    2.556299425
   2     4.49819212    9.716643068
```

Note the following:

1. We called the next-to-last argument of lnbetaden1() S rather than g in accordance with standard statistical jargon. What is being returned is in fact the observation-level scores, which sum to the gradient vector.

2. We called the next-to-last argument S even though that name conflicted with S outside the program, where S is the problem handle. Perhaps we should have renamed the outside S, but there is no confusion on Mata's part.

3. In our program we allocated *S* for ourselves: S = J(rows(x), 2, .). It is worth comparing this with the example of d1 in *Example of d0, d1, and d2*, where we did not need to allocate g. In d1, optimize() preallocates g for us. In v1, optimize() cannot do this because it has no idea how many "observations" we have.

The v2 solution to this problem is

```
: void lnbetaden2(todo, p,  x,  lnf, S, H)
> {
>         a    = p[1]
>         b    = p[2]
>         lnf = lngamma(a+b) :- lngamma(a) :- lngamma(b) :+
>               (a-1)*log(x) :+ (b-1)*log(1:-x)
>         if (todo >= 1) {
>                 S       = J(rows(x), 2, .)
>                 S[.,1]  = log(x) :+ digamma(a+b) :- digamma(a)
>                 S[.,2]  = log(1:-x) :+ digamma(a+b) :- digamma(b)
>                 if (todo==2) {
>                         n = rows(x)
>                         H[1,1] = n*(trigamma(a+b) - trigamma(a))
>                         H[2,1] = n*(trigamma(a+b))
>                         H[2,2] = n*(trigamma(a+b) - trigamma(b))
>                         _makesymmetric(H)
>                 }
>         }
> }
: S = optimize_init()
```

```
: optimize_init_evaluator(S, &lnbetaden2())

: optimize_init_evaluatortype(S, "v2")

: optimize_init_params(S, (1,1))

: optimize_init_argument(S, 1, x)

: p = optimize(S)
Iteration 0:   f(p) =              0
Iteration 1:   f(p) = 5.7297061
Iteration 2:   f(p) = 5.7641349
Iteration 3:   f(p) = 5.7647121
Iteration 4:   f(p) = 5.7647122

: p
                    1             2

    1 |  3.714209343    7.014925751 |


: optimize_result_V_oim(S)
[symmetric]
                    1             2

    1 |  2.556299574                |
    2 |  4.498192412    9.716643651 |
```

Functions

optimize_init( )

> *transmorphic* optimize_init()

optimize_init() is used to begin an optimization problem. Store the returned result in a variable name of your choosing; we have used S in this documentation. You pass S as the first argument to the other optimize*() functions.

optimize_init() sets all optimize_init_*() values to their defaults. You may use the query form of the optimize_init_*() to determine an individual default, or you can use optimize_query() to see them all.

The query form of optimize_init_*() can be used before or after optimization performed by optimize().

optimize_init_which( )

> *void*   optimize_init_which(S, {"max" | "min"})
>
> *string scalar* optimize_init_which(S)

optimize_init_which(S, *which*) specifies whether optimize() is to perform maximization or minimization. The default is maximization if you do not invoke this function.

optimize_init_which(S) returns "max" or "min" according to which is currently set.

optimize_init_evaluator( ) and optimize_init_evaluatortype( )

> *void* optimize_init_evaluator(*S*, *pointer(real function) scalar fptr*)

> *void* optimize_init_evaluatortype(*S*, *evaluatortype*)

> *pointer(real function) scalar* optimize_init_evaluator(*S*)

> *string scalar*   optimize_init_evaluatortype(*S*)

optimize_init_evaluator(*S*, *fptr*) specifies the function to be called to evaluate $f(p)$. Use of this function is required. If your function is named myfcn(), you code optimize_init_evaluator(*S*, &myfcn()).

optimize_init_evaluatortype(*S*, *evaluatortype*) specifies the capabilities of the function that has been set using optimize_init_evaluator(). Alternatives for *evaluatortype* are "d0", "d1", "d2", "d1debug", "d2debug", "v0", "v1", "v2", "v1debug", and "v2debug". The default is "d0" if you do not invoke this function.

optimize_init_evaluator(*S*) returns a pointer to the function that has been set.

optimize_init_evaluatortype(*S*) returns the *evaluatortype* currently set.

optimize_init_negH( )

optimize_init_negH(*S*, { "off" | "on" }) sets whether the evaluator you have written returns H or $-H$, the Hessian or the negative of the Hessian, if it returns a Hessian at all. This is for backward compatibility with prior versions of Stata's ml command (see [R] **ml**). Modern evaluators return H. The default is "off".

optimize_init_params( )

> *void*   optimize_init_params(*S*, *real rowvector initialvalues*)

> *real rowvector* optimize_init_params(*S*)

optimize_init_params(*S*, *initialvalues*) sets the values of p to be used at the start of the first iteration. Use of this function is required.

optimize_init_params(*S*) returns the initial values that will be (or were) used.

optimize_init_nmsimplexdeltas( )

> *void*   optimize_init_nmsimplexdeltas(*S*, *real rowvector delta*)
>
> *real rowvector* optimize_init_nmsimplexdeltas(*S*)

optimize_init_nmsimplexdeltas(*S*, *delta*) sets the values of *delta* to be used, along with the initial parameters, to build the simplex required by technique "nm" (Nelder–Mead). Use of this function is required only in the Nelder–Mead case. The values in *delta* must be at least 10 times larger than *ptol*, which is set by optimize_init_conv_ptol(). The initial simplex will be $\{ p, p + (d_1, 0), \ldots, 0, p + (0, d_2, 0, \ldots, 0), \ldots, p + (0, 0, \ldots, 0, d_k) \}$.

optimize_init_nmsimplexdeltas(*S*) returns the deltas that will be (or were) used.

optimize_init_argument( ) and optimize_init_narguments( )

> *void*   optimize_init_argument(*S*, *real scalar k*, *X*)
>
> *void*   optimize_init_narguments(*S*, *real scalar K*)
>
> *pointer scalar* optimize_init_argument(*S*, *real scalar k*)
>
> *real scalar*   optimize_init_narguments(*S*)

optimize_init_argument(*S*, *k*, *X*) sets the *k*th extra argument of the evaluator function to be *X*, where *k* can only 1, 2, 3, $\ldots$, 9. *X* can be anything, including a view matrix or even a pointer to a function. No copy of *X* is made; it is a pointer to *X* that is stored, so any changes you make to *X* between setting it and *X* being used will be reflected in what is passed to the evaluator function.

optimize_init_narguments(*S*, *K*) sets the number of extra arguments to be passed to the evaluator function. This function is useless and included only for completeness. The number of extra arguments is automatically set as you use optimize_init_argument().

optimize_init_argument(*S*) returns a pointer to the object that was previously set.

optimize_init_nargs(*S*) returns the number of extra arguments that are passed to the evaluator function.

optimize_init_cluster( )

optimize_init_cluster(*S*, *c*) specifies a cluster variable. *c* may be a string scalar containing a Stata variable name, or *c* may be real colvector directly containing the cluster values. The default is "", meaning no clustering. If clustering is specified, the default *vcetype* becomes "robust".

optimize_init_colstripe( )

optimize_init_colstripe(*S* [, stripe]) sets the string matrix to be associated with the parameter vector. See matrix colnames in [P] **matrix rownames**.

optimize_init_technique( )

> *void*   optimize_init_technique(*S*, *string scalar technique*)
>
> *string scalar* optimize_init_technique(*S*)

optimize_init_technique(*S*, *technique*) sets the optimization technique to be used. Current choices are

technique	description
"nr"	modified Newton–Raphson
"dfp"	Davidon–Fletcher–Powell
"bfgs"	Broyden–Fletcher–Goldfarb–Shanno
"bhhh"	Berndt–Hall–Hall–Hausman
"nm"	Nelder–Mead
"gn"	Gauss–Newton (quadratic optimization)

The default is "nr".

optimize_init_technique(*S*) returns the technique currently set.

Aside: All techniques require optimize_init_params() be set. Technique "nm" also requires that optimize_init_nmsimplexdeltas() be set. Parameters (and delta) can be set before or after the technique is set.

You can switch between "nr", "dfp", "bfgs", and "bhhh" by specifying two or more of them in a space-separated list. By default, optimize() will use an algorithm for five iterations before switching to the next algorithm. To specify a different number of iterations, include the number after the technique. For example, specifying optimize_init_technique(*M*, "bhhh 10 nr 1000") requests that optimize() perform 10 iterations using the Berndt–Hall–Hall–Hausman algorithm, followed by 1,000 iterations using the modified Newton–Raphson algorithm, and then switch back to Berndt–Hall–Hall–Hausman for 10 iterations, and so on. The process continues until convergence or until *maxiter* is exceeded.

optimize_init_gnweightmatrix( )

optimize_init_gnweightmatrix(*S*, *W*) sets real matrix W: $L \times L$, which is used only by type q evaluators. The objective function is $r'Wr$. If W is not set and if observation weights w are set using optimize_init_weight(), then $W = \text{diag}(w)$. If w is not set, then W is the identity matrix.

optimize_init_singularHmethod( )

> *void*   optimize_init_singularHmethod(*S*, *string scalar method*)
>
> *string scalar* optimize_init_singularHmethod(*S*)

optimize_init_singularHmethod(*S*, *method*) specifies what the optimizer should do when, at an iteration step, it finds that H is singular. Current choices are

method	description
`"m-marquardt"`	modified Marquardt algorithm
`"hybrid"`	mixture of steepest descent and Newton

The default is `"m-marquardt"` if not set;
`"hybrid"` is equivalent to `ml`'s `difficult` option; see [R] **ml**.

`optimize_init_technique`(*S*) returns the *method* currently set.

optimize_init_conv_maxiter( )

> *void*   `optimize_init_conv_maxiter`(*S*, *real scalar max*)
>
> *real scalar* `optimize_init_conv_maxiter`(*S*)

`optimize_init_conv_maxiter`(*S*, *max*) sets the maximum number of iterations to be performed before `optimization()` is stopped; results are posted to `optimize_result_*()` just as if convergence were achieved, but `optimize_result_converged()` is set to 0. The default *max* if not set is `c(maxiter)`, which is probably 16,000; type `creturn list` in Stata to determine the current default value.

`optimize_init_conv_maxiter`(*S*) returns the *max* currently set.

optimize_init_conv_warning( )

`optimize_init_conv_warning`(*S*, { `"on"` | `"off"` }) specifies whether the warning message "convergence not achieved" is to be displayed when this stopping rule is invoked. The default is `"on"`.

optimize_init_conv_ptol( ), . . . _vtol( ), . . . _nrtol( )

> *void*   `optimize_init_conv_ptol`(*S*, *real scalar ptol*)
>
> *void*   `optimize_init_conv_vtol`(*S*, *real scalar vtol*)
>
> *void*   `optimize_init_conv_nrtol`(*S*, *real scalar nrtol*)
>
> *real scalar* `optimize_init_conv_ptol`(*S*)
>
> *real scalar* `optimize_init_conv_vtol`(*S*)
>
> *real scalar* `optimize_init_conv_nrtol`(*S*)

The two-argument form of these functions set the tolerances that control `optimize()`'s convergence criterion. `optimize()` performs iterations until the convergence criterion is met or until the number of iterations exceeds `optimize_init_conv_maxiter()`. When the convergence criterion is met, `optimize_result_converged()` is set to 1. The default values of *ptol*, *vtol*, and *nrtol* are 1e–6, 1e–7, and 1e–5, respectively.

The single-argument form of these functions return the current values of *ptol*, *vtol*, and *nrtol*.

Optimization criterion: In all cases except `optimize_init_technique`(*S*)==`"nm"`, i.e., in all cases except Nelder–Mead, i.e., in all cases of derivative-based maximization, the optimization criterion is defined as follows:

Define

C_ptol: $\texttt{mreldif}(p, p\_prior) < ptol$

C_vtol: $\texttt{reldif}(v, v\_prior) < vtol$

C_nrtol: $g * \texttt{invsym}(-H) * g' < nrtol$

$C\_concave$: $-H$ is positive semidefinite

The above definitions apply for maximization. For minimization, think of it as maximization of $-f(p)$. $\texttt{optimize()}$ declares convergence when

$$(C\_ptol \mid C\_vtol) \ \& \ C\_concave \ \& \ C\_nrtol$$

For $\texttt{optimize\_init\_technique}(S)==\texttt{"nm"}$ (Nelder–Mead), the criterion is defined as follows:

Let R be the minimum and maximum values on the simplex and define

C_ptol: $\texttt{mreldif}(\text{vertices of } R) < ptol$

C_vtol: $\texttt{reldif}(R) < vtol$

$\texttt{optimize()}$ declares successful convergence when

$$C\_ptol \mid C\_vtol$$

optimize_init_conv_ignorenrtol( )

$\texttt{optimize\_init\_conv\_ignorenrtol}(S, \{\, \texttt{"off"} \mid \texttt{"on"} \,\})$ sets whether C_nrtol should simply be treated as true in all cases, which in effects removes the *nrtol* criterion from the convergence rule. The default is $\texttt{"off"}$.

optimize_init_iterid( )

     void $\texttt{optimize\_init\_iterid}(S, \textit{string scalar id})$

     string scalar $\texttt{optimize\_init\_iterid}(S)$

By default, $\texttt{optimize()}$ shows an iteration log, a line of which looks like

     $\texttt{Iteration 1: f(p) = 5.7641349}$

See *optimize_init_tracelevel()* below.

$\texttt{optimize\_init\_iterid}(S, \textit{id})$ sets the string used to label the iteration in the iteration log. The default is $\texttt{"Iteration"}$.

$\texttt{optimize\_init\_iterid}(S)$ returns the *id* currently in use.

optimize_init_valueid( )

     void $\texttt{optimize\_init\_valueid}(S, \textit{string scalar id})$

     string scalar $\texttt{optimize\_init\_valueid}(S)$

By default, $\texttt{optimize()}$ shows an iteration log, a line of which looks like

     $\texttt{Iteration 1: f(p) = 5.7641349}$

See *optimize_init_tracelevel()* below.

optimize_init_valueid(S, *id*) sets the string used to identify the value. The default is "f(p)".

optimize_init_valueid(S) returns the *id* currently in use.

optimize_init_tracelevel( )

> *void*   optimize_init_tracelevel(S, *string scalar tracelevel*)
>
> *string scalar* optimize_init_tracelevel(S)

optimize_init_tracelevel(S, *tracelevel*) sets what is displayed in the iteration log. Allowed values of *tracelevel* are

tracelevel	To be displayed each iteration
"none"	nothing (suppress the log)
"value"	function value
"tolerance"	previous + convergence values
"params"	previous + parameter values
"step"	previous + stepping information
"gradient"	previous + gradient vector
"hessian"	previous + Hessian matrix

The default is "value" if not reset.

optimize_init_tracelevel(S) returns the value of *tracelevel* currently set.

optimize_init_trace_dots( ), . . . _value( ), . . . _tol( ), . . . _step( ), . . . _gradient( ), . . . _Hessian( )

optimize_init_trace_dots(S, { "off" | "on" }) displays a dot each time your evaluator is called. The default is "off".

optimize_init_trace_value(S, { "on" | "off" }) displays the function value at the start of each iteration. The default is "on".

optimize_init_trace_tol(S, { "off" | "on" }) displays the value of the calculated result that is compared to the effective convergence criterion at the end of each iteration. The default is "off".

optimize_init_trace_params(S, { "off" | "on" }) displays the parameters at the start of each iteration. The default is "off".

optimize_init_trace_step(S, { "off" | "on" }) displays the steps within iteration. Listed are the value of objective function along with the word forward or backward. The default is "off".

optimize_init_trace_gradient(S, { "off" | "on" }) displays the gradient vector at the start of each iteration. The default is "off".

optimize_init_trace_Hessian(S, { "off" | "on" }) displays the Hessian matrix at the start of each iteration. The default is "off".

optimize_init_evaluations( )

optimize_init_evaluations(S, { "off" | "on" }) specifies whether the system is to count the number of times the evaluator is called. The default is "off".

optimize_init_constraints( )

> *void*   optimize_init_constraints(*S*, *real matrix Cc*)
>
> *real matrix* optimize_init_constraints(*S*)

nc linear constraints may be imposed on the *np* parameters in *p* according to $Cp' = c$, *C*: $nc \times np$ and *c*: $nc \times 1$. For instance, if there are four parameters and you wish to impose the single constraint $p_1 = p_2$, then $C = (1, -1, 0, 0)$ and $c = (0)$. If you wish to add the constraint $p_4 = 2$, then $C = (1, -1, 0, 0 \backslash 0, 0, 0, 1)$ and $c = (0 \backslash 2)$.

optimize_init_constraints(*S*, *Cc*) allows you to impose such constraints where $Cc = (C, c)$. Use of this function is optional. If no constraints have been set, then *Cc* is $0 \times (np + 1)$.

optimize_init_constraints(*S*) returns the current *Cc* matrix.

optimize_init_verbose( )

> *void*   optimize_init_verbose(*S*, *real scalar verbose*)
>
> *real scalar* optimize_init_verbose(*S*)

optimize_init_verbose(*S*, *verbose*) sets whether error messages that arise during the execution of optimize() or _optimize() are to be displayed. *verbose*=1 means that they are; 0 means that they are not. The default is 1. Setting *verbose* to 0 is of interest only to users of _optimize(). If you wish to suppress all output, code

        optimize_init_verbose(*S*, 0)
        optimize_init_tracelevel(*S*, "none")

optimize_init_verbose(*S*) returns the current value of *verbose*.

optimize( )

> *real rowvector* optimize(*S*)

optimize(*S*) invokes the optimization process and returns the resulting parameter vector. If something goes wrong, optimize() aborts with error.

Before you can invoke optimize(), you must have defined your evaluator function *evaluator*() and you must have set initial values:

        S = optimize_init()
        optimize_init_evaluator(*S*, &*evaluator*())
        optimize_init_params(*S*, (...))

The above assumes that your evaluator function is d0. Often you will also have coded

        optimize_init_evaluatortype(*S*, "..."))

and you may have coded other optimize_init_*() functions as well.

Once `optimize()` completes, you may use the `optimize_result_*()` functions. You may also continue to use the `optimize_init_*()` functions to access initial settings, and you may use them to change settings and restart optimization (i.e., invoke `optimize()` again) if you wish. If you do that, you will usually want to use the resulting parameter values from the first round of optimization as initial values for the second. If so, do not forget to code

> `optimize_init_params(S, optimize_result_params(S))`

_optimize( )

> *real scalar* `_optimize(S)`

`_optimize(S)` performs the same actions as `optimize(S)` except that, rather than returning the resulting parameter vector, `_optimize()` returns a real scalar and, rather than aborting if numerical issues arise, `_optimize()` returns a nonzero value. `_optimize()` returns 0 if all went well. The returned value is called an error code.

`optimize()` returns the resulting parameter vector *p*. It can work that way because optimization must have gone well. Had it not, `optimize()` would have aborted execution.

`_optimize()` returns an error code. If it is 0, optimization went well and you can obtain the parameter vector by using `optimize_result_param()`. If optimization did not go well, you can use the error code to diagnose what went wrong and take the appropriate action.

Thus, `_optimize(S)` is an alternative to `optimize(S)`. Both functions do the same thing. The difference is what happens when there are numerical difficulties.

`optimize()` and `_optimize()` work around most numerical difficulties. For instance, the evaluator function you write is allowed to return *v* equal to missing if it cannot calculate the $f()$ at the current values of *p*. If that happens during optimization, `optimize()` and `_optimize()` will back up to the last value that worked and choose a different direction. `optimize()`, however, cannot tolerate that happening with the initial values of the parameters because `optimize()` has no value to back up to. `optimize()` issues an error message and aborts, meaning that execution is stopped. There can be advantages in that. The calling program need not include complicated code for such instances, figuring that stopping is good enough because a human will know to address the problem.

`_optimize()`, however, does not stop execution. Rather than aborting, `_optimize()` returns a nonzero value to the caller, identifying what went wrong.

Programmers implementing advanced systems will want to use `_optimize()` instead of `optimize()`. Everybody else should use `optimize()`.

Programmers using `_optimize()` will also be interested in the functions

```
optimize_init_verbose()
optimize_result_errorcode()
optimize_result_errortext()
optimize_result_returncode()
```

If you perform optimization by using `_optimize()`, the behavior of all `optimize_result_*()` functions is altered. The usual behavior is that, if calculation is required and numerical problems arise, the functions abort with error. After `_optimize()`, however, a properly dimensioned missing result is returned and `optimize_result_errorcode()` and `optimize_result_errortext()` are set appropriately.

The error codes returned by _optimize() are listed under the heading *optimize_result_errorcode()* below.

optimize_evaluate( )

> *void* optimize_evaluate(*S*)

optimize_evaluate(*S*) evaluates *f*() at optimize_init_params() and posts results to optimize_result_*() just as if optimization had been performed, meaning that all optimize_result_*() functions are available for use. optimize_result_converged() is set to 1.

The setup for running this function is the same as for running optimize():

> *S* = optimize_init()
> optimize_init_evaluator(*S*, &*evaluator*())
> optimize_init_params(*S*, (...))

Usually, you will have also coded

> optimize_init_evaluatortype(*S*, ...))

The other optimize_init_*() settings do not matter.

_optimize_evaluate( )

> *real scalar* _optimize_evaluate(*S*)

The relationship between _optimize_evaluate() and optimize_evaluate() is the same as that between _optimize() and optimize(); see *_optimize()*.

_optimize_evaluate() returns an error code.

optimize_result_params( )

> *real rowvector* optimize_result_params(*S*)

optimize_result_params(*S*) returns the resulting parameter values. These are the same values that were returned by optimize() itself. There is no computational cost to accessing the results, so rather than coding

> *p* = optimize(*S*)

if you find it more convenient to code

> (void) optimize(*S*)
> ...
> *p* = optimize_result_params(*S*)

then do so.

optimize_result_value( ) and optimize_result_value0( )

> *real scalar* optimize_result_value(*S*)

> *real scalar* optimize_result_value0(*S*)

optimize_result_value(*S*) returns the value of *f*() evaluated at *p* equal to optimize_result_param().

optimize_result_value0(*S*) returns the value of *f*() evaluated at *p* equal to optimize_init_param().

These functions may be called regardless of the evaluator or technique used.

optimize_result_gradient( )

> *real rowvector* optimize_result_gradient(*S*)

optimize_result_gradient(*S*) returns the value of the gradient vector evaluated at *p* equal to optimize_result_param(). This function may be called regardless of the evaluator or technique used.

optimize_result_scores( )

> *real matrix* optimize_result_scores(*S*)

optimize_result_scores(*S*) returns the value of the scores evaluated at *p* equal to optimize_result_param(). This function may be called only if a type v evaluator is used, but regardless of the technique used.

optimize_result_Hessian( )

> *real matrix* optimize_result_Hessian(*S*)

optimize_result_Hessian(*S*) returns the value of the Hessian matrix evaluated at *p* equal to optimize_result_param(). This function may be called regardless of the evaluator or technique used.

optimize_result_V( ) and optimize_result_Vtype( )

> *real matrix*   optimize_result_V(*S*)

> *string scalar* optimize_result_Vtype(*S*)

optimize_result_V(*S*) returns optimize_result_V_oim(*S*) or optimize_result_V_opg(*S*), depending on which is the natural conjugate for the optimization technique used. If there is no natural conjugate, optimize_result_V_oim(*S*) is returned.

optimize_result_Vtype(*S*) returns "oim" or "opg".

optimize_result_V_oim( ), . . . _opg( ), . . . _robust( )

>   *real matrix* optimize_result_V_oim(*S*)

>   *real matrix* optimize_result_V_opg(*S*)

>   *real matrix* optimize_result_V_robust(*S*)

These functions return the variance matrix of p evaluated at p equal to optimize_result_param( ). These functions are relevant only for maximization of log-likelihood functions but may be called in any context, including minimization.

optimize_result_V_oim(*S*) returns invsym($-H$), which is the variance matrix obtained from the observed information matrix. For minimization, returned is invsym(H).

optimize_result_V_opg(*S*) returns invsym($S'S$), where S is the $N \times np$ matrix of scores. This is known as the variance matrix obtained from the outer product of the gradients. optimize_result_V_opg( ) is available only when the evaluator function is type v, but regardless of the technique used.

optimize_result_V_robust(*S*) returns $H * \text{invsym}(S'S) * H$, which is the robust estimate of variance, also known as the sandwich estimator of variance. optimize_result_V_robust( ) is available only when the evaluator function is type v, but regardless of the technique used.

optimize_result_iterations( )

>   *real scalar* optimize_result_iterations(*S*)

optimize_result_iterations(*S*) returns the number of iterations used in obtaining results.

optimize_result_converged( )

>   *real scalar* optimize_result_converged(*S*)

optimize_result_converged(*S*) returns 1 if results converged and 0 otherwise. See *optimize_init_conv_ptol()* for the definition of convergence.

optimize_result_iterationlog( )

>   *real colvector* optimize_result_iterationlog(*S*)

optimize_result_iterationlog(*S*) returns a column vector of the values of $f()$ at the start of the final 20 iterations, or, if there were fewer, however many iterations there were. Returned vector is min(optimize_result_iterations( ), 20) $\times$ 1.

optimize_result_evaluations( )

optimize_result_evaluations(*S*) returns a 1×3 real rowvector containing the number of times the evaluator was called, assuming optimize_init_evaluations( ) was set on. Contents are the number of times called for the purposes of 1) calculating the objective function, 2) calculating the objective function and its first derivative, and 3) calculating the objective function and its first and second derivatives. If optimize_init_evaluations( ) was set to off, returned is $(0, 0, 0)$.

optimize_result_errorcode( ), ..._errortext( ), and ..._returncode( )

> *real scalar*   optimize_result_errorcode(*S*)
>
> *string scalar* optimize_result_errortext(*S*)
>
> *real scalar*   optimize_result_returncode(*S*)

These functions are for use after _optimize().

optimize_result_errorcode(*S*) returns the error code of _optimize(), _optimize_evaluate(), or the last optimize_result_*() run after either of the first two functions. The value will be zero if there were no errors. The error codes are listed directly below.

optimize_result_errortext(*S*) returns a string containing the error message corresponding to the error code. If the error code is zero, the string will be "".

optimize_result_returncode(*S*) returns the Stata return code corresponding to the error code. The mapping is listed directly below.

In advanced code, these functions might be used as

```
(void) _optimize(S)
...
if (ec = optimize_result_code(S)) {
        errprintf("{p}\n")
        errprintf("%s\n", optimize_result_errortext(S))
        errprintf("{p_end}\n")
        exit(optimize_result_returncode(S))
        /*NOTREACHED*/
}
```

(Continued on next page)

The error codes and their corresponding Stata return codes are

Error code	Return code	Error text
1	1400	initial values not feasible
2	412	redundant or inconsistent constraints
3	430	missing values returned by evaluator
4	430	Hessian is not positive semidefinite *or* Hessian is not negative semidefinite
5	430	could not calculate numerical derivatives—discontinuous region with missing values encountered
6	430	could not calculate numerical derivatives—flat or discontinuous region encountered
7	430	could not calculate improvement—discontinuous region encountered
8	430	could not calculate improvement—flat region encountered
10	111	technique unknown
11	111	incompatible combination of techniques
12	111	singular H method unknown
13	198	matrix stripe invalid for parameter vector
14	198	negative convergence tolerance values are not allowed
15	503	invalid starting values
17	111	simplex delta required
18	3499	simplex delta not conformable with parameter vector
19	198	simplex delta value too small (must be greater than $10 \times$ ptol)
20	198	evaluator type requires the nr technique
23	198	evaluator type not allowed with bhhh technique
24	111	evaluator functions required
25	198	starting values for parameters required
26	198	missing parameter values not allowed

NOTES: (1) Error 1 can occur only when evaluating $f()$ at initial parameters.

(2) Error 2 can occur only if constraints are specified.

(3) Error 3 can occur only if the technique is "nm".

(4) Error 9 can occur only if technique is "bfgs" or "dfp".

optimize_query( )

> *void* optimize_query(*S*)

optimize_query(*S*) displays a report on all optimize_init_*() and optimize_result*() values. optimize_query() may be used before or after optimize() and is useful when using optimize() interactively or when debugging a program that calls optimize() or _optimize().

Conformability

All functions have 1×1 inputs and have 1×1 or *void* outputs except the following:

optimize_init_params(*S*, *initialvalues*):

S:	*transmorphic*
initialvalues:	$1 \times np$
result:	*void*

optimize_init_params(*S*):

S:	*transmorphic*
result:	$1 \times np$

optimize_init_argument(*S*, *k*, *X*):

S:	*transmorphic*
k:	1×1
X:	*anything*
result:	*void*

optimize_init_nmsimplexdeltas(*S*, *delta*):

S:	*transmorphic*
delta:	$1 \times np$
result:	*void*

optimize_init_nmsimplexdeltas(*S*):

S:	*transmorphic*
result:	$1 \times np$

optimize_init_constraints(*S*, *Cc*):

S:	*transmorphic*
Cc:	$nc \times (np + 1)$
result:	*void*

optimize_init_constraints(*S*):

S:	*transmorphic*
result:	$nc \times (np + 1)$

optimize(*S*):

S:	*transmorphic*
result:	$1 \times np$

optimize_result_params(*S*):

S:	*transmorphic*
result:	$1 \times np$

optimize_result_gradient(*S*), optimize_result_evaluations(*S*):

> *S*:   *transmorphic*
> *result*: $1 \times np$

optimize_result_scores(*S*):

> *S*:   *transmorphic*
> *result*: $N \times np$

optimize_result_Hessian(*S*):

> *S*:   *transmorphic*
> *result*: $np \times np$

optimize_result_V(*S*), optimize_result_V_oim(*S*), optimize_result_V_opg(*S*), optimize_result_V_robust(*S*):

> *S*:   *transmorphic*
> *result*: $np \times np$

optimize_result_iterationlog(*S*):

> *S*:   *transmorphic*
> *result*: $L \times 1, L \le 20$

For optimize_init_cluster(*S*, *c*), optimize_init_colstripe(*S*), and optimize_init_gnweightmatrix(*S*, *W*), see *Syntax* above.

Diagnostics

All functions abort with error when used incorrectly.

optimize() aborts with error if it runs into numerical difficulties. _optimize() does not; it instead returns a nonzero error code.

optimize_evaluate() aborts with error if it runs into numerical difficulties. _optimize_evaluate() does not; it instead returns a nonzero error code.

The optimize_result_*() functions abort with error if they run into numerical difficulties when called after optimize() or optimize_evaluate(). They do not abort when run after _optimize() or _optimize_evaluate(). They instead return a properly dimensioned missing result and set optimize_result_errorcode() and optimize_result_errortext().

The formula $x_{i+1} = x_i - f(x_i)/f'(x_i)$ and its generalizations for solving $f(x) = 0$ (and its generalizations) are known variously as Newton's method or the Newton–Raphson method. The real history is more complicated than these names imply and has roots in the earlier work of Arabic algebraists and François Viète.

Newton's first formulation dating from about 1669 refers only to solution of polynomial equations and does not use calculus. In his *Philosophiae Naturalis Principia Mathematica*, first published in 1687, the method is used, but not obviously, to solve a nonpolynomial equation. Raphson's work, first published in 1690, also concerns polynomial equations, and proceeds algebraically without using calculus, but lays more stress on iterative calculation and so is closer to present ideas. It was not until 1740 that Thomas Simpson published a more general version explicitly formulated in calculus terms that was applied to both polynomial and nonpolynomial equations and to both single equations and systems of equations. Simpson's work was in turn overlooked in influential later accounts by Lagrange and Fourier, but his contribution also deserves recognition.

Isaac Newton (1643–1727) was an English mathematician, astronomer, physicist, natural philosopher, alchemist, theologian, biblical scholar, historian, politician and civil servant. He was born in Lincolnshire, according to the calendar then in use, in 1642, and studied there and at the University of Cambridge, where he was a fellow of Trinity College and elected Lucasian Professor in 1669. Newton demonstrated the generalized binomial theorem, did major work on power series, and deserves credit with Gottfried Leibniz for the development of calculus: during his lifetime and long afterward, that fact was obscured by a bitter and protracted priority dispute. He described universal gravitation and the laws of motion central to classical mechanics and showed that the motions of objects on Earth and beyond are subject to the same laws. Newton invented the reflecting telescope and developed a theory of color that was based on the fact that a prism splits white light into a visible spectrum. He also studied cooling and the speed of sound and proposed a theory of the origin of stars. Much of his later life was spent in London, including brief spells as member of Parliament and longer periods as master of the Mint and president of the Royal Society. He was knighted in 1705. Although undoubtedly one of the greatest mathematical and scientific geniuses of all time, Newton was also outstandingly contradictory, secretive, and quarrelsome.

Joseph Raphson (1648–1715) was an English or possibly Irish mathematician. No exact dates are known for his birth or death years. He appears to have been largely self-taught and was awarded a degree by the University of Cambridge after the publication of his most notable work, *Analysis Aequationum Universalis* (1690), and his election as a fellow of the Royal Society.

Thomas Simpson (1710–1761) was born in Market Bosworth, Leicestershire, England. He received little formal education and was self-taught in mathematics. Simpson moved to London and worked as a teacher in London coffee houses (as was De Moivre) and then at the Royal Military Academy at Woolwich. He published texts on calculus, astronomy, and probability and is best remembered for his work on interpolation and numerical methods of integration. However, what is now known as "Simpson's rule" was known earlier to Newton. He was also a fellow of the Royal Society.

References

Berndt, E. K., B. H. Hall, R. E. Hall, and J. A. Hausman. 1974. Estimation and inference in nonlinear structural models. *Annals of Economic and Social Measurement* 3/4: 653–665.

Davidon, W. C. 1959. Variable metric method for minimization. Technical Report ANL-5990, Argonne National Laboratory, U.S. Department of Energy.

Fletcher, R. 1970. A new approach to variable metric algorithms. *Computer Journal* 13: 317–322.

——. 1987. *Practical Methods of Optimization.* 2nd ed. New York: Wiley.

Fletcher, R., and M. J. D. Powell. 1963. A rapidly convergent descent method for minimization. *Computer Journal* 6: 163–168.

Gleick, J. 2003. *Isaac Newton.* New York: Pantheon.

Goldfarb, D. 1970. A family of variable-metric methods derived by variational means. *Mathematics of Computation* 24: 23–26.

Marquardt, D. W. 1963. An algorithm for least-squares estimation of nonlinear parameters. *Journal of the Society for Industrial and Applied Mathematics* 11: 431–441.

Nelder, J. A., and R. Mead. 1965. A simplex method for function minimization. *Computer Journal* 7: 308–313.

Newton, I. 1671. *De methodis fluxionum et serierum infinitorum.* Translated by john colson as *the method of fluxions and infinite series* ed. London: Henry Wood Fall, 1736.

Raphson, J. 1690. *Analysis Aequationum Universalis.* Londioni: Prostant venales apud Abelem Swalle.

Shanno, D. F. 1970. Conditioning of quasi-Newton methods for function minimization. *Mathematics of Computation* 24: 647–656.

Westfall, R. S. 1980. *Never at Rest: A Biography of Isaac Newton.* Cambridge: Cambridge University Press.

Ypma, T. J. 1995. Historical development of the Newton–Raphson method. *SIAM Review* 37: 531–551.

Also see

[M-5] **moptimize( )** — Model optimization

[M-4] **mathematical** — Important mathematical functions

[M-4] **statistical** — Statistical functions

Title

Syntax

info = panelsetup(*V*, *idcol*)

info = panelsetup(*V*, *idcol*, *minobs*)

info = panelsetup(*V*, *idcol*, *minobs*, *maxobs*)

real rowvector panelstats(*info*)

real matrix panelsubmatrix(*V*, *i*, *info*)

void panelsubview(*SV*, *V*, *i*, *info*)

where,

V:	*real* or *string matrix*, possibly a view
idcol:	*real scalar*
minobs:	*real scalar*
maxobs:	*real scalar*
info:	*real matrix*
i:	*real scalar*
SV:	*matrix* to be created, possibly as view

Description

These functions assist with the processing of panel data. The idea is to make it easy and fast to write loops like

```
for (i=1; i<=number_of_panels; i++) {
    X = matrix corresponding to panel i
    . . .
    . . . (calculations using X) . . .
    . . .
}
```

Using these functions, this loop could become

```
st_view(Vid, ., "idvar",        "touse")
st_view(V,   ., ("x1", "x2"), "touse")
info = panelsetup(Vid, 1)
for (i=1; i<=rows(info); i++) {
    X = panelsubmatrix(V, i, info)
    . . .
    . . . (calculations using X) . . .
    . . .
}
```

669

panelsetup(V, $idcol$, ...) sets up panel processing. It returns a matrix (*info*) that is passed to other panel-processing functions.

panelstats(*info*) returns a row vector containing the number of panels, number of observations, minimum number of observations per panel, and maximum number of observations per panel.

panelsubmatrix(V, i, *info*) returns a matrix containing the contents of V for panel i.

panelsubview(SV, V, i, *info*) does nearly the same thing. Rather than returning a matrix, however, it places the matrix in SV. If V is a view, then the matrix placed in SV will be a view.

Remarks

Remarks are presented under the following headings:

> *Definition of panel data*
> *Definition of problem*
> *Preparation*
> *Use of panelsetup( )*
> *Using panelstats( )*
> *Using panelsubmatrix( )*
> *Using panelsubview( )*

Definition of panel data

Panel data include multiple observations on subjects, countries, etc.:

subject ID	time ID	x1	x2
1	1	4.2	3.7
1	2	3.2	3.7
1	3	9.2	4.2
2	1	1.7	4.0
2	2	1.9	5.0
3	1	9.5	1.3
⋮	⋮	⋮	⋮

In the above dataset, there are three observations for subject 1, two for subject 2, etc. We labeled the identifier within subject to be time, but that is only suggestive, and in any case, the secondary identifier will play no role in what follows.

If we speak about the first panel, we are discussing the first 3 observations of this dataset. If we speak about the second, that corresponds to observations 4 and 5.

It is common to refer to panel numbers with the letter i. It is common to refer to the number of observations in the ith panel as T_i even when the data within panel have nothing to do with repeated observations over time.

Definition of problem

We want to calculate some statistic on panel data. The calculation amounts to

$$\sum_{i=1}^{K} f(X_i)$$

where the sum is performed across panels, and X_i is the data matrix for panel i. For instance, given the example in the previous section

$$X_1 = \begin{bmatrix} 4.2 & 3.7 \\ 3.2 & 3.7 \\ 9.2 & 4.2 \end{bmatrix}$$

and X_2 is a similarly constructed 2×2 matrix.

Depending on the nature of the calculation, there will be problems for which

1. we want to use all the panels,

2. we want to use only panels for which there are two or more observations, and

3. we want to use the same number of observations in all the panels (balanced panels).

In addition to simple problems of the sort,

$$\sum_{i=1}^{K} f(X_i)$$

you may also need to deal with problems of the form,

$$\sum_{i=1}^{K} f(X_i, Y_i, \ldots)$$

That is, you may need to deal with problems where there are multiple matrices per subject.

We use the sum operator purely for illustration, although it is the most common. Your problem might be

$$F(X_1, Y_1, \ldots, X_2, Y_2, \ldots)$$

Preparation

Before using the functions documented here, create a matrix or matrices containing the data. For illustration, it will be sufficient to create V containing all the data in our example problem:

$$V = \begin{bmatrix} 1 & 1 & 4.2 & 3.7 \\ 1 & 2 & 3.2 & 3.7 \\ 1 & 3 & 9.2 & 4.2 \\ 2 & 1 & 1.7 & 4.0 \\ 2 & 2 & 1.9 & 5.0 \\ 3 & 1 & 9.5 & 1.3 \\ \vdots & \vdots & \vdots & \vdots \end{bmatrix}$$

But you will probably find it more convenient (and we recommend) if you create at least two matrices, one containing the subject identifier and the other containing the x variables (and omit the within-subject "time" identifier altogether):

$$V1 = \begin{bmatrix} 1 \\ 1 \\ 1 \\ 2 \\ 2 \\ 3 \\ \vdots \end{bmatrix} \qquad V2 = \begin{bmatrix} 4.2 & 3.7 \\ 3.2 & 3.7 \\ 9.2 & 4.2 \\ 1.7 & 4.0 \\ 1.9 & 5.0 \\ 9.5 & 1.3 \\ \vdots & \vdots \end{bmatrix}$$

In the above, matrix $V1$ contains the subject identifier, and matrix $V2$ contains the data for all the X_i matrices in

$$\sum_{i=1}^{K} f(X_i)$$

If your calculation is

$$\sum_{i=1}^{K} f(X_i, Y_i, \ldots)$$

create additional V matrices, $V3$ corresponding to Y_i, and so on.

To create these matrices, use [M-5] **st_view( )**

```
st_view(V1,  ., "idvar",        "touse")
st_view(V2,  ., ("x1", "x2"), "touse")
```

although you could use [M-5] **st_data( )** if you preferred. Using st_view() will save memory. You can also construct $V1$, $V2$, ..., however you wish; they are just matrices. Be sure that the matrices align, for example, that row 4 of one matrix corresponds to row 4 of another. We did that above by assuming a *touse* variable had been included (or constructed) in the dataset.

Use of panelsetup( )

panelsetup(V, *idcol*, ...) sets up panel processing, returning a $K \times 2$ matrix that contains a row for each panel. The row records the first and last observation numbers (row numbers in V) that correspond to the panel.

For instance, with our example, `panelsetup()` will return

$$\begin{bmatrix} 1 & 3 \\ 4 & 5 \\ 6 & 9 \\ \vdots & \vdots \end{bmatrix}$$

The first panel is recorded in observations 1 to 3; it contains $3 - 1 + 1 = 3$ observations. The second panel is recorded in observations 4 to 5 and it contains $5 - 4 + 1 = 2$ observations, and so on. We recorded the third panel as being observations 6 to 9, although we did not show you enough of the original data for you to know that 9 was the last observation with ID 3.

`panelsetup()` has many more capabilities in constructing this result, but it is important to appreciate that returning this observation-number matrix is all that `panelsetup()` does. This matrix is all that other panel functions need to know. They work with the information produced by `panelsetup()`, but they will equally well work with any two-column matrix that contains observation numbers. Correspondingly, `panelsetup()` engages in no secret behavior that ties up memory, puts you in a mode, or anything else. `panelsetup()` merely produces this matrix.

The number of rows of the matrix `panelsetup()` returns equals K, the number of panels.

The syntax of `panelsetup()` is

>   *info* = panelsetup(*V*, *idcol*, *minobs*, *maxobs*)

The last two arguments are optional.

The required argument *V* specifies a matrix containing at least the panel identification numbers and required argument *idcol* specifies the column of *V* that contains that ID. Here we will use the matrix *V1*, which contains only the identification number:

>   *info* = panelsetup(*V1*, 1)

The two optional arguments are *minobs* and *maxobs*. *minobs* specifies the minimum number of observations within panel that we are willing to tolerate; if a panel has fewer observations, we want to omit it entirely. For instance, were we to specify

>   *info* = panelsetup(*V1*, 1, 3)

then the matrix `panelsetup()` would contain fewer rows. In our example, the returned *info* matrix would contain

$$\begin{bmatrix} 1 & 3 \\ 6 & 9 \\ \vdots & \vdots \end{bmatrix}$$

Observations 4 and 5 are now omitted because they correspond to a two-observation panel, and we said only panels with three or more observations should be included.

We chose three as a demonstration. In fact, it is most common to code

>   *info* = panelsetup(*V1*, 1, 2)

because that eliminates the singletons (panels with one observation).

The final optional argument is *maxobs*. For example,

>   *info* = panelsetup(*V1*, 1, 2, 5)

means to include only up to five observations per panel. Any observations beyond five are to be trimmed. If we code

> $info$ = panelsetup($V1$, 1, 3, 3)

then all the panels contained in *info* would have three observations. If a panel had fewer than three observations, it would be omitted entirely. If a panel had more than three observations, only the first three would be included.

Panel datasets with the same number of observations per panel are said to be balanced. panelsetup( ) also provides panel-balancing capabilities. If you specify *maxobs* as 0, then

1. panelsetup( ) first calculates the min(T_i) among the panels with *minobs* observations or more. Call that number *m*.

2. panelsetup( ) then returns panelsetup($V1$, *idcol*, *m*, *m*), thus creating balanced panels of size *m* and producing a dataset that has the maximum number of within-panel observations given it has the maximum number of panels.

If we coded

> $info$ = panelsetup($V1$, 1, 2, 0)

then panelsetup( ) would create the maximum number of panels with the maximum number of within-panel observations subject to the constraint of no singletons and the panels being balanced.

Using panelstats( )

panelstats(*info*) can be used on any two-column matrix that contains observation numbers. panelstats( ) returns a row vector containing

> panelstats( )[1] = number of panels (same as rows(*info*))
> panelstats( )[2] = number of observations
> panelstats( )[3] = min(T_i)
> panelstats( )[4] = max(T_i)

Using panelsubmatrix( )

Having created an *info* matrix using panelsetup( ), you can obtain the matrix corresponding to the *i*th panel using

> X = panelsubmatrix(V, i, *info*)

It is not necessary that panelsubmatrix( ) be used with the same matrix that was used to produce *info*. We created matrix *V1* containing the ID numbers, and we created matrix *V2* containing the *x* variables

> st_view($V1$, ., "idvar",   "*touse*")
> st_view($V2$, ., ("x1", "x2"), "*touse*")

and we create *info* using *V1*:

> $info$ = panelsetup($V1$, 1)

We can now create the corresponding X matrix by coding

$$X = \mathtt{panelsubmatrix}(V2, \; i, \; info)$$

and, had we created a $V3$ matrix corresponding to Y_i, we could also code

$$Y = \mathtt{panelsubmatrix}(V3, \; i, \; info)$$

and so on.

Using panelsubview( )

panelsubview() works much like panelsubmatrix(). The difference is that rather than coding

$$X = \mathtt{panelsubmatrix}(V, \; i, \; info)$$

you code

$$\mathtt{panelsubview}(X, \; V, \; i, \; info)$$

The matrix to be defined becomes the first argument of panelsubview(). That is because panel-subview() is designed especially to work with views. panelsubmatrix() will work with views, but panelsubview() does something special. Rather than returning an ordinary matrix (an array, in the jargon), if V is a view, panelsubview() returns a view in its first argument. Views save memory.

Views can save much memory, so it would seem that you would always want to use panelsubview() in place of panelsubmatrix(). What is not always appreciated, however, is that it takes Mata longer to access the data recorded in views, and so there is a tradeoff.

If the panels are likely to be large, you want to use panelsubview(). Conserving memory trumps all other considerations.

In fact, the panels that occur in most datasets are not that large, even when the dataset itself is. If you are going to make many calculations on X, you may wish to use panelsubmatrix().

Both panelsubmatrix() and panelsubview() work with view and nonview matrices. pan-elsubview() produces a regular matrix when the base matrix V is not a view, just as does panelsubmatrix(). The difference is that panelsubview() will produce a view when V is a view, whereas panelsubmatrix() always produces a nonview matrix.

Conformability

panelsetup(V, $idcol$, $minobs$, $maxobs$):

V:	$r \times c$	
$idcol$:	1×1	
$minobs$:	1×1	(optional)
$maxobs$:	1×1	(optional)
$result$:	$K \times 2$,	K = number of panels

panelstats($info$):

$info$:	$K \times 2$
$result$:	1×4

panelsubmatrix(V, i, *info*):

V:	$r \times c$
i:	1×1, $1 \le i \le$ rows(*info*)
info:	$K \times 2$
result:	$t \times c$, $t =$ number of obs. in panel

panelsubview(SV, V, i, *info*):

input:

SV:	irrelevant
V:	$r \times c$
i:	1×1, $1 \le i \le$ rows(*info*)
info:	$K \times 2$
result:	$t \times c$, $t =$ number of obs. in panel

output:

SV:	$t \times c$, $t =$ number of obs. in panel

Diagnostics

panelsubmatrix(V, i, *info*) and panelsubview(SV, V, i, *info*) abort with error if $i < 1$ or $i >$ rows(*info*).

panelsetup( ) can return a 0×2 result.

Also see

[M-4] **utility** — Matrix utility functions

Title

Syntax

string scalar	`pathjoin(`*string scalar path1*, *string scalar path2*`)`
void	`pathsplit(`*string scalar path*, *path1*, *path2*`)`
string scalar	`pathbasename(`*string scalar path*`)`
string scalar	`pathsuffix(`*string scalar path*`)`
string scalar	`pathrmsuffix(`*string scalar path*`)`
real scalar	`pathisurl(`*string scalar path*`)`
real scalar	`pathisabs(`*string scalar path*`)`
real scalar	`pathasciisuffix(`*string scalar path*`)`
real scalar	`pathstatasuffix(`*string scalar path*`)`
string rowvector	`pathlist(`*string scalar dirlist*`)`
string rowvector	`pathlist()`
string rowvector	`pathsubsysdir(`*string rowvector pathlist*`)`
string rowvector	`pathsearchlist(`*string scalar fn*`)`

Description

`pathjoin(`*path1*, *path2*`)` forms, logically speaking, *path1/path2*, but does so in the appropriate style. For instance, *path1* might be a URL and *path2* a Windows *dirname\filename*, and the two paths will, even so, be joined correctly. All issues of whether *path1* ends with a directory separator, *path2* begins with one, etc., are handled automatically.

`pathsplit(`*path*, *path1*, *path2*`)` performs the inverse operation, removing the last element of the path (which is typically a filename) and storing it in *path2* and storing the rest in *path1*.

`pathbasename(`*path*`)` returns the last element of *path*.

`pathsuffix(`*path*`)` returns the file suffix, with leading dot, if there is one, and returns `""` otherwise. For instance, `pathsuffix("this\that.ado")` returns ".ado".

`pathrmsuffix(`*path*`)` returns *path* with the suffix removed, if there was one. For instance, `pathrmsuffix("this\that.ado")` returns "this\that".

`pathisurl(`*path*`)` returns 1 if *path* is a URL and 0 otherwise.

`pathisabs(`*path*`)` returns 1 if *path* is absolute and 0 if relative. `c:\this` is an absolute path. `this\that` is a relative path. URLs are considered to be absolute.

pathasciisuffix(*path*) and pathstatasuffix(*path*) are more for StataCorp use than anything else. pathasciisuffix( ) returns 1 if the file is known to be ASCII, based on its file suffix. StataCorp uses this function in Stata's net command to decide whether end-of-line characters, which differ across operating systems, should be modified during downloads. pathstatasuffix( ) is the function used by Stata's net and update commands to decide whether a file belongs in the official directories. pathstatasuffix("example.ado") is true, but pathstatasuffix("example.do") is false because do-files do not go in system directories.

pathlist(*dirlist*) returns a row vector, each element of which contains an element of a semicolon-separated path list *dirlist*. For instance, pathlist("a;b;c") returns ("a", "b", "c").

pathlist( ) without arguments returns pathlist(c("adopath")), the broken-out elements of the official Stata ado-path.

pathsubsysdir(*pathlist*) returns *pathlist* with any elements that are Stata system directories' shorthands, such as UPDATES, PLUS, PERSONAL, substituted with the actual directory names. For instance, the right way to obtain the official directories over which Stata searches for files is pathsubsysdir(pathlist( )).

pathsearchlist(*fn*) returns a row vector. The elements are full paths/filenames specifying all the locations, in order, where Stata would look for *fn* along the official Stata ado-path.

Remarks

Using these functions, you are more likely to produce code that works correctly regardless of operating system.

Conformability

pathjoin(*path1*, *path2*):

path1:	1×1
path2:	1×1
result:	1×1

pathsplit(*path*, *path1*, *path2*):

input:

path:	1×1

output:

path1:	1×1
path2:	1×1

pathbasename(*path*), pathsuffix(*path*), pathrmsuffix(*path*):

path:	1×1
result:	1×1

pathisurl(*path*), pathisabs(*path*), pathasciisuffix(*path*), pathstatasuffix(*path*):

path:	1×1
result:	1×1

pathlist(*dirlist*):

dirlist:	1×1	(optional)
result:	$1 \times k$	

pathsubsysdir(*pathlist*):

pathlist:	$1 \times k$
result:	$1 \times k$

pathsearchlist(*fn*):

fn:	1×1
result:	$1 \times k$

Diagnostics

All routines abort with error if the path is too long for the operating system; nothing else causes abort with error.

Also see

[M-4] **io** — I/O functions

Title

Syntax

numeric matrix	pinv(*numeric matrix A*)
numeric matrix	pinv(*numeric matrix A, rank*)
numeric matrix	pinv(*numeric matrix A, rank, real scalar tol*)
real scalar	_pinv(*numeric matrix A*)
real scalar	_pinv(*numeric matrix A, real scalar tol*)

where the type of *rank* is irrelevant; the rank of *A* is returned there.

To obtain a generalized inverse of a symmetric matrix with a different normalization, see [M-5] **invsym( )**.

Description

pinv(*A*) returns the unique Moore–Penrose pseudoinverse of real or complex, symmetric or non-symmetric, square or nonsquare matrix *A*.

pinv(*A, rank*) does the same thing, and it returns in *rank* the rank of *A*.

pinv(*A, rank, tol*) does the same thing, and it allows you to specify the tolerance used to determine the rank of *A*, which is also used in the calculation of the pseudoinverse. See [M-5] **svsolve( )** and [M-1] **tolerance** for information on the optional *tol* argument.

_pinv(*A*) and _pinv(*A, tol*) do the same thing as pinv(), except that *A* is replaced with its inverse and the rank is returned.

Remarks

The Moore–Penrose pseudoinverse is also known as the Moore–Penrose inverse and as the generalized inverse. Whatever you call it, the pseudoinverse $A*$ of A satisfies four conditions,

$$A(A*)A \quad = A$$
$$(A*)A(A*) = A*$$
$$(A(A*))' \quad = A(A*)$$
$$((A*)A)' \quad = (A*)A$$

where the transpose operator $'$ is understood to mean the conjugate transpose when A is complex. Also, if A is of full rank, then

$$A* = A^{-1}$$

$\mathtt{pinv}(A)$ is logically equivalent to $\mathtt{svsolve}(A, \mathtt{I}(\mathtt{rows}(A)))$; see [M-5] **svsolve( )** for details and for use of the optional *tol* argument.

Conformability

$\mathtt{pinv}(A, \textit{rank}, \textit{tol})$:

    input:

	A:	$r \times c$	
	tol:	1×1	(optional)

    output:

	rank:	1×1	(optional)
	result:	$c \times r$	

$\_\mathtt{pinv}(A, \textit{tol})$:

    input:

	A:	$r \times c$	
	tol:	1×1	(optional)

    output:

	A:	$c \times r$	
	result:	1×1	(containing rank)

Diagnostics

The inverse returned by these functions is real if *A* is real and is complex if *A* is complex.

$\mathtt{pinv}(A, \textit{rank}, \textit{tol})$ and $\_\mathtt{pinv}(A, \textit{tol})$ return missing results if *A* contains missing values.

$\mathtt{pinv}()$ and $\_\mathtt{pinv}()$ also return missing values if the algorithm for computing the SVD, [M-5] **svd( )**, fails to converge. This is a near zero-probability event. Here *rank* also is returned as missing.

See [M-5] **svsolve( )** and [M-1] **tolerance** for information on the optional *tol* argument.

References

James, I. 2002. *Remarkable Mathematicians: From Euler to von Neumann*. Cambridge: Cambridge University Press.

Moore, E. H. 1920. On the reciprocal of the general algebraic matrix. *Bulletin of the American Mathematical Society* 26: 394–395.

Penrose, R. 1955. A generalized inverse for matrices. *Mathematical Proceedings of the Cambridge Philosophical Society* 51: 406–413.

Also see

[M-5] **invsym( )** — Symmetric real matrix inversion

[M-5] **cholinv( )** — Symmetric, positive-definite matrix inversion

[M-5] **luinv( )** — Square matrix inversion

[M-5] **qrinv( )** — Generalized inverse of matrix via QR decomposition

[M-5] **svd( )** — Singular value decomposition

[M-5] **fullsvd( )** — Full singular value decomposition

[M-4] **matrix** — Matrix functions

[M-4] **solvers** — Functions to solve AX=B and to obtain A inverse

---

Eliakim Hastings Moore (1862–1932) was born in Marietta, Ohio. He studied mathematics and astronomy at Yale and was awarded a Ph.D. for a thesis on n-dimensional geometry. After a year studying in Germany and teaching posts at Northwestern and Yale, he settled at the University of Chicago in 1892. Moore worked on algebra, including fields and groups, the foundations of geometry and the foundations of analysis, algebraic geometry, number theory, and integral equations. He was an inspiring teacher and a great organizer in American mathematics, playing an important part in the early years of the American Mathematical Society.

Roger Penrose (1931–) was born in Colchester in England. His father was a statistically minded medical geneticist and his mother was a doctor. Penrose studied mathematics at University College London and Cambridge and published an article on generalized matrix inverses in 1955. He taught and researched at several universities in Great Britain and the United States before being appointed Rouse Ball Professor of Mathematics at Oxford in 1973. Penrose is perhaps best known for papers ranging from cosmology and general relativity (including work with Stephen Hawking) to pure mathematics (including results on tilings of the plane) and for semipopular and wide-ranging books making controversial connections between physics, computers, mind, and consciousness. He was knighted in 1994.

Title

[M-5] **polyeval( )** — Manipulate and evaluate polynomials

Syntax

numeric vector	`polyeval(`*numeric rowvector c*, *numeric vector x*`)`
numeric rowvector	`polysolve(`*numeric vector y*, *numeric vector x*`)`
numeric rowvector	`polytrim(`*numeric vector c*`)`
numeric rowvector	`polyderiv(`*numeric rowvector c*, *real scalar i*`)`
numeric rowvector	`polyinteg(`*numeric rowvector c*, *real scalar i*`)`
numeric rowvector	`polyadd(`*numeric rowvector c_1*, *numeric rowvector c_2*`)`
numeric rowvector	`polymult(`*numeric rowvector c_1*, *numeric rowvector c_2*`)`
void	`polydiv(`*numeric rowvector c_1*, *numeric rowvector c_2*, c_q, c_r`)`
complex rowvector	`polyroots(numeric rowvector c)`

In the above, row vector c contains the coefficients for a `cols(`c`)` -1 degree polynomial. For instance,

$$c = (4, 2, 1)$$

records the polynomial

$$4 + 2x + x^2$$

Description

`polyeval(`c, x`)` evaluates polynomial c at each value recorded in x, returning the results in a p-conformable-with-x vector. For instance, `polyeval((4,2,1), (3\5))` returns `(4+2*3+3^2 \ 4+2*5+5^2) = (19\39)`.

`polysolve(`y, x`)` returns the minimal-degree polynomial c fitting $y = $ `polyeval(`c, x`)`. Solution is via Lagrange's interpolation formula.

`polytrim(`c`)` returns polynomial c with trailing zeros removed. For instance, `polytrim((1,2,3,0))` returns `(1,2,3)`. `polytrim((0,0,0,0))` returns `(0)`. Thus if $n = $ `cols(polytrim(`c`))`, then c records an $(n-1)$th degree polynomial.

`polyderiv(`c, i`)` returns the polynomial that is the ith derivative of polynomial c. For instance, `polyderiv((4,2,1), 1)` returns `(2,2)` (the derivative of $4 + 2x + x^2$ is $2 + 2x$). The value of the first derivative of polynomial c at x is `polyeval(polyderiv(`c`,1),`x`)`.

`polyinteg(`c, i`)` returns the polynomial that is the ith integral of polynomial c. For instance, `polyinteg((4,2,1), 1)` returns `(0,4,1,.3333)` (the integral of $4+2x+x^2$ is $0+4x+x^2+.3333x^3$). The value of the integral of polynomial c at x is `polyeval(polyinteg(`c`,1),` x`)`.

polyadd(c_1, c_2) returns the polynomial that is the sum of the polynomials c_1 and c_2. For instance, polyadd((2,1), (3,5,1)) is (5,6,1) (the sum of $2 + x$ and $3 + 5x + x^2$ is $5 + 6x + x^2$).

polymult(c_1, c_2) returns the polynomial that is the product of the polynomials c_1 and c_2. For instance, polymult((2,1), (3,5,1)) is (6,13,7,1) (the product of $2 + x$ and $3 + 5x + x^2$ is $6 + 13x + 7x^2 + x^3$).

polydiv(c_1, c_2, c_q, c_r) calculates polynomial c_1/c_2, storing the quotient polynomial in c_q and the remainder polynomial in c_r. For instance, polydiv((3,5,1), (2,1), c_q, c_r) returns c_q=(3,1) and c_r=(-3); i.e.,

$$\frac{3 + 5x + x^2}{2 + x} = 3 + x \text{ with a remainder of } -3$$

or

$$3 + 5x + x^2 = (3 + x)(2 + x) - 3$$

polyroots(c) find the roots of polynomial c and returns them in complex row vector (complex even if c is real). For instance, polyroots((3,5,1)) returns (-4.303+0i, -.697+0i) (the roots of $3 + 5x + x^2$ are -4.303 and $-.697$).

Remarks

Given the real or complex coefficients c that define an $n - 1$ degree polynomial in x, polyroots(c) returns the $n - 1$ roots for which

$$0 = c_1 + c_2 x^1 + c_3 x^2 + \cdots + c_n x^{n-1}$$

polyroots(c) obtains the roots by calculating the eigenvalues of the companion matrix. The $(n - 1) \times (n - 1)$ companion matrix for the polynomial defined by c is

$$C = \begin{bmatrix} -c_{n-1}s & -c_{n-2}s & \cdots & -c_2 s & -c_1 s \\ 1 & 0 & \cdots & 0 & 0 \\ 0 & 1 & \cdots & 0 & 0 \\ \vdots & \vdots & \ddots & \vdots & \vdots \\ 0 & 0 & \cdots & 1 & 0 \\ 0 & 0 & \cdots & 0 & 1 \end{bmatrix}$$

where $s = 1/c_n$ if c is real and

$$s = C\left(\frac{\text{Re}(c_n)}{\text{Re}(c_n)^2 + \text{Im}(c_n)^2}, \frac{-\text{Im}(c_n)}{\text{Re}(c_n)^2 + \text{Im}(c_n)^2} \right)$$

otherwise.

As in all nonsymmetric eigenvalue problems, the returned roots are complex and sorted from largest to smallest, see [M-5] **eigensystem( )**.

Conformability

polyeval(c, x):

c:	$1 \times n, n > 0$		
x:	$r \times 1$	or	$1 \times c$
result:	$r \times 1$	or	$1 \times c$

polysolve(y, x):

y:	$n \times 1$	or	$1 \times n, n \geq 1$
x:	$n \times 1$	or	$1 \times n$
result:	$1 \times k$,	$1 \leq k \leq n$	

polytrim(c):

c:	$1 \times n$
result:	$1 \times k$, $1 \leq k \leq n$

polyderiv(c, i):

c:	$1 \times n$, $n > 0$
i:	1×1, i may be negative
result:	$1 \times \max(1, n - i)$

polyinteg(c, i):

c:	$1 \times n, n > 0$
i:	1×1, i may be negative
result:	$1 \times \max(1, n + i)$

polyadd(c_1, c_2):

c_1:	$1 \times n_1$, $n_1 > 0$
c_2:	$1 \times n_2$, $n_2 > 0$
result:	$1 \times \max(n_1, \quad n_2)$

polymult(c_1, c_2):

c_1:	$1 \times n_1$, $n_1 > 0$
c_2:	$1 \times n_2$, $n_2 > 0$
result:	$1 \times n_1 + n_2 - 1$

polydiv(c_1, c_2, c_q, c_r):

input:		
c_1:	$1 \times n_1$, $n_1 > 0$	
c_2:	$1 \times n_2$, $n_2 > 0$	
output:		
c_q:	$1 \times k_1$,	$1 \leq k_1 \leq \max(n_1 - n_2 + 1, 1)$
c_r:	$1 \times k_2$,	$1 \leq k_2 \leq \max(n_1 - n_2, 1)$

polyroots(c):

c:	$1 \times n_i$, $n > 0$
result:	$1 \times k - 1$, $k = \text{cols}(\text{polytrim}(c))$

Diagnostics

All functions abort with an error if a polynomial coefficient row vector is void, but they do not necessarily give indicative error messages as to the problem. Polynomial coefficient row vectors may contain missing values.

polyderiv(c, i) returns c when $i = 0$. It returns polyinteg(c, $-i$) when $i < 0$. It returns (0) when i is missing (think of missing as positive infinity).

polyinteg(c, i) returns c when $i = 0$. It returns polyderiv(c, $-i$) when $i < 0$. It aborts with error if i is missing (think of missing as positive infinity).

polyroots(c) returns a vector of missing values if any element of c equals missing.

Also see

[M-4] **mathematical** — Important mathematical functions

Title

[M-5] printf( ) — Format output

Syntax

void   printf(*string scalar fmt*, *r1*, *r2*, ... , *rN*)

string scalar sprintf(*string scalar fmt*, *r1*, *r2*, ... , *rN*)

where *fmt* may contain a mix of text and %*fmts*, such as

```
printf("The result is %9.2f, adjusted\n", result)
printf("%s = %9.0g\n", name, value)
```

There must be a one-to-one correspondence between the %*fmts* in *fmt* and the number of results to be displayed.

Along with the usual %*fmts* that Stata provides (see [D] **format**), also provided are

format	meaning
%f	%11.0g, compressed
%g	%11.0g, compressed
%e	%11.8e, compressed
%s	%#s, # = whatever necessary

Compressed means that, after the indicated format is applied, all leading and trailing blanks are removed.

C programmers, be warned: %d is Stata's (old) calendar date format (equivalent to modern Stata's %td format) and not an integer format; use %f for formatting integers.

The following character sequences are given a special meaning when contained within *fmt*:

character sequence	meaning
%%	one %
\n	newline
\r	carriage return
\t	tab
\\	one \

Description

printf() displays output at the terminal.

687

`sprintf()` returns a string that can then be displayed at the terminal, written to a file, or used in any other way a string might be used.

Remarks

Remarks are presented under the following headings:

> *printf( )*
> *sprintf( )*

printf( )

`printf()` displays output at the terminal. A program might contain the line

```
printf("the result is %f\n", result)
```

and display the output

```
the result is 5.213
```

or it might contain the lines

```
printf("{txt}{space 13}{c |}      Coef.    Std. Err.\n")
printf("{hline 13}{c +}{hline 24}\n")
printf("{txt}%12s {c |} {res}%10.0g  %10.0g\n",
    varname[i], coef[i], se[i])
```

and so display the output

	Coef.	Std. Err.
mpg	−.0059541	.0005921

Do not forget to include \n at the end of lines. When \n is not included, the line continues. For instance, the code

```
printf("{txt}{space 13}{c |}      Coef.    Std. Err.\n")
printf("{hline 13}{c +}{hline 24}\n")
printf("{txt}%12s {c |} {res}", varname[i])
printf("%10.0g", coef[i])
printf(" ")
printf("%10.0g", se[i])
printf("\n")
```

produces the same output as shown above.

Although users are unaware of it, Stata buffers output. This makes Stata faster. A side effect of the buffering, however, is that output may not appear when you want it to appear. Consider the code fragment

```
for (n=1; !converged(b, b0); n++) {
    printf("iteration %f:  diff = %12.0g\n", n, b-b0)
    b0 = b
    ... new calculation of b ...
}
```

One of the purposes of the iteration output is to keep the user informed that the code is indeed working, yet as the above code is written, the user probably will not see the iteration messages as they occur. Instead, nothing will appear for a while, and then, unexpectedly, many iteration messages will appear as Stata, buffers full, decides to send to the terminal the waiting output.

To force output to be displayed, use [M-5] **displayflush( )**:

```
for (n=1; !converged(b, b0); n++) {
        printf("iteration %f:  diff = %12.0g\n", n, b-b0)
        displayflush()
        b0 = b
        ... new calculation of b ...
}
```

It is only in situations like the above that use of displayflush() is necessary. In other cases, it is better to let Stata decide when output buffers should be flushed. (Ado-file programmers: you have never had to worry about this because, at the ado-level, all output is flushed as it is created. Mata, however, is designed to be fast and so printf() does not force output to be flushed until it is efficient to do so.)

sprintf( )

The difference between sprintf() and printf() is that, whereas printf() sends the resulting string to the terminal, sprintf() returns it. Since Mata displays the results of expressions that are not assigned to variables, sprintf() used by itself also displays output:

```
: sprintf("the result is %f\n", result)
  the result is 5.2130a
```

The outcome is a little different from that produced by printf() because the output-the-unassigned-expression routine indents results by 2 and displays all the characters in the string (the 0a at the end is the \n newline character). Also, the output-the-unassigned-expression routine does not honor SMCL, electing instead to display the codes:

```
: sprintf("{txt}the result is {res}%f", result)
  {txt}the result is {res}5.213
```

The purpose of sprintf() is to create strings that will then be used with printf(), with [M-5] **display( )**, with fput() (see [M-5] **fopen( )**), or with some other function.

Pretend that we are creating a dynamically formatted table. One of the columns in the table contains integers, and we want to create a %fmt that is exactly the width required. That is, if the integers to appear in the table are 2, 9, and 20, we want to create a %2.0f format for the column. We assume the integers are in the column vector dof in what follows:

```
max = 0
for (i=1; i<=rows(dof); i++) {
        len = strlen(sprintf("%f", dof[i])
        if (len>max) max = len
}
fmt = sprintf("%%%f.0f", max)
```

We used `sprintf()` twice in the above. We first used `sprintf()` to produce the string representation of the integer `dof[i]`, and we used the `%f` format so that the length would be whatever was necessary, and no more. We obtained in `max` the maximum length. If `dof` contained 2, 9, and 20, by the end of our loop, `max` will contain 2. We finally used `sprintf()` to create the `%#.0f` format that we wanted: `%2.0f`.

The format string `%%%f.0f` in the final `sprintf()` is a little difficult to read. The first two percent signs amount to one real percent sign, so in the output we now have `%` and we are left with `%f.0f`. The `%f` is a format—it is how we are to format `max`—and so in the output we now have `%2`, and we are left with `.0f`. `.0f` is just a string, so the final output is `%2.0f`.

Conformability

$$\text{printf}(\mathit{fmt}, r_1, r_2, \ldots, r_N)$$

$$
\begin{array}{rl}
\mathit{fmt}: & 1 \times 1 \\
r_1: & 1 \times 1 \\
r_2: & 1 \times 1 \\
\ldots & \\
r_N: & 1 \times 1 \\
\mathit{result}: & \mathit{void}
\end{array}
$$

$$\text{sprintf}(\mathit{fmt}, r_1, r_2, \ldots, r_N)$$

$$
\begin{array}{rl}
\mathit{fmt}: & 1 \times 1 \\
r_1: & 1 \times 1 \\
r_2: & 1 \times 1 \\
\ldots & \\
r_N: & 1 \times 1 \\
\mathit{result}: & 1 \times 1
\end{array}
$$

Diagnostics

`printf()` and `sprintf()` abort with error if a `%`*fmt* is misspecified, if a numeric `%`*fmt* corresponds to a string result or a string `%`*fmt* to a numeric result, or there are too few or too many `%`*fmts* in *fmt* relative to the number of *results* specified.

Also see

[M-5] **displayas( )** — Set display level

[M-5] **displayflush( )** — Flush terminal-output buffer

[M-4] **io** — I/O functions

Title

[M-5] qrd( ) — QR decomposition

Syntax

void	qrd(*numeric matrix A*, *Q*, *R*)
void	hqrd(*numeric matrix A*, *H*, *tau*, R_1)
void	_hqrd(*numeric matrix A*, *tau*, R_1)
numeric matrix	hqrdmultq(*numeric matrix H*, *rowvector tau*, *numeric matrix X*, *real scalar transpose*)
numeric matrix	hqrdmultq1t(*numeric matrix H*, *rowvector tau*, *numeric matrix X*)
numeric matrix	hqrdq(*numeric matrix H*, *numeric matrix tau*)
numeric matrix	hqrdq1(*numeric matrix H*, *numeric matrix tau*)
numeric matrix	hqrdr(*numeric matrix H*)
numeric matrix	hqrdr1(*numeric matrix H*)
void	qrdp(*numeric matrix A*, *Q*, *R*, *real rowvector p*)
void	hqrdp(*numeric matrix A*, *H*, *tau*, R_1, *real rowvector p*)
void	_hqrdp(*numeric matrix A*, *tau*, R_1, *real rowvector p*)
void	_hqrdp_la(*numeric matrix A*, *tau*, *real rowvector p*)

Description

qrd(*A*, *Q*, *R*) calculates the QR decomposition of *A*: $m \times n$, $m \geq n$, returning results in *Q* and *R*.

hqrd(*A*, *H*, *tau*, R_1) calculates the QR decomposition of *A*: $m \times n$, $m \geq n$, but rather than returning *Q* and *R*, returns the Householder vectors in *H* and the scale factors *tau*—from which *Q* can be formed—and returns an upper-triangular matrix in R_1 that is a submatrix of *R*; see *Remarks* below for its definition. Doing this saves calculation and memory, and other routines allow you to manipulate these matrices:

1. hqrdmultq(*H*, *tau*, *X*, *transpose*) returns QX or $Q'X$ on the basis of the *Q* implied by *H* and *tau*. QX is returned if *transpose* $= 0$, and $Q'X$ is returned otherwise.

2. hqrdmultq1t(*H*, *tau*, *X*) returns $Q_1'X$ on the basis of the *Q*1 implied by *H* and *tau*.

3. hqrdq(*H*, *tau*) returns the *Q* matrix implied by *H* and *tau*. This function is rarely used.

4. hqrdq1(*H*, *tau*) returns the Q_1 matrix implied by *H* and *tau*. This function is rarely used.

5. hqrdr(H) returns the full R matrix. This function is rarely used. (It may surprise you that hqrdr( ) is a function of H and not R_1. R_1 also happens to be stored in H, and there is other useful information there, as well.)

6. hqrdr1(H) returns the R_1 matrix. This function is rarely used.

_hqrd(A, tau, R_1) does the same thing as hqrd(A, H, tau, R_1), except that it overwrites H into A and so conserves even more memory.

qrdp(A, Q, R, p) is similar to qrd(A, Q, R): it returns the QR decomposition of A in Q and R. The difference is that this routine allows for pivoting. New argument p specifies whether a column is available for pivoting and, on output, p is overwritten with a permutation vector that records the pivoting actually performed. On input, p can be specified as . (missing)—meaning all columns are available for pivoting—or p can be specified as an $n \times 1$ column vector containing 0s and 1s, with 1 meaning the column is fixed and so may not be pivoted.

hqrdp(A, H, tau, R_1, p) is a generalization of hqrd(A, H, tau, R_1) just as qrdp( ) is a generalization of qrd( ).

_hqrdp(A, tau, R_1, p) does the same thing as hqrdp(A, H, tau, R_1, p), except that _hqrdp( ) overwrites H into A.

_hqrdp_la( ) is the interface into the [M-1] **LAPACK** routine that performs the QR calculation; it is used by all the above routines. Direct use of _hqrdp_la( ) is not recommended.

Remarks

Remarks are presented under the following headings:

> *QR decomposition*
> *Avoiding calculation of Q*
> *Pivoting*
> *Least-squares solutions with dropped columns*

QR decomposition

The decomposition of square or nonsquare matrix A can be written

$$A = QR \tag{1}$$

where Q is an orthogonal matrix ($Q'Q = I$), and R is upper triangular. qrd(A, Q, R) will make this calculation:

```
: A
          1    2

     1    7    4
     2    9    6
     3    9    6
     4    7    2
     5    3    1

: Q = R = .

: qrd(A, Q, R)

: Ahat = Q*R

: mreldif(Ahat, A)
  3.55271e-16
```

Avoiding calculation of Q

In fact, you probably do not want to use qrd(). Calculating the necessary ingredients for Q is not too difficult, but going from those necessary ingredients to form Q is devilish. The necessary ingredients are usually all you need, which are the Householder vectors and their scale factors, known as H and *tau*. For instance, one can write down a mathematical function $f(H, tau, X)$ that will calculate QX or $Q'X$ for some matrix X.

Also, QR decomposition is often carried out on violently nonsquare matrices A: $m \times n$, $m \gg n$. We can write

$$\underset{m \times n}{A} = \begin{bmatrix} \underset{m \times n}{Q_1} & \underset{m \times m-n}{Q_2} \end{bmatrix} \begin{bmatrix} \underset{n \times n}{R_1} \\ \underset{m-n \times n}{R_2} \end{bmatrix} = \underset{m \times n}{Q_1 R_1} + \underset{m \times n}{Q_2 R_2}$$

R_2 is zero, and thus

$$\underset{m \times n}{A} = \begin{bmatrix} \underset{m \times n}{Q_1} & \underset{m \times m-n}{Q_2} \end{bmatrix} \begin{bmatrix} \underset{n \times n}{R_1} \\ \underset{m-n \times n}{0} \end{bmatrix} = \underset{m \times n}{Q_1 R_1}$$

Thus it is enough to know Q_1 and R_1. Rather than defining QR decomposition as

$$A = QR, \qquad Q : m \times m, \quad R : m \times n \tag{1}$$

it is better to define it as

$$A = Q_1 R_1 \qquad Q_1 : m \times n \quad R_1 : n \times n \tag{1'}$$

To appreciate the savings, consider the reasonable case where $m = 4{,}000$ and $n = 3$:

$$A = QR, \qquad Q : 4{,}000 \times 4{,}000, \quad R : 4{,}000 \times 3$$

versus,

$$A = Q_1 R_1 \qquad Q_1 : 4{,}000 \times 3 \qquad R_1 : 3 \times 3$$

Memory consumption is reduced from 125,094 kilobytes to 94 kilobytes, a 99.92% saving!

Combining the arguments, we need not save Q because Q_1 is sufficient, we need not calculate Q_1 because H and *tau* are sufficient, and we need not store R because R_1 is sufficient.

That is what hqrd(A, H, *tau*, R_1) does. Having used hqrd(), if you need to multiply the full Q by some matrix X, you can use hqrdmultq(). Having used hqrd(), if you need the full Q, you can use hqrdq() to obtain it, but by that point you will be making the devilish calculation you sought to avoid and so you might as well have used qrd() to begin with. If you want Q_1, you can use hqrdq1(). Finally, having used hqrd(), if you need R or R_1, you can use hqrdr() and hqrdr1():

```
: A
        1    2

  1     7    4
  2     9    6
  3     9    6
  4     7    2
  5     3    1

: H = tau = R1 = .
: hqrd(A, H, tau, R1)
: Ahat = hqrdq1(H, tau) * R1                    // i.e.,  Q1*R1
: mreldif(Ahat, A)
  3.55271e-16
```

Pivoting

The QR decomposition with column pivoting solves

$$AP = QR \tag{2}$$

or, if you prefer,

$$AP = Q_1 R_1 \tag{2'}$$

where P is a permutation matrix; see [M-1] **permutation**. We can rewrite this as

$$A = QRP' \tag{3}$$

and

$$A = Q_1 R_1 P' \tag{3'}$$

Column pivoting can improve the numerical accuracy. The functions qrdp(A, Q, R, p) and hqrdp(A, H, *tau*, R_1, p) perform pivoting and return the permutation matrix P in permutation vector form:

```
: A
        1    2

  1     7    4
  2     9    6
  3     9    6
  4     7    2
  5     3    1

: Q = R = p = .
: qrdp(A, Q, R, p)
: Ahat = (Q*R)[., invorder(p)]                  // i.e., QRP'
: mreldif(Ahat, A)
  3.55271e-16
: H = tau = R1 = p = .
: hqrdp(A, H, tau, R1, p)
: Ahat = (hqrdq1(H, tau)*R1)[., invorder(p)]    // i.e., Q1*R1*P'
: mreldif(Ahat, A)
  3.55271e-16
```

Before calling qrdp() or hqrdp(), we set p equal to missing, specifying that all columns could be pivoted. We could just as well have set p equal to (0, 0), which would have stated that both columns were eligible for pivoting.

When pivoting is disallowed, and when A is not of full column rank, the order in which columns appear affects the kind of generalized solution produced; later columns are, in effect, dropped. When pivoting is allowed, the columns are reordered based on numerical accuracy considerations. In the rank-deficient case, you no longer know ahead of time which columns will be dropped, because you do not know in what order the columns will appear. Generally, you do not care, but there are occasions when you do.

In such cases, you can specify which columns are eligible for pivoting and which are not—you specify p as a vector and if p_i==1, the ith column may not be pivoted. The p_i==1 columns are (conceptually) moved to appear first in the matrix, and the remaining columns are ordered optimally after that. The permutation vector that is returned in p accounts for all of this.

Least-squares solutions with dropped columns

Least-square solutions are one popular use of QR decomposition. We wish to solve for x

$$Ax = b \qquad (A : m \times n, \quad m \geq n) \qquad (4)$$

The problem is that there is no solution to (4) when $m > n$ because we have more equations than unknowns. Then we want to find x such that $(Ax - b)'(Ax - b)$ is minimized.

If A is of full column rank then it is well known that the least-squares solution for x is given by solveupper(R_1, $Q_1'b$) where solveupper() is an upper-triangular solver; see [M-5] **solvelower( )**.

If A is of less than full column rank and we do not care which columns are dropped, then we can use the same solution: solveupper(R_1, $Q_1'b$).

Adding pivoting to the above hardly complicates the issue; the solution becomes solveupper(R_1, $Q_1'b$)[invorder(p)].

For both cases, the full details are

```
        : A
             1    2    3

        1    3    9    1
        2    3    8    1
        3    3    7    1
        4    3    6    1

        : b
             1

        1    7
        2    3
        3    12
        4    0

        : H = tau = R1 = p = .
        : hqrdp(A, H, tau, R1, p)
        : q1b = hqrdmultq1t(H, tau, b)            // i.e., Q1'b
        : xhat = solveupper(R1, q1b)[invorder(p)]
```

```
: xhat
                  1

    1 | -1.166666667
    2 |          1.2
    3 |            0
```

The *A* matrix in the above example has less than full column rank; the first column contains a variable with no variation and the third column contains the data for the intercept. The solution above is correct, but we might prefer a solution that included the intercept. To do that, we need to specify that the third column cannot be pivoted:

```
: p = (0, 0, 1)
: H = tau = R1 = .
: hqrdp(A, H, tau, R1, p)
: q1b = hqrdmultq1t(H, tau, b)
: xhat = solveupper(R1, q1b)[invorder(p)]
: xhat
            1

    1 |    0
    2 |  1.2
    3 | -3.5
```

Conformability

qrd(*A*, *Q*, *R*):

   input:

           A: $m \times n$, $m \geq n$

   output:

           Q: $m \times m$

           R: $m \times n$

hqrd(*A*, *H*, *tau*, R_1):

   input:

           A: $m \times n$, $m \geq n$

   output:

           H: $m \times n$

        tau: $1 \times n$

         R_1: $n \times n$

_hqrd(*A*, *tau*, R_1):

   input:

           A: $m \times n$, $m \geq n$

   output:

           A: $m \times n$ (contains *H*)

        tau: $1 \times n$

         R_1: $n \times n$

hqrdmultq(H, *tau*, X, *transpose*):

H:	$m \times n$
tau:	$1 \times n$
X:	$m \times c$
transpose:	1×1
result:	$m \times c$

hqrdmultq1t(H, *tau*, X):

H:	$m \times n$
tau:	$1 \times n$
X:	$m \times c$
result:	$n \times c$

hqrdq(H, *tau*):

H:	$m \times n$
tau:	$1 \times n$
result:	$m \times m$

hqrdq1(H, *tau*):

H:	$m \times n$
tau:	$1 \times n$
result:	$m \times n$

hqrdr(H):

H:	$m \times n$
result:	$m \times n$

hqrdr1(H):

H:	$m \times n$
result:	$n \times n$

qrdp(A, Q, R, p):

input:

A:	$m \times n, \quad m \geq n$
p:	$1 \times 1 \quad$ or $\quad 1 \times n$

output:

Q:	$m \times m$
R:	$m \times n$
p:	$1 \times n$

hqrdp(A, H, *tau*, R_1, p):

input:

A:	$m \times n, \quad m \geq n$
p:	$1 \times 1 \quad$ or $\quad 1 \times n$

output:

H:	$m \times n$
tau:	$1 \times n$
R_1:	$n \times n$
p:	$1 \times n$

_hqrdp(A, *tau*, R_1, p):

   input:

         A: $m \times n$, $m \geq n$

         p: 1×1 or $1 \times n$

   output:

         A: $m \times n$ (contains H)

         tau: $1 \times n$

         R_1: $n \times n$

         p: $1 \times n$

_hqrdp_la(A, *tau*, p):

   input:

         A: $m \times n$, $m \geq n$

         p: 1×1 or $1 \times n$

   output:

         A: $m \times n$ (contains H)

         tau: $1 \times n$

         p: $1 \times n$

Diagnostics

qrd(A, ...), hqrd(A, ...), _hqrd(A, ...), qrdp(A, ...), hqrdp(A, ...), and _hqrdp(A, ...) return missing results if A contains missing values. That is, Q will contain all missing values. R will contain missing values on and above the diagonal. p will contain the integers 1, 2, ...

_hqrd(A, ...) and _hqrdp(A, ...) abort with error if A is a view.

hqrdmultq(H, *tau*, X, *transpose*) and hqrdmultq1t(H, *tau*, X) return missing results if X contains missing values.

> Alston Scott Householder (1904–1993) was born in Rockford, Illinois, and grew up in Alabama. He studied philosophy at Northwestern and Cornell, and then mathematics, earning a doctorate in the calculus of variations from the University of Chicago. Householder worked on mathematical biology for several years at Chicago, but in 1946 he joined the Mathematics Division of Oak Ridge National Laboratory. There he moved into numerical analysis, specializing in linear equations and eigensystems and helping to unify the field through reviews and symposia. His last post was at the University of Tennessee.

Also see

[M-5] **qrsolve( )** — Solve AX=B for X using QR decomposition

[M-5] **qrinv( )** — Generalized inverse of matrix via QR decomposition

[M-4] **matrix** — Matrix functions

Title

[M-5] **qrinv( )** — Generalized inverse of matrix via QR decomposition

Syntax

numeric matrix	qrinv(numeric matrix A)
numeric matrix	qrinv(numeric matrix A, *rank*)
numeric matrix	qrinv(numeric matrix A, *rank*, *real scalar tol*)
real scalar	_qrinv(numeric matrix A)
real scalar	_qrinv(numeric matrix A, *real scalar tol*)

where the type of *rank* is irrelevant; the rank of A is returned there.

Description

qrinv(A, ...) returns the inverse or generalized inverse of real or complex matrix A: $m \times n$, $m \geq n$. If optional argument *rank* is specified, the rank of A is returned there.

_qrinv(A, ...) does the same thing except that, rather than returning the result, it overwrites the original matrix A with the result. _qrinv() returns the rank of A.

In both cases, optional argument *tol* specifies the tolerance for determining singularity; see *Remarks* below.

Remarks

qrinv() and _qrinv() are most often used on square and possibly rank-deficient matrices but may be used on nonsquare matrices that have more rows than columns. Also see [M-5] **pinv( )** for an alternative. See [M-5] **luinv( )** for a more efficient way to obtain the inverse of full-rank, square matrices, and see [M-5] **invsym( )** for inversion of real, symmetric matrices.

When A is of full rank, the inverse calculated by qrinv() is essentially the same as that computed by the faster luinv(). When A is singular, qrinv() and _qrinv() compute a generalized inverse, $A*$, which satisfies

$$A(A*)A = A$$
$$(A*)A(A*) = A*$$

This generalized inverse is also calculated for nonsquare matrices that have more rows than columns and, then returned is a least-squares solution. If A is $m \times n$, $m \geq n$, and if the rank of A is equal to n, then $(A*)A = I$, ignoring roundoff error.

qrinv(A) is implemented as qrsolve(A, I(rows(A))); see [M-5] **qrsolve( )** for details and for use of the optional *tol* argument.

Conformability

qrinv(*A*, *rank*, *tol*):

> *input*:

A:	$m \times n$, $m \geq n$	
tol:	1×1	(optional)

> *output*:

rank:	1×1	(optional)
result:	$n \times m$	

_qrinv(*A*, *tol*):

> *input*:

A:	$m \times n$, $m \geq n$	
tol:	1×1	(optional)

> *output*:

A:	$n \times m$	
result:	1×1	(containing rank)

Diagnostics

The inverse returned by these functions is real if *A* is real and is complex if *A* is complex.

qrinv(*A*, ...) and _qrinv(*A*, ...) return a result containing missing values if *A* contains missing values.

_qrinv(*A*, ...) aborts with error if *A* is a view.

See [M-5] **qrsolve( )** and [M-1] **tolerance** for information on the optional *tol* argument.

Also see

[M-5] **invsym( )** — Symmetric real matrix inversion

[M-5] **cholinv( )** — Symmetric, positive-definite matrix inversion

[M-5] **luinv( )** — Square matrix inversion

[M-5] **pinv( )** — Moore–Penrose pseudoinverse

[M-5] **qrsolve( )** — Solve AX=B for X using QR decomposition

[M-5] **solve_tol( )** — Tolerance used by solvers and inverters

[M-4] **matrix** — Matrix functions

[M-4] **solvers** — Functions to solve AX=B and to obtain A inverse

Title

[M-5] **qrsolve( )** — Solve AX=B for X using QR decomposition

Syntax

numeric matrix   qrsolve(*A*, *B*)

numeric matrix   qrsolve(*A*, *B*, *rank*)

numeric matrix   qrsolve(*A*, *B*, *rank*, *tol*)

real scalar   _qrsolve(*A*, *B*)

real scalar   _qrsolve(*A*, *B*, *tol*)

where

A:	*numeric matrix*
B:	*numeric matrix*
rank:	irrelevant; *real scalar* returned
tol:	*real scalar*

Description

qrsolve(*A*, *B*, ...) uses QR decomposition to solve $AX = B$ and returns X. When *A* is singular or nonsquare, qrsolve() computes a least-squares generalized solution. When *rank* is specified, in it is placed the rank of *A*.

_qrsolve(*A*, *B*, ...), does the same thing, except that it destroys the contents of *A* and it overwrites *B* with the solution. Returned is the rank of *A*.

In both cases, *tol* specifies the tolerance for determining whether *A* is of full rank. *tol* is interpreted in the standard way—as a multiplier for the default if *tol* > 0 is specified and as an absolute quantity to use in place of the default if *tol* ≤ 0 is specified; see [M-1] **tolerance**.

Remarks

qrsolve(*A*, *B*, ...) is suitable for use with square and possibly rank-deficient matrix *A*, or when *A* has more rows than columns. When *A* is square and full rank, qrsolve() returns the same solution as lusolve() (see [M-5] **lusolve( )**), up to roundoff error. When *A* is singular, qrsolve() returns a generalized (least-squares) solution.

Remarks are presented under the following headings:

> Derivation
> Relationship to inversion
> Tolerance

701

Derivation

We wish to solve for X

$$AX = B \qquad (1)$$

Perform QR decomposition on A so that we have $A = QRP'$. Then (1) can be rewritten

$$QRP'X = B$$

Premultiplying by Q' and remembering that $Q'Q = QQ' = I$, we have

$$RP'X = Q'B \qquad (2)$$

Define

$$Z = P'X \qquad (3)$$

Then (2) can be rewritten

$$RZ = Q'B \qquad (4)$$

It is easy to solve (4) for Z because R is upper triangular. Having Z, we can obtain X via (3), because $Z = P'X$, premultiplied by P (and if we remember that $PP' = I$), yields

$$X = PZ$$

For more information on QR decomposition, see [M-5] **qrd( )**.

Relationship to inversion

For a general discussion, see *Relationship to inversion* in [M-5] **lusolve( )**.

For an inverse based on QR decomposition, see [M-5] **qrinv( )**. $qrinv(A)$ amounts to $qrsolve(A, I(rows(A)))$, although it is not actually implemented that way.

Tolerance

The default tolerance used is

$$eta = \texttt{1e-13} * \texttt{trace(abs}(R))/\texttt{rows}(R)$$

where R is the upper-triangular matrix of the QR decomposition; see *Derivation* above. When A is less than full rank, by, say, d degrees of freedom, then R is also rank deficient by d degrees of freedom and the bottom d rows of R are essentially zero. If the ith diagonal element of R is less than or equal to *eta*, then the ith row of Z is set to zero. Thus if the matrix is singular, $qrsolve()$ provides a generalized solution.

If you specify *tol* > 0, the value you specify is used to multiply *eta*. You may instead specify *tol* ≤ 0, and then the negative of the value you specify is used in place of *eta*; see [M-1] **tolerance**.

Conformability

qrsolve(*A*, *B*, *rank*, *tol*):

input:

A:	$m \times n$,	$m \geq n$
B:	$m \times k$	
tol:	1×1	(optional)

output:

rank:	1×1	(optional)
result:	$n \times k$	

_qrsolve(*A*, *B*, *tol*):

input:

A:	$m \times n$,	$m \geq n$
B:	$m \times k$	
tol:	1×1	(optional)

output:

A:	0×0
B:	$n \times k$
result:	1×1

Diagnostics

qrsolve(*A*, *B*, ...) and _qrsolve(*A*, *B*, ...) return a result containing missing if *A* or *B* contain missing values.

_qrsolve(*A*, *B*, ...) aborts with error if *A* or *B* are views.

Also see

[M-5] **qrinv( )** — Generalized inverse of matrix via QR decomposition

[M-5] **qrd( )** — QR decomposition

[M-5] **solvelower( )** — Solve AX=B for X, A triangular

[M-5] **cholsolve( )** — Solve AX=B for X using Cholesky decomposition

[M-5] **lusolve( )** — Solve AX=B for X using LU decomposition

[M-5] **svsolve( )** — Solve AX=B for X using singular value decomposition

[M-5] **solve_tol( )** — Tolerance used by solvers and inverters

[M-4] **matrix** — Matrix functions

[M-4] **solvers** — Functions to solve AX=B and to obtain A inverse

Title

Syntax

real matrix   quadcross(X, Z)

real matrix   quadcross(X, w, Z)

real matrix   quadcross(X, xc, Z, zc)

real matrix   quadcross(X, xc, w, Z, zc)

real matrix   quadcrossdev(X, x, Z, z)

real matrix   quadcrossdev(X, x, w, Z, z)

real matrix   quadcrossdev(X, xc, x, Z, zc, z)

real matrix   quadcrossdev(X, xc, x, w, Z, zc, z)

where

X:	*real matrix X*
xc:	*real scalar xc*
x:	*real rowvector x*
w:	*real vector w*
Z:	*real matrix Z*
zc:	*real scalar zc*
z:	*real rowvector z*

Description

quadcross( ) makes calculations of the form

$X'X$

$X'Z$

$X'\mathrm{diag}(w)X$

$X'\mathrm{diag}(w)Z$

This function mirrors cross( ) (see [M-5] **cross( )**), the difference being that sums are formed in quad precision rather than in double precision, so quadcross( ) is more accurate.

quadcrossdev() makes calculations of the form

$$(X: -x)'(X: -x)$$

$$(X: -x)'(Z: -z)$$

$$(X: -x)'\text{diag}(w)(X: -x)$$

$$(X: -x)'\text{diag}(w)(Z: -z)$$

This function mirrors crossdev() (see [M-5] **crossdev( )**), the difference being that sums are formed in quad precision rather than in double precision, so quadcrossdev() is more accurate.

Remarks

The returned result is double precision, but the sum calculations made in creating that double-precision result were made in quad precision.

Conformability

quadcross() has the same conformability requirements as cross(); see [M-5] **cross( )**.

quadcrossdev() has the same conformability requirements as crossdev(); see [M-5] **crossdev( )**.

Diagnostics

See *Diagnostics* under [M-5] **cross( )** and [M-5] **crossdev( )**.

Also see

[M-5] **cross( )** — Cross products

[M-5] **crossdev( )** — Deviation cross products

[M-4] **statistical** — Statistical functions

[M-4] **utility** — Matrix utility functions

Title

Syntax

> *numeric colvector* range(*a*, *b*, *numeric scalar delta*)

> *numeric colvector* rangen(*a*, *b*, *real scalar n*)

where *a* and *b* are numeric scalars.

Description

range(*a*, *b*, *delta*) returns a column vector going from *a* to *b* in steps of abs(*delta*) ($b \geq a$) or −abs(*delta*) ($b < a$).

rangen(*a*, *b*, *n*) returns a round(*n*) × 1 column vector going from *a* to *b* in round(*n*)-1 steps. *a* may be less than, equal to, or greater than *b*.

Remarks

range(0, 1, .25) returns (0 \ .25 \ .5 \ .75 \ 1). The sign of the third argument does not matter; range(0, 1, -.25) returns the same thing. range(1, 0, .25) and range(1, 0, -.25) return (1 \ .75 \ .5 \ .25 \ 0).

rangen(0, .5, 6) returns (0 \ .1 \ .2 \ .3 \ .4 \ .5). rangen(.5, 0, 6) returns (.5 \ .4 \ .3 \ .2 \ .1 \ 0).

range( ) and rangen( ) may be used with complex arguments. range(1, 1i, .4) returns (1 \ .75+.25i \ .5+.5i \ .25+.75i \ 1i). rangen(1, 1i, 5) returns the same thing. For range( ), only the distance of *delta* from zero matters, so range(1, 1i, .4i) would produce the same result, as would range(1, 1i, .25+.312i).

Conformability

range(*a*, *b*, *delta*):
>| | |
>|---|---|
>| *a*: | 1 × 1 |
>| *b*: | 1 × 1 |
>| *delta*: | 1 × 1 |
>| *result*: | 1 × 1, if $a = b$ |
>| | max(1+abs(*b−a*)/abs(*delta*),2) × 1, otherwise |

rangen(*a*, *b*, *n*):
>| | |
>|---|---|
>| *a*: | 1 × 1 |
>| *b*: | 1 × 1 |
>| *n*: | n × 1 |

Diagnostics

range(*a*, *b*, *delta*) aborts with error if *a*, *b*, or *delta* contains missing, if abs(*b*−*a*)/abs(*delta*) results in overflow, or if 1+abs(*b*−*a*)/abs(*delta*) results in a vector that is too big given the amount of memory available.

range(*a*, *b*, *delta*) returns a 1×1 result when *a* = *b*. In all other cases, the result is 2×1 or longer.

rangen(*a*, *b*, *n*) aborts with error if round(*n*) is less than 0 or missing.

Also see

[M-4] **standard** — Functions to create standard matrices

Title

[M-5] rank( ) — Rank of matrix

Syntax

> *real scalar*   rank(*numeric matrix A*)
>
> *real scalar*   rank(*numeric matrix A*, *real scalar tol*)

Description

rank(*A*) and rank(*A*, *tol*) return the rank of *A*: $m \times n$.

Remarks

The row rank of a matrix *A*: $m \times n$ is the number of rows of *A* that are linearly independent. The column rank is the number of columns that are linearly independent. The terms row rank and column rank, however, are used merely for emphasis. The ranks are equal, and the result is simply called the rank of *A*.

rank( ) calculates the rank by counting the number of nonzero singular values of the SVD of *A*, where nonzero is interpreted relative to a tolerance. rank( ) uses the same tolerance as pinv( ) (see [M-5] **pinv( )**) and as svsolve( ) (see [M-5] **svsolve( )**), and optional argument *tol* is specified in the same way as with those functions.

Thus if you were going to use rank( ) before calculating an inverse using pinv( ), it would be better to skip rank( ) altogether and proceed to the pinv( ) step, because pinv( ) will return the rank, calculated as a by-product of calculating the inverse. Using rank( ) ahead of time, the SVD would be calculated twice.

rank( ) in general duplicates calculations; and, worse, if you are not planning on using pinv( ) or svsolve( ) but rather are planning on using some other function, the rank returned by rank( ) may disagree with the implied rank of whatever numerical method you subsequently use because each numerical method has its own precision and tolerances.

All that said, rank( ) is useful in interactive and pedagogical situations.

Conformability

rank(*A*, *tol*):

A:	$m \times n$	
tol:	1×1	(optional)
result:	1×1	

Diagnostics

rank(*A*) returns missing if *A* contains missing values.

Also see

[M-5] **svd( )** — Singular value decomposition

[M-5] **fullsvd( )** — Full singular value decomposition

[M-5] **pinv( )** — Moore–Penrose pseudoinverse

[M-4] **matrix** — Matrix functions

Title

[M-5] Re( ) — Extract real or imaginary part

Syntax

>   *real matrix*   Re(*numeric matrix Z*)

>   *real matrix*   Im(*numeric matrix Z*)

Description

Re(Z) returns a real matrix containing the real part of Z. Z may be real or complex.

Im(Z) returns a real matrix containing the imaginary part of Z. Z may be a real or complex. If Z is real, Im(Z) returns a matrix of zeros.

Conformability

Re(Z), Im(Z):
>   Z: $r \times c$
>   *result*: $r \times c$

Diagnostics

Re(Z), if Z is real, literally returns Z and not a copy of Z. This makes execution of Re() applied to real arguments instant.

Also see

[M-5] **C( )** — Make complex

[M-4] **scalar** — Scalar mathematical functions

[M-4] **utility** — Matrix utility functions

Title

[M-5] reldif( ) — Relative/absolute difference

Syntax

real matrix   reldif (*numeric matrix X , numeric matrix Y*)

real scalar   mreldif (*numeric matrix X , numeric matrix Y*)

real scalar   mreldifsym (*numeric matrix X*)

real scalar   mreldifre (*numeric matrix X*)

Description

reldif (*X, Y*) returns the relative difference defined by

$$r = \frac{|X - Y|}{|Y| + 1}$$

calculated element by element.

mreldif (*X, Y*) returns the maximum relative difference and is equivalent to max (reldif (*X, Y*)).

mreldifsym (*X*) is equivalent to mreldif (*X'*, *X*) and so is a measure of how far the matrix is from being symmetric (Hermitian).

mreldifre (*X*) is equivalent to mreldif (Re(*X*), *X*) and so is a measure of how far the matrix is from being real.

Conformability

reldif (*X, Y*):
X:	$r \times c$
Y:	$r \times c$
result:	$r \times c$

mreldif (*X, Y*):
X:	$r \times c$
Y:	$r \times c$
result:	1×1

mreldifsym (*X*):
X:	$n \times n$
result:	1×1

mreldifre (*X*):
X:	$r \times c$
result:	1×1

Diagnostics

The relative difference function treats equal missing values as having a difference of 0 and different missing values as having a difference of missing (.):

$$\text{reldif}(.,\ .) == \text{reldif}(.a,\ .a) == \cdots == \text{reldif}(.z,\ .z) == 0$$

$$\text{reldif}(.,\ .a) == \text{reldif}(.,\ .z) == \cdots == \text{reldif}(.y,\ .z) == .$$

Also see

[M-4] **utility** — Matrix utility functions

Title

[M-5] **rows( )** — Number of rows and number of columns

Syntax

real scalar   rows (*transmorphic matrix P*)

real scalar   cols (*transmorphic matrix P*)

real scalar   length (*transmorphic matrix P*)

Description

rows(*P*) returns the number of rows of *P*.

cols(*P*) returns the number of columns of *P*.

length(*P*) returns rows(*P*)*cols(*P*).

Remarks

length(*P*) is typically used with vectors, as in

```
for (i=1; i<=length(x); i++) {
        ... x[i] ...
}
```

Conformability

rows(*P*), cols(*P*), length(*P*):

$$
\begin{array}{rl}
P\text{:} & r \times c \\
result\text{:} & 1 \times 1
\end{array}
$$

Diagnostics

rows(*P*), cols(*P*), and length(*P*) return a result that is greater than or equal to zero.

Also see

[M-4] **utility** — Matrix utility functions

Title

[M-5] **rowshape( )** — Reshape matrix

Syntax

transmorphic matrix   rowshape(*transmorphic matrix T*, *real scalar r*)

transmorphic matrix   colshape(*transmorphic matrix T*, *real scalar c*)

Description

rowshape(T, r) returns T transformed into a matrix with trunc(r) rows.

colshape(T, c) returns T having trunc(c) columns.

In both cases, elements are assigned sequentially with the column index varying more rapidly. See [M-5] **vec( )** for a function that varies the row index more rapidly.

Remarks

Remarks are presented under the following headings:

> *Example of rowshape( )*
> *Example of colshape( )*

Example of rowshape( )

```
: A
        1     2     3     4

  1    11    12    13    14
  2    21    22    23    24
  3    31    32    33    34
  4    41    42    43    44

: rowshape(A,2)
        1     2     3     4     5     6     7     8

  1    11    12    13    14    21    22    23    24
  2    31    32    33    34    41    42    43    44
```

Example of colshape( )

```
: colshape(A, 2)
        1     2

  1    11    12
  2    13    14
  3    21    22
  4    23    24
  5    31    32
  6    33    34
  7    41    42
  8    43    44
```

Conformability

$\texttt{rowshape}(T, r)$:

$\qquad\qquad T$: $r_0 \times c_0$
$\qquad\qquad r$: 1×1
$\qquad result$: $r \times r_0 c_0 / r$

$\texttt{colshape}(T, c)$:

$\qquad\qquad T$: $r_0 \times c_0$
$\qquad\qquad c$: 1×1
$\qquad result$: $r_0 c_0 / c \times c$

Diagnostics

Let r_0 and c_0 be the number of rows and columns of T.

$\texttt{rowshape}()$ aborts with error if $r_0 \times c_0$ is not evenly divisible by $\texttt{trunc}(r)$.

$\texttt{colshape}()$ aborts with error if $r_0 \times c_0$ is not evenly divisible by $\texttt{trunc}(c)$.

Also see

[M-4] **manipulation** — Matrix manipulation

Title

[M-5] runiform( ) — Uniform and nonuniform pseudorandom variates

Syntax

real matrix   runiform(*real scalar r, real scalar c*)

string scalar rseed( )

void   rseed(*string scalar newseed*)

void   rseed(*real scalar newseed*)

real matrix   rbeta(*real scalar r, real scalar c, real matrix a, real matrix b*)

real matrix   rbinomial(*real scalar r, real scalar c, real matrix n, real matrix p*)

real matrix   rchi2(*real scalar r, real scalar c, real matrix df*)

real matrix   rdiscrete(*real scalar r, real scalar c, real colvector p*)

real matrix   rgamma(*real scalar r, real scalar c, real matrix a, real matrix b*)

real matrix   rhypergeometric(*real scalar r, real scalar c, real matrix N,*
                              real matrix K, real matrix n)

real matrix   rnbinomial(*real scalar r, real scalar c, real matrix n, real matrix p*)

real matrix   rnormal(*real scalar r, real scalar c, real matrix m, real matrix s*)

real matrix   rpoisson(*real scalar r, real scalar c, real matrix m*)

real matrix   rt(*real scalar r, real scalar c, real matrix df*)

Description

runiform(r, c) returns an $r \times c$ real matrix containing uniformly distributed random variates on [0,1). runiform( ) is the same function as Stata's runiform( ) function.

rseed( ) returns the current random-variate seed in an encrypted string form. rseed( ) returns the same thing as Stata's c(seed); see [R] **set seed** and [P] **creturn**.

rseed(*newseed*) sets the seed: a string previously obtained from rseed( ) can be specified for the argument or an integer number can be specified. rseed( ) has the same effect as Stata's set seed command; see [R] **set seed**.

rbeta(r, c, a, b) returns an $ir \times jc$ real matrix containing beta random variates. The real-valued matrices a and b contain the beta shape parameters. The matrices a and b must be r-conformable, where $i = \max(\text{rows}(a), \text{rows}(b))$ and $j = \max(\text{cols}(a), \text{cols}(b))$.

rbinomial(r, c, n, p) returns an $ir \times jc$ real matrix containing binomial random variates. The real-valued matrices n and p contain the number of trials and the probability parameters. The matrices n and p must be r-conformable, where $i = \max(\text{rows}(n), \text{rows}(p))$ and $j = \max(\text{cols}(n), \text{cols}(p))$.

rchi2(r, c, df) returns an $ir \times jc$ real matrix containing chi-squared random variates. The real-valued matrix df contains the degrees of freedom parameters, where $i = \text{rows}(df)$ and $j = \text{cols}(df)$.

rdiscrete(r, c, p) returns an $r \times c$ real matrix containing random variates from the discrete distribution specified by the probabilities in the vector p of length k. The range of the discrete variates is 1, 2, ..., k. The alias method of Walker (1977) is used to sample from the discrete distribution.

rgamma(r, c, a, b) returns an $ir \times jc$ real matrix containing gamma random variates. The real-valued matrices a and b contain the gamma shape and scale parameters, respectively. The matrices a and b must be r-conformable, where $i = \max(\text{rows}(a), \text{rows}(b))$ and $j = \max(\text{cols}(a), \text{cols}(b))$.

rhypergeometric(r, c, N, K, n) returns an $ir \times jc$ real matrix containing hypergeometric random variates. The integer-valued matrix N contains the population sizes, the integer-valued matrix K contains the number of elements in each population that have the attribute of interest, and the integer-valued matrix n contains the sample size. The matrices N, K, and n must be r-conformable, where $i = \max(\text{rows}(N), \text{rows}(K), \text{rows}(n))$ and $j = \max(\text{cols}(N), \text{cols}(K), \text{cols}(n))$.

rnbinomial(r, c, n, p) returns an $ir \times jc$ real matrix containing negative binomial random variates. When the elements of the matrix n are integer-valued, rnbinomial() returns the number of failures before the nth success, where the probability of success on a single draw is contained in the real-valued matrix p. The elements of n can also be nonintegral but must be positive. The matrices n and p must be r-conformable, where $i = \max(\text{rows}(n), \text{rows}(p))$ and $j = \max(\text{cols}(n), \text{cols}(p))$.

rnormal(r, c, m, s) returns an $ir \times jc$ real matrix containing normal (Gaussian) random variates. The real-valued matrices m and s contain the mean and standard deviation parameters, respectively. The matrices m and s must be r-conformable, where $i = \max(\text{rows}(m), \text{rows}(s))$ and $j = \max(\text{cols}(m), \text{cols}(s))$.

rpoisson(r, c, m) returns an $ir \times jc$ real matrix containing Poisson random variates. The real-valued matrix m contains the Poisson mean parameters, where $i = \text{rows}(m)$ and $j = \text{cols}(m)$.

rt(r, c, df) returns an $ir \times jc$ real matrix containing Student's t random variates. The real-valued matrix df contains the degrees-of-freedom parameters, where $i = \text{rows}(df)$ and $j = \text{cols}(df)$.

Remarks

The functions described here generate random variates. The parameter limits for each generator are the same as those documented for Stata's random-number functions, except for rdiscrete(), which has no Stata equivalent.

In the example below, we generate and summarize 1,000 random normal deviates with a mean of 3 and standard deviation of 1.

```
: rseed(13579)
: x = rnormal(1000, 1, 3, 1)
: meanvariance(x)
                 1

    1 │ 3.002713574
    2 │ .9843730019
```

The next example uses a 1×3 vector of gamma shape parameters to generate a 1000×3 matrix of gamma random variates, X.

```
: a = (0.5,1.5,2.5)
: rseed(13579)
: X = rgamma(1000,1,a,1)
: mean(X)
              1              2              3

    1 │ .5339343609    1.510028772    2.451447187

: diagonal(variance(X))'
              1              2              3

    1 │ .6129729256    1.669457192    2.284915684
```

The first column of X contains gamma variates with shape parameter 0.5, the second column contains gamma variates with shape parameter 1.5, and the third column contains gamma variates with shape parameter 2.5.

Below we generate a 4×3 matrix of beta variates where we demonstrate the use of two r-conformable parameter matrices, a and b.

```
: a = (0.5,1.5,2.5)
: b = (0.5,0.75,1.0\1.25,1.5,1.75)
: rseed(13579)
: rbeta(2,1,a,b)
              1              2              3

    1 │ .668359305     .3238859912    .7785175363
    2 │ .266459731     .7665943496    .634730294
    3 │ .0373430126    .9246702534    .851879254
    4 │ .1903514438    .5012842811    .8759050005
```

The 4×3 shape-parameter matrices used to generate these beta variates are given below:

```
: J(2,1,J(rows(b),1,a))
          1    2    3

    1 │  .5   1.5  2.5
    2 │  .5   1.5  2.5
    3 │  .5   1.5  2.5
    4 │  .5   1.5  2.5
```

```
: J(2,1,b)
            1      2      3
      ┌────────────────────────┐
   1  │    .5    .75      1    │
   2  │  1.25    1.5    1.75   │
   3  │    .5    .75      1    │
   4  │  1.25    1.5    1.75   │
      └────────────────────────┘
```

Conformability

runiform(r, c):
    r: 1×1
    c: 1×1
   result: $r \times c$

rseed( ):
   result: 1×1

rseed(*newseed*):
   newseed: 1×1
   result: *void*

rbeta(r, c, a, b):
    r: 1×1
    c: 1×1
    a: 1×1 or $i \times 1$ or $1 \times j$ or $i \times j$
    b: 1×1 or $i \times 1$ or $1 \times j$ or $i \times j$
   result: $r \times c$ or $ir \times c$ or $r \times jc$ or $ir \times jc$

rbinomial(r, c, n, p):
    r: 1×1
    c: 1×1
    n: 1×1 or $i \times 1$ or $1 \times j$ or $i \times j$
    p: 1×1 or $i \times 1$ or $1 \times j$ or $i \times j$
   result: $r \times c$ or $ir \times c$ or $r \times jc$ or $ir \times jc$

rchi2(r, c, df):
    r: 1×1
    c: 1×1
    df: $i \times j$
   result: $ir \times jc$

rdiscrete(r,c,p):
    r: 1×1
    c: 1×1
    p: $k \times 1$
   result: $r \times c$

rgamma(r, c, a, b):
  r: 1×1
  c: 1×1
  a: 1×1 or $i \times 1$ or $1 \times j$ or $i \times j$
  b: 1×1 or $i \times 1$ or $1 \times j$ or $i \times j$
  result: $r \times c$ or $ir \times c$ or $r \times jc$ or $ir \times jc$

rhypergeometric(r, c, N, K, n):
  r: 1×1
  c: 1×1
  N: 1×1
  K: 1×1 or $i \times 1$ or $1 \times j$ or $i \times j$
  n: 1×1 or $i \times 1$ or $1 \times j$ or $i \times j$
  result: $r \times c$ or $ir \times c$ or $r \times jc$ or $ir \times jc$

rnbinomial(r, c, n, p):
  r: 1×1
  c: 1×1
  n: 1×1 or $i \times 1$ or $1 \times j$ or $i \times j$
  p: 1×1 or $i \times 1$ or $1 \times j$ or $i \times j$
  result: $r \times c$ or $ir \times c$ or $r \times jc$ or $ir \times jc$

rnormal(r, c, m, s):
  r: 1×1
  c: 1×1
  m: 1×1 or $i \times 1$ or $1 \times j$ or $i \times j$
  s: 1×1 or $i \times 1$ or $1 \times j$ or $i \times j$
  result: $r \times c$ or $ir \times c$ or $r \times jc$ or $ir \times jc$

rpoisson(r, c, m):
  r: 1×1
  c: 1×1
  m: $i \times j$
  result: $ir \times jc$

rt(r, c, df):
  r: 1×1
  c: 1×1
  df: 1×1 or $i \times 1$ or $1 \times j$ or $i \times j$
  result: $r \times c$ or $ir \times c$ or $r \times jc$ or $ir \times jc$

Diagnostics

All random-variate generators abort with an error if $r < 0$ or $c < 0$.

rseed(*seed*) aborts with error if a string seed is specified and it is malformed (was not obtained from rseed( )).

rnormal(r, c, m, s), rbeta(r, c, a, b), rbinomial(r, c, n, p), rhypergeometric(r, c, N, K, n), and rnbinomial(r, c, k, p) abort with an error if the parameter matrices do not conform. See *r-conformability* in [M-6] **Glossary** for rules on matrix conformability.

rdiscrete( ) aborts with error if the probabilities in p are not in [0,1] or do not sum to 1.

Reference

Walker, A. J. 1977. An efficient method for generating discrete random variables with general distributions. *ACM Transactions on Mathematical Software* 3: 253–256.

Also see

[M-4] **standard** — Functions to create standard matrices

[M-4] **statistical** — Statistical functions

Title

Syntax

numeric vector   runningsum(*numeric vector x* [, *missing*])

numeric vector   quadrunningsum(*numeric vector x* [, *missing*])

void   _runningsum(*y*, *numeric vector x* [, *missing*])

void   _quadrunningsum(*y*, *numeric vector x* [, *missing*])

where optional argument *missing* is a *real scalar* that determines how missing values in *x* are treated:

1. Specifying *missing* as 0 is equivalent to not specifying the argument; missing values in *x* are treated as contributing 0 to the sum.

2. Specifying *missing* as 1 specifies that missing values in *x* are to be treated as missing values and turn the sum to missing.

Description

runningsum(*x*) returns a vector of the same dimension as *x* containing the running sum of *x*. Missing values are treated as contributing zero to the sum.

runningsum(*x*, *missing*) does the same but lets you specify how missing values are treated. runningsum(*x*, 0) is the same as runningsum(*x*). runningsum(*x*, 1) specifies that missing values are to turn the sum to missing where they occur.

quadrunningsum(*x*) and quadrunningsum(*x*, *missing*) do the same but perform the accumulation in quad precision.

_runningsum(*y*, *x* [, *missing*]) and _quadrunningsum(*y*, *x* [, *missing*]) work the same way, except that rather than returning the running-sum vector, they store the result in *y*. This method is slightly more efficient when *y* is a view.

Remarks

The running sum of (1, 2, 3) is (1, 3, 6).

All functions return the same type as the argument, real if argument is real, complex if complex.

Conformability

runningsum(x, *missing*), quadrunningsum(x, *missing*):

x:	$r \times 1$ or $1 \times c$	
missing:	1×1	(optional)
result:	$r \times 1$ or $1 \times c$	

_runningsum(y, x, *missing*), _quadrunningsum(y, x, *missing*):

input:

x:	$r \times 1$ or $1 \times c$	
y:	$r \times 1$ or $1 \times c$	(contents irrelevant)
missing:	1×1	(optional)

output:

y:	$r \times 1$ or $1 \times c$

Diagnostics

If *missing* $= 0$, missing values are treated as contributing zero to the sum; they do not turn the sum to missing. Otherwise, missing values turn the sum to missing.

_runningsum(y, x, *missing*) and _quadrunningsum(y, x, *missing*) abort with error if y is not p-conformable with x and of the same eltype. The contents of y are irrelevant.

Also see

[M-5] **sum( )** — Sums

[M-4] **mathematical** — Important mathematical functions

[M-4] **utility** — Matrix utility functions

Title

[M-5] **schurd( )** — Schur decomposition

Syntax

void schurd(X, T, Q)

void _schurd(X, Q)

void schurdgroupby(X, f, T, Q, w, m)

void _schurdgroupby(X, f, Q, w, m)

where inputs are

X: *numeric matrix*
f: *pointer scalar*   (points to a function used to group eigenvalues)

and outputs are

T: *numeric matrix* (Schur-form matrix)
Q: *numeric matrix* (orthogonal or unitary)
w: *numeric vector* of eigenvalues
m: *real scalar*   (the number of eigenvalues satisfy the grouping condition)

Description

schurd(X, T, Q) computes the Schur decomposition of a square, numeric matrix, X, returning the Schur-form matrix, T, and the matrix of Schur vectors, Q. Q is orthogonal if X is real and unitary if X is complex.

_schurd(X, Q) does the same thing as schurd(), except that it returns T in X.

schurdgroupby(X, f, T, Q, w, m) computes the Schur decomposition and the eigenvalues of a square, numeric matrix, X, and groups the results according to whether a condition on each eigenvalue is satisfied. schurdgroupby() returns the Schur-form matrix in T, the matrix of Schur vectors in Q, the eigenvalues in w, and the number of eigenvalues for which the condition is true in m. f is a pointer of the function that implements the condition on each eigenvalue, as discussed below.

_schurdgroupby(X, f, Q, w, m) does the same thing as schurdgroupby() except that it returns T in X.

_schurd_la() and _schurdgroupby_la() are the interfaces into the LAPACK routines used to implement the above functions; see [M-1] **LAPACK**. Their direct use is not recommended.

Remarks

Remarks are presented under the following headings:

> *Schur decomposition*
> *Grouping the results*

Schur decomposition

Many algorithms begin by obtaining the Schur decomposition of a square matrix.

The Schur decomposition of matrix $\mathbf{X}$ can be written as

$$\mathbf{Q}' \times \mathbf{X} \times \mathbf{Q} = \mathbf{T}$$

where $\mathbf{T}$ is in Schur form, $\mathbf{Q}$, the matrix of Schur vectors, is orthogonal if $\mathbf{X}$ is real or unitary if $\mathbf{X}$ is complex.

A real, square matrix is in Schur form if it is block upper triangular with 1×1 and 2×2 diagonal blocks. Each 2×2 diagonal block has equal diagonal elements and opposite sign off-diagonal elements. A complex, square matrix is in Schur form if it is upper triangular. The eigenvalues of $\mathbf{X}$ are obtained from the Schur form by a few quick computations.

In the example below, we define X, obtain the Schur decomposition, and list T.

```
: X=(.31,.69,.13,.56\.31,.5,.72,.42\.68,.37,.71,.8\.09,.16,.83,.9)
: schurd(X, T=., Q=.)
: T
```

	1	2	3	4
1	2.10742167	.1266712792	.0549744934	.3329112999
2	0	-.0766307549	.3470959084	.1042286546
3	0	-.4453774705	-.0766307549	.3000409803
4	0	0	0	.4658398402

Grouping the results

In many applications, there is a stable solution if the modulus of an eigenvalue is less than one and an explosive solution if the modulus is greater than or equal to one. One frequently handles these cases differently and would group the Schur decomposition results into a block corresponding to stable solutions and a block corresponding to explosive solutions.

In the following example, we use `schurdgroupby()` to put the stable solutions first. One of the arguments to `schurdgroupby()` is a pointer to a function that accepts a complex scalar argument, an eigenvalue, and returns 1 to select the eigenvalue and 0 otherwise. Here `isstable()` returns 1 if the eigenvalue is less than 1:

```
: real scalar isstable(scalar p)
> {
>         return((abs(p)<1))
> }
```

Using this function to group the results, we see that the Schur-form matrix has been reordered.

```
: schurdgroupby(X, &isstable(), T=., Q=., w=., m=.)
: T
```

	1	2	3	4
1	-.0766307549	.445046622	.3029641608	-.0341867415
2	-.3473539401	-.0766307549	-.1036266286	.0799058566
3	0	0	.4658398402	-.3475944606
4	0	0	0	2.10742167

Listing the moduli of the eigenvalues reveals that they are grouped into stable and explosive groups.

```
: abs(w)
```

	1	2	3	4
1	.4005757984	.4005757984	.4658398402	2.10742167

m contains the number of stable solutions

```
: m
  3
```

Conformability

schurd(X, T, Q):

>   *input*:
>> X: $n \times n$
>   *output*:
>> T: $n \times n$
>> Q: $n \times n$

_schurd(X, Q):

>   *input*:
>> X: $n \times n$
>   *output*:
>> X: $n \times n$
>> Q: $n \times n$

schurdgroupby(X, f, T, Q, w, m):

>   *input*:
>> X: $n \times n$
>> f: 1×1
>   *output*:
>> T: $n \times n$
>> Q: $n \times n$
>> w: $1 \times n$
>> m: 1×1

_schurdgroupby(X, f, Q, w, m):

>   *input*:
>> X: $n \times n$
>> f: 1×1
>   *output*:
>> X: $n \times n$
>> Q: $n \times n$
>> w: $1 \times n$
>> m: 1×1

Diagnostics

_schurd( ) and _schurdgroupby( ) abort with error if X is a view.

schurd(), _schurd(), schurdgroupby(), and _schurdgroupby() return missing results if X contains missing values.

schurdgroupby() groups the results via a matrix transform. If the problem is very ill conditioned, applying this matrix transform can cause changes in the eigenvalues. In extreme cases, the grouped eigenvalues may no longer satisfy the condition used to perform the grouping.

Also see

[M-1] **LAPACK** — The LAPACK linear-algebra routines

[M-5] **hessenbergd( )** — Hessenberg decomposition

[M-4] **matrix** — Matrix functions

Title

[M-5] **select( )** — Select rows or columns

Syntax

> *transmorphic matrix* select(*transmorphic matrix X*, *real vector v*)
>
> *void* st_select(*A*, *transmorphic matrix X*, *real vector v*)

Description

select(*X*, *v*) returns *X*

1. omitting the rows for which *v*[*i*]==0 (*v* a column vector) or

2. omitting the columns for which *v*[*j*]==0 (*v* a row vector).

st_select(*A*, *X*, *v*) does the same thing, except that the result is placed in *A* and, if *X* is a view, *A* will be a view.

Remarks

Remarks are presented under the following headings:

> *Examples*
> *Using st_select()*

Examples

1. To select rows 1, 2, and 4 of 5 × *c* matrix X,

   submat = select(X, (1\1\0\1\0))

   See [M-2] **subscripts** for another solution, submat = X[(1\2\4), .].

2. To select columns 1, 2, and 4 of *r* × 5 matrix X,

   submat = select(X, (1,1,0,1,0))

   See [M-2] **subscripts** for another solution, submat = X[., (1,2,4)].

3. To select rows of X for which the first element is positive,

   submat = select(X, X[.,1]:>0)

4. To select columns of X for which the first element is positive,

   submat = select(X, X[1,.]:>0)

5. To select rows of X for which there are no missing values,

   submat = select(X, rowmissing(X):==0)

728

6. To select rows and columns of square matrix X for which the diagonal elements are positive,

```
pos     = diagonal(X):>0
submat = select(X, pos)
submat = select(submat, pos')
```

or, equivalently,

```
pos     = diagonal(X):>0
submat = select(select(X, pos), pos')
```

Using st_select()

Coding

```
st_select(submat, X, v)                    (1)
```

produces the same result as coding

```
submat = st_select(X, v)                    (2)
```

The difference is in how the result is stored. If X is a view (it need not be), then (1) will produce submat as a view or, if you will, a subview, whereas in (2), submat will always be a regular (nonview) matrix.

When X is a view, (1) executes more quickly than (2) and produces a result that consumes less memory.

See [M-5] **st_view( )** for a description of views.

Conformability

select(X, v):

X:	$r_1 \times c_1$			
v:	$r_1 \times 1$	or	$1 \times c_1$	
result:	$r_2 \times c_1$	or	$r_1 \times c_2$,	$r_2 \le r_1, c_2 \le c_1$

st_select(A, X, v):

input:

X:	$r_1 \times c_1$		
v:	$r_1 \times 1$	or	$1 \times c_1$

output:

A:	$r_2 \times c_1$	or	$r_1 \times c_2$, $r_2 \le r_1, c_2 \le c_1$

Diagnostics

None.

Also see

[M-5] **st_subview( )** — Make view from view

[M-2] **op_colon** — Colon operators

[M-2] **subscripts** — Use of subscripts

[M-4] **utility** — Matrix utility functions

Title

[M-5] setbreakintr( ) — Break-key processing

Syntax

real scalar	setbreakintr(*real scalar val*)
real scalar	querybreakintr()
real scalar	breakkey()
void	breakkeyreset()

Description

setbreakintr(*val*) turns the break-key interrupt off (*val*==0) or on (*val*!=0) and returns the value of the previous break-key mode, 1, it was on, or 0, it was off.

querybreakintr() returns 1 if the break-key interrupt is on and 0 otherwise.

breakkey() (for use in setbreakintr(0) mode) returns 1 if the break key has been pressed since it was last reset.

breakkeyreset() (for use in setbreakintr(0) mode) resets the break key.

Remarks

Remarks are presented under the following headings:

> *Default break-key processing*
> *Suspending the break-key interrupt*
> *Break-key polling*

Default break-key processing

By default, if the user presses *Break*, Mata stops execution and returns control to the console, setting the return code to 1.

To obtain this behavior, there is nothing you need do. You do not need to use these functions.

Suspending the break-key interrupt

The default behavior is known as interrupt-on-break and is also known as setbreakintr(1) mode.

The alternative is break-key suspension, also known as setbreakintr(0) mode.

For instance, you have several steps that must be performed in their entirety or not at all. The way to do this is

```
val = setbreakintr(0)
...
... (critical code) ...
...
(void) setbreakintr(val)
```

The first line stores in *val* the current break-key processing mode and then sets the mode to break-key suspension. The critical code then runs. If the user presses *Break* during the execution of the critical code, that will be ignored. Finally, the code restores the previous break-key processing mode.

Break-key polling

In coding large, interactive systems, you may wish to adopt the break-key polling style of coding rather than interrupt-on-break. In this alternative style of coding, you turn off interrupt-on-break:

```
val = setbreakintr(0)
```

and, from then on in your code, wherever you are willing to interrupt your code, you ask (poll whether) the break key has been pressed:

```
...
if (breakkey()) {
    ...
}
...
```

In this style of coding, you must decide where and when you are going to reset the break key, because once the break key has been pressed, breakkey() will continue to return 1 every time it is called. To reset the break key, code,

```
breakkeyreset()
```

You can also adopt a mixed style of coding, using interrupt-on-break in some places and polling in others. Function querybreakintr() can then be used to determine the current mode.

Conformability

```
setbreakintr(val):
          val:      1 × 1
        result:      1 × 1

querybreakintr(), breakkey():
        result:      1 × 1

breakkeyreset():
        result:      void
```

Diagnostics

setbreakintr(1) aborts with break if the break key has been pressed since the last setbreak-intr(0) or breakkeyreset(). Code breakkeyreset() before setbreakintr(1) if you do not want this behavior.

After coding setbreakintr(1), remember to restore setbreakintr(0) mode. It is not, however, necessary, to restore the original mode if exit() or _error() is about to be executed.

breakkey(), once the break key has been pressed, continues to return 1 until breakkeyreset() is executed.

There is absolutely no reason to use breakkey() in setbreakintr(0) mode, because the only value it could return is 0.

Also see

[M-5] **error( )** — Issue error message

[M-4] **programming** — Programming functions

Title

[M-5] **sign( )** — Sign and complex quadrant functions

Syntax

real matrix   sign(*real matrix R*)

real matrix   quadrant(*complex matrix Z*)

Description

sign(*R*) returns the elementwise sign of *R*. sign() is defined

argument range	sign(*arg*)
$arg \geq .$	.
$arg < 0$	-1
$arg = 0$	0
$arg > 0$	1

quadrant(*Z*) returns a real matrix recording the quadrant of each complex entry in *Z*. quadrant() is defined

argument range		quadrant(*arg*)
Re(*arg*)	Im(*arg*)	
$Re \geq .$		.
$Re = 0$	$Im = 0$	.
$Re > 0$	$Im \geq 0$	1
$Re \leq 0$	$Im > 0$	2
$Re < 0$	$Im \leq 0$	3
$Re \geq 0$	$Im < 0$	4

quadrant(1+0i)==1, quadrant(-1+0i)==3
quadrant(0+1i)==2, quadrant(0-1i)==4

Conformability

sign(*R*):

R:	$r \times c$
result:	$r \times c$

quadrant(*Z*):

Z:	$r \times c$
result:	$r \times c$

Diagnostics

sign(R) returns missing when R is missing.

quadrant(Z) returns missing when Z is missing.

Also see

[M-5] **dsign( )** — FORTRAN-like DSIGN( ) function

[M-4] **scalar** — Scalar mathematical functions

Title

[M-5] **sin( )** — Trigonometric and hyperbolic functions

Syntax

numeric matrix	sin(*numeric matrix Z*)
numeric matrix	cos(*numeric matrix Z*)
numeric matrix	tan(*numeric matrix Z*)
numeric matrix	asin(*numeric matrix Z*)
numeric matrix	acos(*numeric matrix Z*)
numeric matrix	atan(*numeric matrix Z*)
real matrix	atan2(*real matrix X*, *real matrix Y*)
real matrix	arg(*complex matrix Z*)
numeric matrix	sinh(*numeric matrix Z*)
numeric matrix	cosh(*numeric matrix Z*)
numeric matrix	tanh(*numeric matrix Z*)
numeric matrix	asinh(*numeric matrix Z*)
numeric matrix	acosh(*numeric matrix Z*)
numeric matrix	atanh(*numeric matrix Z*)
real scalar	pi()

Description

$\sin(Z)$, $\cos(Z)$, and $\tan(Z)$ return the appropriate trigonometric functions. Angles are measured in radians. All return real if the argument is real and complex if the argument is complex.

$\sin(x)$, x real, returns the sine of x. $\sin()$ returns a value between -1 and 1.

$\sin(z)$, z complex, returns the complex sine of z, mathematically defined as $\{\exp(i*z) - \exp(-i*z)\}/2i$.

$\cos(x)$, x real, returns the cosine of x. $\cos()$ returns a value between -1 and 1.

$\cos(z)$, z complex, returns the complex cosine of z, mathematically defined as $\{\exp(i*z) + \exp(-i*z)\}/2$.

$\tan(x)$, x real, returns the tangent of x.

tan(z), z complex, returns the complex tangent of z, mathematically defined as $\sin(z)/\cos(z)$.

asin(Z), acos(Z), and atan(Z) return the appropriate inverse trigonometric functions. Returned results are in radians. All return real if the argument is real and complex if the argument is complex.

asin(x), x real, returns arcsine in the range $[-\pi/2, \pi/2]$. If $x < -1$ or $x > 1$, missing (.) is returned.

asin(z), z complex, returns the complex arcsine, mathematically defined as $-i * \ln\{i * z + \text{sqrt}(1 - z * z)\}$. Re(asin( )) is chosen to be in the interval $[-\pi/2, \pi/2]$.

acos(x), x real, returns arccosine in the range $[0, \pi]$. If $x < -1$ or $x > 1$, missing (.) is returned.

acos(z), z complex, returns the complex arccosine, mathematically defined as $-i * \ln\{z + \text{sqrt}(z * z - 1)\}$. Re(acos( )) is chosen to be in the interval $[0, \pi]$.

atan(x), x real, returns arctangent in the range $(-\pi/2, \pi/2)$.

atan(z), z complex, returns the complex arctangent, mathematically defined as $\ln\{(1 + iz)/(1 - iz)\}/(2i)$. Re(atan( )) is chosen to be in the interval $[0, \pi]$.

atan2(X, Y) returns the radian value in the range $(-\pi, \pi]$ of the angle of the vector determined by (X,Y), the result being in the range $[0, \pi]$ for quadrants 1 and 2 and $[0, -\pi)$ for quadrants 4 and 3. X and Y must be real. atan2(X, Y) is equivalent to arg(C(X, Y)).

arg(Z) returns the arctangent of Im(Z)/Re(Z) in the correct quadrant, the result being in the range $(-\pi, \pi]$; $[0, \pi]$ in quadrants 1 and 2 and $[0, -\pi)$ in quadrants 4 and 3. arg(Z) is equivalent to atan2(Re(Z), Im(Z)).

sinh(Z), cosh(Z), and tanh(Z) return the hyperbolic sine, cosine, and tangent, respectively. The returned value is real if the argument is real and complex if the argument is complex.

sinh(x), x real, returns the inverse hyperbolic sine of x, mathematically defined as $\{\exp(x) - \exp(-x)\}/2$.

sinh(z), z complex, returns the complex hyperbolic sine of z, mathematically defined as $\{\exp(z) - \exp(-z)\}/2$.

cosh(x), x real, returns the inverse hyperbolic cosine of x, mathematically defined as $\{\exp(x) + \exp(-x)\}/2$.

cosh(z), z complex, returns the complex hyperbolic cosine of z, mathematically defined as $\{\exp(z) + \exp(-z)\}/2$.

tanh(x), x real, returns the inverse hyperbolic tangent of x, mathematically defined as $\sinh(x)/\cosh(x)$.

tanh(z), z complex, returns the complex hyperbolic tangent of z, mathematically defined as $\sinh(z)/\cosh(z)$.

asinh(Z), acosh(Z), and atanh(Z) return the inverse hyperbolic sine, cosine, and tangent, respectively. The returned value is real if the argument is real and complex if the argument is complex.

asinh(x), x real, returns the inverse hyperbolic sine.

asinh(z), z complex, returns the complex inverse hyperbolic sine, mathematically defined as $\ln\{z + \text{sqrt}(z * z + 1)\}$. $\text{Im}(\text{asinh}(\ ))$ is chosen to be in the interval $[-\pi/2, \pi/2]$.

acosh(x), x real, returns the inverse hyperbolic cosine. If $x < 1$, missing (.) is returned.

acosh(z), z complex, returns the complex inverse hyperbolic cosine, mathematically defined as $\ln\{z + \text{sqrt}(z * z - 1)\}$. $\text{Im}(\text{acosh}(\ ))$ is chosen to be in the interval $[-\pi, \pi]$; $\text{Re}(\text{acosh}(\ ))$ is chosen to be nonnegative.

atanh(x), x real, returns the inverse hyperbolic tangent. If $|x| > 1$, missing (.) is returned.

atanh(z), z complex, returns the complex inverse hyperbolic tangent, mathematically defined as $\ln\{(1 + z)/(1 - z)\}/2$.

pi( ) returns the value of π.

Conformability

atan2$(X,\ Y)$:

X:	$r_1 \times c_1$	
Y:	$r_2 \times c_2$,	X and Y r-conformable
result:	$\max(r_1, r_2) \times \max(c_1, c_2)$	

pi( ) returns a 1×1 scalar.

All other functions return a matrix of the same dimension as input containing element-by-element calculated results.

Diagnostics

All functions return missing for real arguments when the result would be complex. For instance, acos(2) = ., whereas acos(2+0i) = -1.317i.

Also see

[M-4] **scalar** — Scalar mathematical functions

Title

[M-5] **sizeof( )** — Number of bytes consumed by object

Syntax

real scalar sizeof (*transmorphic matrix A*)

Description

sizeof (*A*) returns the number of bytes consumed by *A*.

Remarks

sizeof (*A*) returns the same number as shown by mata describe; see [M-3] **mata describe**.

A 500×5 real matrix consumes 20,000 bytes:

```
: sizeof(mymatrix)
  20000
```

A 500×5 view matrix, however, consumes only 24 bytes:

```
: sizeof(myview)
  24
```

To obtain the number of bytes consumed by a function, pass a dereferenced function pointer:

```
: sizeof(*&myfcn())
  320
```

Conformability

sizeof (*A*):

A:	$r \times c$
result:	1×1

Diagnostics

The number returned by sizeof (*A*) does not include any overhead, which usually amounts to 64 bytes, but can be less (as small as zero in the case of recently used scalars).

If *A* is a pointer matrix, the number returned reflects the amount of memory required to store *A* itself and does not include the memory consumed by its siblings.

Also see

[M-4] **programming** — Programming functions

Title

[M-5] **solve_tol( )** — Tolerance used by solvers and inverters

Syntax

> *real scalar* `solve_tol(`*numeric matrix Z, real scalar usertol*`)`

Description

`solve_tol(`*Z, usertol*`)` returns the tolerance used by many Mata solvers to solve $AX = B$ and by many Mata inverters to obtain A^{-1}. *usertol* is the tolerance specified by the user or is missing value if the user did not specify a tolerance.

Remarks

The tolerance used by many Mata solvers to solve $AX = B$ and by many Mata inverters to obtain A^{-1} is

$$eta = s * \frac{\texttt{trace(abs}(Z))}{n} \tag{1}$$

where $s = $ 1e–13 or a value specified by the user, n is the `min(rows(`Z`)`, `cols(`Z`))`, and Z is a matrix related to A, usually by some form of decomposition, but could be A itself (for instance, if A were triangular). See, for instance, [M-5] **solvelower( )** and [M-5] **cholsolve( )**.

When *usertol* > 0 and *usertol* $< .$ is specified, `solvetol()` returns *eta* calculated with $s = usertol$.

When *usertol* ≤ 0 is specified, `solvetol()` returns $-usertol$.

When *usertol* $\ge .$ is specified, `solvetol()` returns a default result, calculated as

1. If external real scalar `_solvetolerance` does not exist, as is usually the case, the value of *eta* is returned using $s = $ 1e–13.

2. If external real scalar `_solvetolerance` does exist,

    a. If `_solvetolerance` > 0, the value of *eta* is returned using $s = $ `solvetolerance`.

    b. If `_solvetolerance` ≤ 0, $-$`_solvetolerance` is returned.

Conformability

`solve_tol(`*Z, usertol*`)`:
$$
\begin{array}{rl}
Z: & r \times c \\
usertol: & 1 \times 1 \\
result: & 1 \times 1
\end{array}
$$

Diagnostics

solve_tol(Z, *usertol*) skips over missing values in Z in calculating (1); n is defined as the number of nonmissing elements on the diagonal.

Also see

[M-4] **utility** — Matrix utility functions

Title

[M-5] **solvelower( )** — Solve AX=B for X, A triangular

Syntax

numeric matrix	solvelower$(A, B [, rank [, tol [, d]]])$
numeric matrix	solveupper$(A, B [, rank [, tol [, d]]])$
real scalar	_solvelower$(A, B [, tol [, d]])$
real scalar	_solveupper$(A, B [, tol [, d]])$

where

A:	*numeric matrix*
B:	*numeric matrix*
rank:	irrelevant; *real scalar* returned
tol:	*real scalar*
d:	*numeric scalar*

Description

These functions are used in the implementation of the other solve functions; see [M-5] **lusolve( )**, [M-5] **qrsolve( )**, and [M-5] **svsolve( )**.

solvelower$(A, B, ...)$ and _solvelower$(A, B, ...)$ solve lower-triangular systems.

solveupper$(A, B, ...)$ and _solveupper$(A, B, ...)$ solve upper-triangular systems.

Functions without a leading underscore—solvelower() and solveupper()—return the solution; *A* and *B* are unchanged.

Functions with a leading underscore—_solvelower() and _solveupper()—return the solution in *B*.

All four functions produce a generalized solution if *A* is singular. The functions without an underscore place the rank of *A* in *rank*, if the argument is specified. The underscore functions return the rank.

Determination of singularity is made via *tol*. *tol* is interpreted in the standard way—as a multiplier for the default if *tol* > 0 is specified and as an absolute quantity to use in place of the default if *tol* ≤ 0 is specified.

All four functions allow *d* to be optionally specified. Specifying $d = .$ is equivalent to not specifying *d*.

If $d \neq .$ is specified, that value is used as if it appeared on the diagonal of *A*. The four functions do not in fact require that *A* be triangular; they merely look at the lower or upper triangle and pretend that the opposite triangle contains zeros. This feature is useful when a decomposition utility has stored both the lower and upper triangles in one matrix, because one need not take apart the combined matrix. In such cases, it sometimes happens that the diagonal of the matrix corresponds to one matrix but not the other, and that for the other matrix, one merely knows that the diagonal elements are, say, 1. Then you can specify $d = 1$.

Remarks

The triangular-solve functions documented here exploit the triangular structure in A and solve for X by recursive substitution.

When A is of full rank, these functions provide the same solution as the other solve functions, such as [M-5] **lusolve( )**, [M-5] **qrsolve( )**, and [M-5] **svsolve( )**. The solvelower( ) and solveupper( ) functions, however, will produce the answer more quickly because of the large computational savings.

When A is singular, however, you may wish to consider whether you want to use these triangular-solve functions. The triangular-solve functions documented here reach a generalized solution by setting $B_{ij} = 0$, for all j, when A_{ij} is zero or too small (as determined by *tol*). The method produces a generalized inverse, but there are many generalized inverses, and this one may not have the other properties you want.

Remarks are presented under the following headings:

> *Derivation*
> *Tolerance*

Derivation

We wish to solve

$$AX = B \tag{1}$$

when A is triangular. Let us consider the lower-triangular case first. solvelower( ) is up to handling full matrices for B and X, but let us assume X: $n \times 1$ and B: $m \times 1$:

$$\begin{bmatrix} a_{11} & 0 & 0\dots & 0 \\ a_{21} & 0 & 0\dots & 0 \\ \vdots & \vdots & \ddots & \vdots \\ a_{m1} & a_{m2} & a_{m3} & a_{mn} \end{bmatrix} \begin{bmatrix} x_1 \\ x_2 \\ \vdots \\ x_n \end{bmatrix} = \begin{bmatrix} b_1 \\ b_2 \\ \vdots \\ b_m \end{bmatrix}$$

The first equation to be solved is

$$a_{11}x_1 = b_1$$

and the solution is simply

$$x_1 = \frac{b_1}{a_{11}} \tag{2}$$

The second equation to be solved is

$$a_{21}x_1 + a_{22}x_2 = b_2$$

and because we have already solved for x_1, the solution is simply

$$x_2 = \frac{b_2 - a_{21}x_1}{a_{22}} \tag{3}$$

We proceed similarly for the remaining rows of A. If there are additional columns in B and X, we can then proceed to handling each remaining column just as we handled the first column above.

In the upper-triangular case, the formulas are similar except that you start with the last row of A.

Tolerance

In (2) and (3), we divide by the diagonal elements of A. If element a_{ii} is less than *eta* in absolute value, the corresponding x_i is set to zero. *eta* is given by

$$eta = \text{1e-13} * \texttt{trace(abs}(A))/\texttt{rows}(A)$$

If you specify *tol* > 0, the value you specify is used to multiply *eta*. You may instead specify *tol* $\leq$ 0, and then the negative of the value you specify is used in place of *eta*; see [M-1] **tolerance**.

Conformability

solvelower(A, B, *rank*, *tol*, d), solveupper(A, B, *rank*, *tol*, d):

> *input*:

A:	$n \times n$	
B:	$n \times k$	
tol:	1×1	(optional)
d:	1×1	(optional)

> *output*:

rank:	1×1	(optional)
result:	$n \times k$	

_solvelower(A, B, *tol*, d), _solveupper(A, B, *tol*, d):

> *input*:

A:	$n \times n$	
B:	$n \times k$	
tol:	1×1	(optional)
d:	1×1	(optional)

> *output*:

B:	$n \times k$	
result:	1×1	(contains rank)

Diagnostics

solvelower(A, B, ...), _solvelower(A, B, ...), solveupper(A, B, ...), and _solveupper(A, B, ...) do not verify that the upper (lower) triangle of A contains zeros; they just use the lower (upper) triangle of A.

_solvelower(A, B, ...) and _solveupper(A, B, ...) do not abort with error if B is a view but can produce results subject to considerable roundoff error.

Also see

[M-5] **cholsolve( )** — Solve AX=B for X using Cholesky decomposition

[M-5] **lusolve( )** — Solve AX=B for X using LU decomposition

[M-5] **qrsolve( )** — Solve AX=B for X using QR decomposition

[M-5] **svsolve( )** — Solve AX=B for X using singular value decomposition

[M-5] **solve_tol( )** — Tolerance used by solvers and inverters

[M-4] **matrix** — Matrix functions

Title

Syntax

transmorphic matrix	sort (*transmorphic matrix X*, *real rowvector idx*)
void	_sort (*transmorphic matrix X*, *real rowvector idx*)
transmorphic matrix	jumble (*transmorphic matrix X*)
void	_jumble (*transmorphic matrix X*)
real colvector	order (*transmorphic matrix X*, *real rowvector idx*)
real colvector	unorder (*real scalar n*)
void	_collate (*transmorphic matrix X*, *real colvector p*)

where

1. *X* may not be a pointer matrix.

2. *p* must be a permutation column vector, a $1 \times c$ vector containing the integers 1, 2, ..., *c* in some order.

Description

sort (*X*, *idx*) returns *X* with rows in ascending or descending order of the columns specified by *idx*. For instance, sort (*X*, 1) sorts *X* on its first column; sort (*X*, (1,2)) sorts *X* on its first and second columns (meaning rows with equal values in their first column are ordered on their second column). In general, the *i*th sort key is column abs (*idx*[*i*]). Order is ascending if *idx*[*i*] > 0 and descending otherwise. Ascending and descending are defined in terms of [M-5] **abs( )** (length of elements) for complex.

_sort (*X*, *idx*) does the same as sort (*X*, *idx*), except that *X* is sorted in place.

jumble (*X*) returns *X* with rows in random order. For instance, to shuffle a deck of cards numbered 1 to 52, one could code jumble (1::52). See rseed ( ) in [M-5] **runiform( )** for information on setting the random-number seed.

_jumble (*X*) does the same as jumble (*X*), except that *X* is jumbled in place.

order (*X*, *idx*) returns the permutation vector—see [M-1] **permutation**—that would put *X* in ascending (descending) order of the columns specified by *idx*. A row-permutation vector is a $1 \times c$ column vector containing the integers 1, 2, ..., *c* in some order. Vectors (1\2\3), (1\3\2), (2\1\3), (2\3\1), (3\1\2), and (3\2\1) are examples. Row-permutation vectors are used to specify the order in which the rows of a matrix *X* are to appear. If *p* is a row-permutation vector, *X*[*p*, .] returns *X* with its rows in the order of *p*; *p* = (3\2\1) would reverse the rows of *X*. order (*X*, *idx*) returns the row-permutation vector that would sort *X* and, as a matter of fact, sort (*X*, *idx*) is implemented as *X*[order (*X*, *idx*), .].

unorder(n) returns a $1 \times n$ permutation vector for placing the rows in random order. Random numbers are calculated by runiform(); see rseed() in [M-5] **runiform( )** for information on setting the random-number seed. jumble() is implemented in terms of unorder(): jumble(X) is equivalent to X[unorder(rows(X)), .].

_collate(X, p) is equivalent to $X = X[p, .]$; it changes the order of the rows of X. _collate() is used by _sort() and _jumble() and has the advantage over subscripting in that no extra memory is required when the result is to be assigned back to itself. Consider

$$X = X[p, .]$$

There will be an instant after $X[p, .]$ has been calculated but before the result has been assigned back to X when two copies of X exist. _collate(X, p) avoids that. _collate() is not a substitute for subscripting in all cases; _collate() requires p be a permutation vector.

Remarks

If X is complex, the ordering is defined in terms of [M-5] **abs( )** of its elements.

Also see inorder() and revorder() in [M-5] **inorder( )**. Let p be the permutation vector returned by order():

$$p = order(X, \ldots)$$

Then $X[p,.]$ are the sorted rows of X. revorder() can be used to reverse sort order: X[revorder(p),.] are the rows of X in the reverse of the order of $X[p,.]$. inorder() provides the inverse transform: If $Y = X[p,.]$, then $X = Y$[inorder(p),.].

Conformability

sort(X, idx), jumble(X):

X:	$r_1 \times c_1$
idx:	$1 \times c_2, c_2 \le c_1$
result:	$r_1 \times c_1$

_sort(X, idx), _jumble(X):

X:	$r_1 \times c_1$
idx:	$1 \times c_2, c_2 \le c_1$
result:	void; X row order modified

order(X, idx):

X:	$r_1 \times c_1$
idx:	$1 \times c_2, c_2 \le c_1$
result:	$r_1 \times 1$

unorder(n):

n:	1×1
result:	$n \times 1$

_collate(X, p):

X:	$r \times c$
p:	$r \times 1$
result:	void; X row order modified

Diagnostics

sort(X, idx) aborts with error if any element of abs(idx) is less than 1 or greater than rows(X).

_sort(X, idx) aborts with error if any element of abs(idx) is less than 1 or greater than rows(X), or if X is a view.

_jumble(X) aborts with error if X is a view.

order(X, idx) aborts with error if any element of abs(idx) is less than 1 or greater than rows(X).

unorder(n) aborts with error if $n < 1$.

_collate(X, p) aborts with error if p is not a permutation vector or if X is a view.

Also see

[M-5] **invorder( )** — Permutation vector manipulation

[M-5] **uniqrows( )** — Obtain sorted, unique values

[M-4] **manipulation** — Matrix manipulation

Title

> **[M-5] soundex( )** — Convert string to soundex code

Syntax

> *string matrix*   soundex(*string matrix s*)
>
> *string matrix* soundex_nara(*string matrix s*)

Description

soundex(*s*) returns the soundex code for a string, *s*. The soundex code consists of a letter followed by three numbers: the letter is the first letter of the name and the numbers encode the remaining consonants. Similar sounding consonants are encoded by the same number.

soundex_nara(*s*) returns the U.S. Census soundex code for a string, *s*. The soundex code consists of a letter followed by three numbers: the letter is the first letter of the name and the numbers encode the remaining consonants. Similar sounding consonants are encoded by the same number.

When *s* is not a scalar, these functions return element-by-element results.

Remarks

soundex("Ashcraft") returns "A226".

soundex_nara("Ashcraft") returns "A261".

Conformability

soundex(*s*), soundex_nara(*s*):

s:	$r \times c$
result:	$r \times c$

Diagnostics

None.

Also see

[M-4] **string** — String manipulation functions

Title

[M-5] **spline3( )** — Cubic spline interpolation

Syntax

$real\ matrix$ spline3($real\ vector\ x$, $real\ vector\ y$)

$real\ vector$ spline3eval($real\ matrix\ spline\_info$, $real\ vector\ x$)

Description

spline3(x, y) returns the coefficients of a cubic natural spline $S(x)$. The elements of x must be strictly monotone increasing.

spline3eval($spline\_info$, x) uses the information returned by spline3( ) to evaluate and return the spline at the abscissas x. Elements of the returned result are set to missing if outside the range of the spline. x is assumed to be monotonically increasing.

Remarks

spline3( ) and spline3eval( ) is a translation into Mata of Herriot and Reinsch (CUBNATSPLINE)

For xx in $[x_i, x_{i+1})$:

$$S(xx) = \{(d_i t + c_i)t + b_i\}t + y_i$$

with $t = xx - x_i$.

spline3( ) returns (b, c, d, x, y) or, if x and y are row vectors, (b, c, d, x', y').

Conformability

spline3(x, y):

x:	$n \times 1$	or	$1 \times n$
y:	$n \times 1$	or	$1 \times n$
$result$:	$n \times 5$		

spline3eval($spline\_info$, x):

$spline\_info$:	$n \times 5$		
x:	$m \times 1$	or	$1 \times m$
$result$:	$m \times 1$	or	$1 \times m$

Diagnostics

spline3(x, y) requires that x be in ascending order.

spline3eval($spline\_info$, x) requires that x be in ascending order.

Reference

Herriot, J. G., and C. H. Reinsch. 1973. Algorithm 472: Procedures for natural spline interpolation [E1]. *Communications of the ACM* 16: 763–768.

Also see

[M-4] **mathematical** — Important mathematical functions

Title

[M-5] **sqrt( )** — Square root

Syntax

>*numeric matrix*   sqrt(*numeric matrix Z*)

Description

sqrt(*Z*) returns the elementwise square root of *Z*.

Conformability

sqrt(*Z*)
>>*Z*: $r \times c$
>*result*: $r \times c$

Diagnostics

sqrt(*Z*) returns missing when Z is real and $Z < 0$; i.e., sqrt(-4) = . but sqrt(-4+0i) = 2i.

Also see

[M-5] **cholesky( )** — Cholesky square-root decomposition

[M-4] **scalar** — Scalar mathematical functions

Title

> **[M-5] st_addobs( )** — Add observations to current Stata dataset

Syntax

void	st_addobs(*real scalar n*)
void	st_addobs(*real scalar n*, *real scalar nofill*)
real scalar	_st_addobs(*real scalar n*)
real scalar	_st_addobs(*real scalar n*, *real scalar nofill*)

Description

st_addobs(*n*) adds *n* observations to the current Stata dataset.

st_addobs(*n*, *nofill*) does the same thing but saves computer time by not filling in the additional observations with the appropriate missing-value code if *nofill* $\neq$ 0. st_addobs(*n*, 0) is equivalent to st_addobs(*n*). Use of st_addobs( ) with *nofill* $\neq$ 0 is not recommended. If you specify *nofill* $\neq$ 0, it is your responsibility to ensure that the added observations ultimately are filled in or removed before control is returned to Stata.

_st_addobs(*n*) and _st_addobs(*n*, *nofill*) perform the same action as st_addobs(*n*) and st_addobs(*n*, *nofill*), except that they return 0 if successful and the appropriate Stata return code otherwise (otherwise usually being caused by insufficient memory). Where _st_addobs( ) would return nonzero, st_addobs( ) aborts with error.

Remarks

There need not be any variables defined to add observations. If you are attempting to create a dataset from nothing, you can add the observations first and then add the variables, or you can add the variables and then add the observations. Use st_addvar( ) (see [M-5] **st_addvar( )**) to add variables.

Conformability

st_addobs(*n*, *nofill*):
      n: 1 × 1
      nofill: 1 × 1   (optional)
      result: *void*

_st_addobs(*n*, *nofill*):
      n: 1 × 1
      nofill: 1 × 1   (optional)
      result: 1 × 1

Diagnostics

st_addobs(n[, *nofill*]) and _st_addobs(n[, *nofill*]) abort with error if $n < 0$. They do nothing if $n = 0$.

st_addobs() aborts with error if there is insufficient memory to add the requested number of observations.

_st_addobs() aborts with error if $n < 0$ but otherwise returns the appropriate Stata return code if the observations cannot be added. If they are added, 0 is returned.

st_addobs() and _st_addobs() do not set st_updata() (see [M-5] **st_updata( )**); you must set it if you want it set.

Also see

[M-4] **stata** — Stata interface functions

Title

> **[M-5] st_addvar( )** — Add variable to current Stata dataset

Syntax

real rowvector	st_addvar(*type*, *name*)
real rowvector	st_addvar(*type*, *name*, *nofill*)
real rowvector	_st_addvar(*type*, *name*)
real rowvector	_st_addvar(*type*, *name*, *nofill*)

where

> *type*:   *string scalar* or *rowvector* containing "byte", "int", "long", "float", "double", or "str#"
>
>   or
>
>   *real scalar* or *rowvector* containing # (interpreted as str#)
>
> *name*: *string rowvector* containing new variable names
>
> *nofill*: *real scalar* containing 0 or non-0

Description

st_addvar(*type*, *name*) adds new variable *name*(s) of type *type* to the Stata dataset. Returned are the variable indices of the new variables. st_addvar() aborts with error (and adds no variables) if any of the variables already exist or cannot be added for other reasons.

st_addvar(*type*, *name*, *nofill*) does the same thing. *nofill* ≠ 0 specifies that the variables' values are not to be filled in with missing values. st_addvar(*type*, *name*, 0) is the same as st_addvar(*type*, *name*). Use of *nofill* ≠ 0 is not, in general, recommended. See *Remarks* below.

_st_addvar() does the same thing as st_addvar() except that, rather than aborting with error if the new variable cannot be added, returned is a 1×1 scalar containing the negative of the appropriate Stata return code.

Remarks

Remarks are presented under the following headings:

> *Creating a new variable*
> *Creating new variables*
> *Creating new string variables*
> *Creating a new temporary variable*
> *Creating temporary variables*
> *Handling errors*
> *Using nofill*

Creating a new variable

To create new variable `myvar` as a `double`, code

```
idx = st_addvar("double", "myvar")
```

or

```
(void) st_addvar("double", "myvar")
```

You use the first form if you will subsequently need the variable's index number, or you use the second form otherwise.

Creating new variables

You can add more than one variable. For instance,

```
idx = st_addvar("double", ("myvar1","myvar2"))
```

adds two new variables, both of type `double`.

```
idx = st_addvar(("double","float") ("myvar1","myvar2"))
```

also adds two new variables, but this time, `myvar1` is `double` and `myvar2` is `float`.

Creating new string variables

Creating string variables is no different from any other type:

```
idx = st_addvar(("str10","str5"), ("myvar1","myvar2"))
```

creates `myvar1` as a `str10` and `myvar2` as a `str5`.

There is, however, another way to specify the types.

```
idx = st_addvar((10,5), ("myvar1","myvar2"))
```

also creates `myvar1` as a `str10` and `myvar2` as a `str5`.

```
idx = st_addvar(10, ("myvar1","myvar2"))
```

creates both variables as `str10`s.

Creating a new temporary variable

Function `st_tempname()` (see [M-5] **st_tempname( )**) returns temporary variable names. To create a temporary variable as a `double`, code

```
idx = st_addvar("double", st_tempname())
```

or code

```
(void) st_addvar("double", name=st_tempname())
```

You use the first form if you will subsequently need the variable's index, or you use the second form if you will subsequently need the variable's name. You will certainly need one or the other. If you will need both, code

```
idx = st_addvar("double", name=st_tempname())
```

Creating temporary variables

st_tempname() can return a vector of temporary variable names.

>   idx = st_addvar("double", st_tempname(5))

creates five temporary variables, each of type double.

Handling errors

There are three common reasons why st_addvar() might fail: the variable name is invalid or a variable under that name already exists or there is insufficient memory to add another variable. If there is a problem adding a variable, st_addvar() will abort with error. If you wish to avoid the traceback log and just have Stata issue an error, use _st_addvar() and code

>   if ((idx = _st_addvar("double", "myvar"))<0) exit(error(-idx))

If you are adding multiple variables, look at the first element of what _st_addvar() returns:

>   if ((idx = _st_addvar(types, names))[1]<0) exit(error(-idx))

Using nofill

The three-argument versions of st_addvar() and _st_addvar() allow you to avoid filling in the values of the newly created variable. Filling in those values with missing really is a waste of time if the next thing you are going to do is fill in the values with something else. On the other hand, it is important that all the observations be filled in on the new variable before control is returned to Stata, and this includes returning to Stata because of subsequent error or the user pressing *Break*. Thus use of *nofill* $\neq 0$ is not, in general, recommended. Filling in values really does not take that long.

If you are determined to save the computer time, however, see [M-5] **setbreakintr( )**. To do things right, you need to set the break key off, create your variable, fill it in, and turn break-key processing back on.

There is, however, a case in which use of *nofill* $\neq 0$ is acceptable and such effort is not required: when you are creating a temporary variable. Temporary variables vanish in any case, and it does not matter whether they are filled in before they vanish.

Temporary variables in fact vanish not when Mata ends but when the ado-file calling Mata ends, if there is an ado-file. We will assume there is an ado-file because that is the only case in which you would be creating a temporary variable anyway. Because they do not disappear until later, there is the possibility of there being an issue if the variable is not filled in. If we assume, however, that your Mata program is correctly written and does fill in the variable ultimately, then the chances of a problem are minimal. If the user presses *Break* or there is some other problem in your program that causes Mata to abort, the ado-file will be aborted, too, and the variable will vanish.

Let us add that Stata will not crash if a variable is not filled in, even if it regains control. The danger is that the user will look at the variable or, worse, use it and be baffled by what he or she sees, which might concern not only odd values but also NaNs and worse.

Conformability

st_addvar(*type*, *name*, *nofill*):

type:	1×1	or $1 \times k$
name:	$1 \times k$	
nofill:	1×1	(optional)
result:	$1 \times k$	

_st_addvar(*type*, *name*, *nofill*):

type:	1×1	or $1 \times k$
name:	$1 \times k$	
nofill:	1×1	(optional)
result:	$1 \times k$	or, if error, 1×1

Diagnostics

st_addvar(*type*, *name*, *nofill*) aborts with error if

1. *type* is not equal to a valid Stata variable type and it is not a number that would form a valid str# variable type;

2. *name* is not a valid variable name;

3. a variable named *name* already exists;

4. there is insufficient memory to add another variable.

_st_addvar(*type*, *name*, *nofill*) aborts with error for reason 1 above, but otherwise, it returns the negative value of the appropriate Stata return code.

Both functions, when creating multiple variables, create either all the variables or none of them. Whether creating one variable or many, if variables are created, st_updata( ) (see [M-5] **st_updata( )**) is set unless all variables are temporary; see [M-5] **st_tempname( )**.

Reference

Gould, W. W. 2006. Mata Matters: Creating new variables—sounds boring, isn't. *Stata Journal* 6: 112–123.

Also see

[M-5] **st_store( )** — Modify values stored in current Stata dataset

[M-5] **st_tempname( )** — Temporary Stata names

[M-4] **stata** — Stata interface functions

Title

[M-5] **st_data( )** — Load copy of current Stata dataset

Syntax

real scalar	_st_data(*real scalar i, real scalar j*)

real matrix	st_data(*real matrix i, j*)	(1,2)
real matrix	st_data(*real matrix i, j, scalar selectvar*)	(1,2,3)

string scalar	_st_sdata(*real scalar i, real scalar j*)

string matrix	st_sdata(*real matrix i, j*)	(1,2)
string matrix	st_sdata(*real matrix i, j, scalar selectvar*)	(1,2,3)

where

1. *i* may be specified as a 1×1 scalar, as a 1×1 scalar containing missing, as a column vector of observation numbers, as a row vector specifying an observation range, or as a $k \times 2$ matrix specifying both.

   a. st_data(1, 2) returns the first observation on the second variable.

   b. st_data(., 2) returns all observations on the second variable.

   c. st_data((1\2\5), 2) returns observations 1, 2, and 5 on the second variable.

   d. st_data((1,5), 2) returns observations 1 through 5 on the second variable.

   e. st_data((1,5\7,9), 2) returns observations 1 through 5 and observations 7 through 9 on the second variable.

   When a range is specified, any element of the range (i_1, i_2) may be specified to contribute zero observations if $i_2 = i_1 - 1$.

2. *j* may be specified as a real row vector or as a string scalar or string row vector.

   a. st_data(., .) returns the values of all variables, all observations of the Stata dataset.

   b. st_data(., 1) returns the value of the first variable, all observations.

   c. st_data(., (3,1,9)) returns the values of the third, first, and ninth variables of all observations.

   d. st_data(., ("mpg", "weight")) returns the values of variables mpg and weight, all observations.

   e. st_data(., ("mpg weight")) does the same as d above.

   f. st_data(., ("gnp", "l.gnp")) returns the values of gnp and the lag of gnp, all observations.

     g. st_data(., ("gnp l.gnp")) does the same as f above.

     h. st_data(., ("mpg i.rep78")) returns the value of mpg and the 5 pseudovariables associated with i.rep78. There are 5 pseudovariables because we are imagining that auto.dta is in memory; the actual number is a function of the values taken on by the variable in the sample specified. Factor variables can be specified only with string scalars; specifying ("mpg", "i.rep78") will not work.

3. *selectvar* may be specified as real or as a string. Observations for which *selectvar* $\neq 0$ will be selected. If *selectvar* is real, it is interpreted as a variable number. If string, *selectvar* should contain the name of a Stata variable.

Specifying *selectvar* as "" or as missing (.) has the same result as not specifying *selectvar*; no observations are excluded.

Specifying *selectvar* as 0 means that observations with missing values of the variables specified by *j* are to be excluded.

Description

_st_data(i, j) returns the numeric value of the *i*th observation of the *j*th Stata variable. Observations are numbered 1 through st_nobs(). Variables are numbered 1 through st_nvar().

st_data(i, j) is similar to _st_data(i, j) except

    1. *i* may be specified as a vector or matrix to obtain multiple observations simultaneously,

    2. *j* may be specified using names or indices (indices are faster), and

    3. *j* may be specified to obtain multiple variables simultaneously.

The net effect is that st_data() can return a scalar (the value of one variable in one observation), a row vector (the value of many variables in an observation), a column vector (the value of a variable in many observations), or a matrix (the value of many variables in many observations).

st_data(i, j, *selectvar*) works like st_data(i, j) except that only observations for which *selectvar* $\neq 0$ are returned.

_st_sdata() and st_sdata() are the string variants of _st_data() and st_data(). _st_data() and st_data() are for use with numeric variables; they return missing (.) when used with string variables. _st_sdata() and st_sdata() are for use with string variables; they return empty string ("") when used with numeric variables.

Remarks

Remarks are presented under the following headings:

        Description of _st_data( ) and _st_sdata( )
        Description of st_data( ) and st_sdata( )
        Details of observation subscripting using st_data( ) and st_sdata( )

Description of _st_data( ) and _st_sdata( )

_st_data( ) returns one variable's value in one observation. You refer to variables and observations by their numbers. The first variable in the Stata dataset is 1; the first observation is 1.

_st_data(1, 1)	value of 1st obs., 1st variable
_st_data(1, 2)	value of 1st obs., 2nd variable
_st_data(2, 1)	value of 2nd obs., 1st variable

_st_sdata( ) works the same way. _st_data( ) is for use with numeric variables, and _st_sdata( ) is for use with string variables.

_st_data( ) and _st_sdata( ) are the fastest way to obtain the value of a variable in one observation.

Description of st_data( ) and st_sdata( )

st_data( ) can be used just like _st_data( ), and used that way, it produces the same result.

Variables, however, can be referred to by their names or their numbers:

st_data(1, 1)	value of 1st obs., 1st variable
st_data(1, 2)	value of 1st obs., 2nd variable
st_data(2, 1)	value of 2nd obs., 1st variable
st_data(1, "mpg")	value of 1st obs, variable mpg
st_data(2, "mpg")	value of 2nd obs, variable mpg

Also, you may specify more than one variable:

st_data(2, (1,2,3))	value of 2nd obs., variables 1, 2, and 3
st_data(2, ("mpg","weight","displ"))	value of 2nd obs., variables mpg, weight, and displ
st_data(2, "mpg weight displ")	(same as previous)

Used this way, st_data( ) returns a row vector.

Similarly, you may obtain multiple observations:

st_data((1\2\3), 10)	values of obs. 1, 2, and 3, variable 10
st_data((1,5), 10)	values of obs. 1 through 5, variable 10
st_data((1,5)\(7,9), 10)	values of obs. 1 through 5 and 7 through 9, variable 10

st_sdata( ) works the same way as st_data( ).

Details of observation subscripting using st_data( ) and st_sdata( )

1. i may be specified as a scalar: the specified, single observation is returned. i must be between 1 and st_nobs( ).

2. i may be specified as a scalar containing missing value: all observations are returned.

3. i may be specified as a column vector: the specified observations are returned. Each element of i must be between 1 and st_nobs( ) or may be missing. Missing is interpreted as st_nobs( ).

4. i may be specified as a 1×2 row vector: the specified range of observations is returned; (c_1, c_2) returns the $c_2 - c_1 + 1$ observations c_1 through c_2.

   $c_2 - c_1 + 1$ must evaluate to a number greater than or equal to 0. In general, c_1 and c_2 must be between 1 and st_nobs( ), but if $c_2 - c_1 + 1 = 0$, then c_1 may be between 1 and st_nobs( ) + 1 and c_2 may be between 0 and st_nobs( ). Regardless, $c_1 ==$. or $c_2 ==$. is interpreted as st_nobs( ).

5. i may be specified as a $k \times 2$ matrix: $((1,5)\backslash(7,7)\backslash(20,30))$ specifies observations 1 through 5, 7, and 20 through 30.

Conformability

_st_data(i, j), _st_sdata(i, j):

i:	1×1
j:	1×1
result:	1×1

st_data(i, j), st_sdata(i, j):

i:	$n \times 1$ or	$n_2 \times 2$
j:	$1 \times k$ or	1×1 containing k elements when expanded
result:	$n \times k$	

st_data($i, j, selectvar$), st_sdata($i, j, selectvar$):

i:	$n \times 1$ or	$n_2 \times 2$
j:	$1 \times k$ or	1×1 containing k elements when expanded
selectvar:	1×1	
result:	$(n - e) \times k$,	where e is number of observations excluded by *selectvar*

Diagnostics

_st_data(i, j) returns missing (.) if i or j is out of range; it does not abort with error.

_st_sdata(i, j) returns "" if i or j is out of range; it does not abort with error.

st_data(i, j) and st_sdata(i, j) abort with error if any element of i or j is out of range. j may be specified as variable names or variable indices. If names are specified, abbreviations are allowed. If you do not want this and no factor variables nor time-series–operated variables are specified, use st_varindex( ) (see [M-5] **st_varindex( )**) to translate variable names into variable indices.

Also see

[M-5] **st_view( )** — Make matrix that is a view onto current Stata dataset

[M-5] **st_store( )** — Modify values stored in current Stata dataset

[M-4] **stata** — Stata interface functions

Title

[M-5] **st_dir( )** — Obtain list of Stata objects

Syntax

> *string colvector*   st_dir(*cat*, *subcat*, *pattern*)

> *string colvector*   st_dir(*cat*, *subcat*, *pattern*, *adorn*)

where

cat:	*string scalar* containing "local", "global", "r()", "e()", "s()", or "char"
subcat:	*string scalar* containing "macro", "numscalar", "strscalar", "matrix", or, if *cat*=="char", "_dta" or a name.
pattern:	*string scalar* containing a pattern as defined in [M-5] **strmatch( )**
adorn:	*string scalar* containing 0 or non-0

The valid *cat–subcat* combinations and their meanings are

cat	subcat	meaning
"local"	"macro"	Stata's local macros
"global"	"macro"	Stata's global macros
"global"	"numscalar"	Stata's numeric scalars
"global"	"strscalar"	Stata's string scalars
"global"	"matrix"	Stata's matrices
"r()"	"macro"	macros in r()
"r()"	"numscalar"	numeric scalars in r()
"r()"	"matrix"	matrices in r()
"e()"	"macro"	macros in e()
"e()"	"numscalar"	numeric scalars in e()
"e()"	"matrix"	matrices in e()
"s()"	"macro"	macros in s()
"char"	"_dta"	characteristics in _dta[]
"char"	"*name*"	characteristics in variable *name*[]

st_dir() returns an empty list if an invalid *cat–subcat* combination is specified.

Description

st_dir(*cat*, *subcat*, *pattern*) and st_dir(*cat*, *subcat*, *pattern*, *adorn*) return a column vector containing the names matching *pattern* of the Stata objects described by *cat–subcat*.

Argument *adorn* is optional; not specifying it is equivalent to specifying *adorn* = 0. By default, simple names are returned. If *adorn* ≠ 0 is specified, the name is adorned in the standard Stata way used to describe the object. Say that one is listing the macros in e() and one of the elements is e(cmd). By default, the returned vector will contain an element equal to "cmd". With *adorn* ≠ 0, the element will be "e(cmd)".

For many objects, the adorned and unadorned forms of the names are the same.

Conformability

st_dir(*cat*, *subcat*, *pattern*, *adorn*):

cat:	1×1	
subcat:	1×1	
pattern:	1×1	
adorn:	1×1	(optional)
result:	$k \times 1$	

Diagnostics

st_dir(*cat*, *subcat*, *pattern*) and st_dir(*cat*, *subcat*, *pattern*, *adorn*) abort with error if *cat* or *subcat* is invalid. If the combination is invalid, however, J(0,1,"") is returned. *subcat*==*name* is considered invalid unless *cat*=="char".

st_dir() aborts with error if any of its arguments are views.

Also see

[M-4] **stata** — Stata interface functions

Title

[M-5] **st_dropvar( )** — Drop variables or observations

Syntax

> *void*   st_dropvar(*transmorphic rowvector vars*)
>
> *void*   st_dropobsin(*real matrix range*)
>
> *void*   st_dropobsif(*real colvector select*)
>
> *void*   st_keepvar(*transmorphic rowvector vars*)
>
> *void*   st_keepobsin(*real matrix range*)
>
> *void*   st_keepobsif(*real colvector select*)

Description

st_dropvar(*vars*) drops the variables specified. *vars* is a row vector that may contain either variable names or variable indices. st_dropvar(.) drops all variables and observations.

st_dropobsin() and st_dropobsif() have to do with dropping observations.

st_dropobsin(*range*) specifies the observations to be dropped:

> st_dropobsin(5) drops observation 5.
>
> st_dropobsin((5,9)) drops observations 5 through 9.
>
> st_dropobsin((5\8\12)) drops observations 5 and 8 and 12.
>
> st_dropobsin((5,7\8,11\13,13)) drops observations 5 through 7, 8 through 11, and 13.
>
> st_dropobsin(.) drops all observations (but not the variables).
>
> st_dropobsin(J(0,1,.)) drops no observations (or variables).

st_dropobsif(*select*) specifies a st_nobs() $\times$ 1 vector. Observations i for which $select_i \neq 0$ are dropped.

st_keepvar(), st_keepobsin(), and st_keepobsif() do the same thing, except that the variables and observations to be kept are specified.

Remarks

To drop all variables and observations, code any of the following:

```
st_dropvar(.)
st_keepvar(J(1,0,.))
st_keepvar(J(1,0,""))
```

All do the same thing. Dropping all the variables clears the dataset.

Dropping all the observations, however, leaves the variables in place.

Conformability

st_dropvar(*vars*), st_keepvar(*vars*):
>   *vars*: $1 \times k$
>   *result*:   *void*

st_dropobsin(*range*), st_keepobsin(*range*):
>   *range*: $k \times 1$ or $k \times 2$
>   *result*:   *void*

st_dropobsif(*select*), st_keepobsif(*select*):
>   *select*:   st_nobs() $\times 1$
>   *result*:   *void*

Diagnostics

st_dropvar(*vars*) and st_keepvar(*vars*) abort with error if any element of *vars* is missing unless *vars* is 1×1, in which case they drop or keep all the variables.

st_dropvar(*vars*) and st_keepvar(*vars*) abort with error if any element of *vars* is not a valid variable index or name, or if *vars* is a view. If *vars* is specified as names, abbreviations are not allowed.

st_dropvar() and st_keepvar() set st_updata() (see [M-5] **st_updata( )**) unless all variables dropped are temporary; see [M-5] **st_tempname( )**.

st_dropobsin(*range*) and st_keepobsin(*range*) abort with error if any element of *range* is missing unless *range* is 1×1, in which case they drop or keep all the observations.

st_dropobsin(*range*) and st_keepobsin(*range*) abort with error if any element of *range* is not a valid observation number (is not between 1 and st_nobs() [see [M-5] **st_nvar( )**] inclusive) or if *range* is a view.

st_dropobsif(*select*) and st_keepobsif(*select*) abort with error if *select* is a view.

st_dropobsin(), st_dropobsif(), st_keepobsin(), and st_keepobsif() set st_updata() if any observations are removed from the data.

Be aware that, after dropping any variables or observations, any previously constructed views (see [M-5] **st_view( )**) are probably invalid because views are internally stored in terms of variable and observation numbers. Subsequent use of an invalid view may lead to unexpected results or an abort with error.

Also see

[M-4] **stata** — Stata interface functions

Title

> **[M-5] st_global( )** — Obtain strings from and put strings into global macros

Syntax

> *string scalar*   st_global(*string scalar name*)
>
> *void*   st_global(*string scalar name*, *string scalar contents*)

where

1. *name* is to contain

   a. global macro such as "myname"

   b. r() macro such as "r(names)"

   c. e() macro such as "e(cmd)"

   d. s() macro such as "s(vars)"

   e. c() macro such as "c(current_date)"

   f. dataset characteristic such as "_dta[date]"

   g. variable characteristic such as "mpg[note]"

2. st_global(*name*, "") deletes, and it deletes even if the *name* is not a macro. For instance, perhaps r(N) is a numeric scalar; st_global(r(N), "") will delete it. Perhaps e(X) is matrix; st_global(e(X), "") will delete it.

Description

st_global(*name*) returns the contents of the specified Stata global.

st_global(*name*, *contents*) sets or resets the contents of the specified Stata global. If the Stata global did not previously exist, a new global is created. If the global did exist, the new contents replace the old.

Remarks

Mata provides a suite of functions for obtaining and setting the contents of global macros, local macros, saved results, etc. It can sometimes be confusing to know which you should use. The table on the following page will help.

Stata component/action	function call
Local macro	
obtain contents	*contents* = st_local("*name*")
create/set/replace	st_local("*name*", *contents*)
delete	st_local("*name*", "")
Global macro	
obtain contents	*contents* = st_global("*name*")
create/set/replace	st_global("*name*", *contents*)
delete	st_global("*name*", "")
Global numeric scalar	
obtain contents	*value* = st_numscalar("*name*")
create/set/replace	st_numscalar("*name*", *value*)
delete	st_numscalar("*name*", J(0,0,.))
Global string scalar	
obtain contents	*contents* = st_strscalar("*name*")
create/set/replace	st_strscalar("*name*", *contents*)
delete	st_strscalar("*name*", J(0,0,""))
Global matrix	
obtain contents	*matrix* = st_matrix("*name*") *rowlabel* = st_matrixrowstripe("*name*") *collabel* = st_matrixcolstripe("*name*")
create/set/replace	st_matrix("*name*", *matrix*) st_matrixrowstripe("*name*", *rowlabel*) st_matrixcolstripe("*name*", *collabel*)
replace	st_replacematrix("*name*", *matrix*)
delete	st_matrix("name", J(0,0,.))
Characteristic	
obtain contents	*contents* = st_global("*name[name]*")
create/set/replace	st_global("*name[name]*", *contents*)
delete	st_global("*name[name]*", "")

Stata component/action	function call
r( ) results	
macro	
obtain contents	*contents* = st_global("r(*name*)")
create/set/replace	st_global("r(*name*)", *contents*)
numeric scalar	
obtain contents	*value* = st_numscalar("r(*name*)")
create/set/replace	st_numscalar("r(*name*)", *value*)
matrix	
obtain contents	*matrix* = st_matrix("r(*name*)")
	rowlabel = st_matrixrowstripe("r(*name*)")
	collabel = st_matrixcolstripe("r(*name*)")
create/set/replace	st_matrix("r(name)", *matrix*)
	st_matrixrowstripe("r(*name*)", *rowlabel*)
	st_matrixcolstripe("r(*name*)", *collabel*)
replace	st_replacematrix("r(*name*)", *matrix*)
IN ALL CASES	
delete	st_global("r(*name*)", "")
to delete all of r( )	st_rclear( )
e( ) results	same as r( ) results, but code e(*name*) and st_eclear( )
s( ) results	
macro	
obtain contents	*contents* = st_global("s(*name*)")
create/set/replace	st_global("s(*name*)", *contents*)
delete	st_global("s(*name*)", "")
to delete all of s( )	st_sclear( )
c( ) results	
macro	
obtain contents	*contents* = st_global("c(*name*)")
numeric scalar	
obtain contents	*value* = st_numscalar("c(*name*)")

See [M-5] **st_local( )**, [M-5] **st_numscalar( )**, [M-5] **st_matrix( )**, and [M-5] **st_rclear( )**.

Conformability

st_global(*name*):
        name: 1 × 1
        result: 1 × 1

st_global(*name*, *contents*):
        name: 1 × 1
    contents: 1 × 1
        result: *void*

Diagnostics

st_global(*name*) returns "" if the name contained in *name* is not defined. st_global(*name*) aborts with error if the name is malformed, such as st_global("invalid name").

st_global(*name*, *contents*) aborts with error if the name contained in *name* is malformed. The maximum length of strings in Mata is significantly longer than in Stata. st_global() truncates what is stored at the appropriate maximum length if that is necessary.

Reference

Gould, W. W. 2008. Mata Matters: Macros. *Stata Journal* 8: 401–412.

Also see

[M-5] **st_rclear( )** — Clear r( ), e( ), or s( )

[M-4] **stata** — Stata interface functions

Title

[M-5] st_isfmt( ) — Whether valid %fmt

Syntax

> *real scalar*   st_isfmt(*string scalar s*)
>
> *real scalar*   st_isnumfmt(*string scalar s*)
>
> *real scalar*   st_isstrfmt(*string scalar s*)

Description

st_isfmt(*s*) returns 1 if *s* contains a valid Stata %*fmt* and 0 otherwise.

st_isnumfmt(*s*) returns 1 if *s* contains a valid Stata numeric %*fmt* and 0 otherwise.

st_isstrfmt(*s*) returns 1 if *s* contains a valid Stata string %*fmt* and 0 otherwise.

Conformability

st_isfmt(*s*), st_isnumfmt(*s*), st_isstrfmt(*s*):
>   *s*: 1×1
> *result*: 1×1

Diagnostics

st_isfmt(*s*), st_isnumfmt(*s*), and st_isstrfmt(*s*) abort with error if *s* is a view.

Also see

[M-4] **stata** — Stata interface functions

Title

[M-5] st_isname( ) — Whether valid Stata name

Syntax

> *real scalar*   st_isname(*string scalar s*)
>
> *real scalar*   st_islmname(*string scalar s*)

Description

st_isname(*s*) returns 1 if *s* contains a valid Stata name and 0 otherwise.

st_islmname(*s*) returns 1 if *s* contains a valid Stata local-macro name and 0 otherwise.

Conformability

st_isname(*s*), st_islmname(*s*):
>   *s*: 1×1
> *result*: 1×1

Diagnostics

st_isname(*s*) aborts with error if *s* is a view (but st_islmname() does not).

Also see

[M-4] **stata** — Stata interface functions

Title

[M-5] st_local( ) — Obtain strings from and put strings into Stata macros

---

Syntax

> *string scalar*   st_local(*string scalar name*)
>
> *void*   st_local(*string scalar name*, *string scalar contents*)

Note: st_local(*name*, "") deletes.

Description

st_local(*name*) returns the contents of the specified local macro.

st_local(*name*, *contents*) sets or resets the contents of the specified local macro. If the macro did not previously exist, a new macro is created. If it did previously exist, the new contents replace the old.

Remarks

See [M-5] **st_global( )** and [M-5] **st_rclear( )**.

Conformability

st_local(*name*):
>   *name*: 1×1
>   *result*: 1×1

st_local(*name*, *contents*):
>   *name*: 1×1
>   *contents*: 1×1
>   *result*:   *void*

Diagnostics

st_local(*name*) returns "" if the name contained in *name* is not defined. st_local(*name*) aborts with error if the name is malformed.

st_local(*name*, *contents*) aborts with error if the name contained in *name* is malformed.

Reference

Gould, W. W. 2008. Mata Matters: Macros. *Stata Journal* 8: 401–412.

Also see

[M-4] **stata** — Stata interface functions

Title

[M-5] **st_macroexpand( )** — Expand Stata macros in string

Syntax

>*string scalar*   st_macroexpand(*string scalar s*)
>
>*real scalar*   _st_macroexpand(*S, string scalar s*)

Note: the type of *S* does not matter; it is replaced and becomes a string scalar.

Description

st_macroexpand(*s*) returns *s* with any quoted or dollar sign–prefixed macros expanded.

_st_macroexpand(*S, s*) places in *S* the contents of *s* with any quoted or dollar sign–prefixed macros expanded and returns a Stata return code (it returns 0 if all went well).

Remarks

Be careful coding string literals containing quoted or prefixed macros because macros are also expanded at compile time. For instance, consider

>s = st_macroexpand("regress `varlist'")

'varlist' will be substituted with its value at compile time. What you probably want is

>s = st_macroexpand("regress " + "`" + "varlist" + "'")

Conformability

st_macroexpand(*s*):

>| | |
>|---:|---|
>| *s*: | 1×1 |
>| *result*: | 1×1 |

_st_macroexpand(*S, s*):

>*input*:

>| | |
>|---:|---|
>| *s*: | 1×1 |

>*output*:

>| | |
>|---:|---|
>| *S*: | 1×1 |
>| *result*: | 1×1 |

Diagnostics

st_macroexpand(*s*) aborts with error if *s* is too long (exceedingly unlikely) or if macro expansion fails (also unlikely).

_st_macroexpand(*S, s*) aborts with error if *s* is too long.

Also see

[M-4] **stata** — Stata interface functions

Title

Syntax

real matrix	st_matrix(*string scalar name*)
string matrix	st_matrixrowstripe(*string scalar name*)
string matrix	st_matrixcolstripe(*string scalar name*)
void	st_matrix(*string scalar name*, *real matrix X*)
void	st_matrixrowstripe(*string scalar name*, *string matrix s*)
void	st_matrixcolstripe(*string scalar name*, *string matrix s*)
void	st_replacematrix(*string scalar name*, *real matrix X*)

where

1. All functions allow *name* to be

    a. global matrix name such as "mymatrix",

    b. r() matrix such as "r(Z)", or

    c. e() matrix such as "e(V)".

2. st_matrix(*name*) returns J(0,0,.) if the matrix does not exist.

3. The contents of r() and e() may be replaced except for the estimation results e(b) and e(V).

4. st_matrix(*name*, X) deletes the specified Stata matrix if X==J(0,0,.).

Description

st_matrix(*name*) returns the contents of Stata's matrix *name*, or it returns J(0,0,.) if the matrix does not exist.

st_matrixrowstripe(*name*) returns the row stripe associated with the matrix *name*, or it returns J(0,2,"") if the matrix does not exist.

st_matrixcolstripe(*name*) returns the column stripe associated with the matrix *name*, or it returns J(0,2,"") if the matrix does not exist.

st_matrix(*name*, X) sets or resets the contents of the Stata matrix *name* to be X. If the matrix did not previously exist, a new matrix is created. If the matrix did exist, the new contents replace the old. Either way, the row and column stripes are also reset to contain "r1", "r2", ..., and "c1", "c2", ...

st_matrix(*name*, X) deletes the Stata matrix *name* when X is 0×0: st_matrix(*name*, J(0,0,.)) deletes Stata matrix *name* or does nothing if *name* does not exist.

st_matrixrowstripe(*name*, s) and st_matrixcolstripe(*name*, s) change the contents to be s of the row and column stripe associated with the already existing Stata matrix *name*. In either case, s must be $n \times 2$, where $n =$ the number of rows (columns) of the underlying matrix.

st_matrixrowstripe(*name*, s) and st_matrixcolstripe(*name*, s) reset the row and column stripe to be "r1", "r2", ..., and "c1", "c2", ..., when s is 0×2 (i.e., J(0,2,"")).

st_replacematrix(*name*, X) resets the contents of the Stata matrix *name* to be X. The existing Stata matrix must have the same number of rows and columns as X. The row stripes and column stripes remain unchanged.

Remarks

Remarks are presented under the following headings:

> *Processing Stata's row and column stripes*
> *Stata's matsize is irrelevant*

Also see [M-5] **st_global( )** and [M-5] **st_rclear( )**.

Processing Stata's row and column stripes

Both row stripes and column stripes are presented in the same way: each row of s represents the *eq:op.name* associated with a row or column of the underlying matrix. The first column records *eq*, and the second column records *op.name*. For instance, given the following Stata matrix

	turn	L. turn	eq2: turn	eq2: L. turn
mpg	1	2	3	4
L.mpg	5	6	7	8
eq2:mpg	9	10	11	12
eq2:L.mpg	13	14	15	16

st_matrixrowstripe(*name*) returns the 4×2 string matrix

""	"mpg"
""	"L.mpg"
"eq2"	"mpg"
"eq2"	"L.mpg"

and st_matrixcolstripe(*name*) returns

""	"turn"
""	"L.turn"
"eq2"	"turn"
"eq2"	"L.turn"

Stata's matsize is irrelevant

Matrices in Stata are limited to `matsize` (see [R] **matsize**), a number between 10 and 11,000. Mata matrices have no such limits.

When getting a matrix, the `matsize` limit plays no role.

When putting a matrix, the `matsize` limit is ignored; meaning that, to use the matrix in Stata, the user may have to reset `matsize` or, if the matrix is too large, the user may not be able to use the matrix at all.

Conformability

st_matrix(*name*):
    name: 1×1
    result: $m \times n$ (0×0 if not found)

st_matrixrowstripe(*name*):
    name: 1×1
    result: $m \times 2$ (0×2 if not found)

st_matrixcolstripe(*name*):
    name: 1×1
    result: $n \times 2$ (0×2 if not found)

st_matrix(*name*, *X*):
    name: 1×1
    X: $r \times c$ (0×0 means delete)
    result: *void*

st_matrixrowstripe(*name*, *s*):
    name: 1×1
    s: $r \times 2$ (0×2 means default "r1", "r2", ...)
    result: *void*

st_matrixcolstripe(*name*, *s*):
    name: 1×1
    s: $c \times 2$ (0×2 means default "c1", "c2", ...)
    result: *void*

st_replacematrix(*name*, *X*):
    name: 1×1
    X: $m \times n$ (0×0 means delete)
    result: *void*

Diagnostics

st_matrix(*name*), st_matrixrowstripe(*name*), and st_matrixcolstripe(*name*) abort with error if *name* is malformed. Also,

1. st_matrix(*name*) returns J(0,0,.) if Stata matrix *name* does not exist.

2. st_matrixrowstripe(*name*) and st_matrixcolstripe(*name*) return J(0,2,"") if Stata matrix *name* does not exist. There is no possibility that matrix *name* might exist and not have row and column stripes.

st_matrix(*name*, *X*), st_matrixrowstripe(*name*, *s*), and st_matrixcolstripe(*name*, *s*) abort with error if *name* is malformed. Also,

1. st_matrixrowstripe(*name*, *s*) aborts with error if rows(*s*) is not equal to the number of rows of Stata matrix *name* and rows(*s*)!=0, or if cols(*s*)!=2.

2. st_matrixcolstripe(*name*, *s*) aborts with error if cols(*s*) is not equal to the number of columns of Stata matrix *name* and cols(*s*)!=0, or if cols(*s*)!=2.

st_replacematrix(*name*, *X*) aborts with error if Stata matrix *name* does not have the same number of rows and columns as *X*. st_replacematrix() also aborts with error if Stata matrix *name* does not exist and *X*!=J(0,0,.); st_replacematrix() does nothing if the matrix does not exist and *X*==J(0,0,.). st_replacematrix() aborts with error if *name* is malformed.

Also see

[M-5] **st_rclear( )** — Clear r( ), e( ), or s( )

[M-4] **stata** — Stata interface functions

Title

[M-5] st_numscalar( ) — Obtain values from and put values into Stata scalars

Syntax

real   st_numscalar(*string scalar name*)

void   st_numscalar(*string scalar name*, *real value*)

string   st_strscalar(*string scalar name*)

void   st_strscalar(*string scalar name*, *string value*)

where

1. All functions allow *name* to be

    a. global scalar such as "myname",

    b. r() scalar such as "r(mean)",

    c. e() scalar such as "e(N)", or

    d. c() scalar such as "c(namelen)".

2. st_numscalar(*name*) and st_strscalar(*name*) return a 1×1 result containing the value of *name* or they return a 0×0 if *name* does not exist.

3. st_numscalar(*name*, *value*) and st_strscalar(*name*, *value*) allow *value* to be 1×1 containing the new value with which *name* is to be created or replaced, or 0×0, specifying that *name* is to be deleted.

Description

st_numscalar(*name*) returns the value of the specified Stata numeric scalar, or it returns J(0,0,.) if the scalar does not exist.

st_numscalar(*name*, *value*) sets or resets the value of the specified numeric scalar, assuming *value* != J(0,0,.). st_numscalar(*name*, *value*) deletes the specified scalar if *value* == J(0,0,.). st_numscalar("x", J(0,0,.)) erases the scalar x, or it does nothing if scalar x did not exist.

st_strscalar(*name*) returns the value of the specified Stata string scalar, or it returns J(0,0,"") if the scalar does not exist.

st_strscalar(*name*, *value*) sets or resets the value of the specified scalar, assuming *value* != J(0,0,""). st_strscalar(*name*, *value*) deletes the specified scalar if *value* == J(0,0,""). st_strscalar("x", J(0,0,"")) erases the scalar x, or it does nothing if scalar x did not exist.

Concerning deletion of a scalar, it does not matter whether you code st_numscalar(*name*, J(0,0,.)) or st_strscalar(*name*, J(0,0,"")); both yield the same result.

Remarks

See [M-5] **st_global**( ) and [M-5] **st_rclear**( ).

Conformability

st_numscalar(*name*), st_strscalar(*name*):

    name: 1×1
    result: 1×1 or 0×0

st_numscalar(*name*, *value*), st_strscalar(*name*, *value*):

    name: 1×1
    value: 1×1 or 0×0
    result: *void*

Diagnostics

st_numscalar(*name*) and st_strscalar(*name*) return $J(0,0,.)$ or $J(0,0,"")$ if Stata scalar *name* does not exist. They abort with error, however, if the name is malformed.

st_numscalar(*name*, *value*) and st_strscalar(*name*, *value*) abort with error if the name is malformed.

Also see

[M-5] **st_rclear**( ) — Clear r( ), e( ), or s( )

[M-4] **stata** — Stata interface functions

Title

[M-5] st_nvar( ) — Numbers of variables and observations

Syntax

>*real scalar*   st_nvar()
>
>*real scalar*   st_nobs()

Description

st_nvar() returns the number of variables defined in the dataset currently loaded in Stata.

st_nobs() returns the number of observations defined in the dataset currently loaded in Stata.

Conformability

st_nvar(), st_nobs():
>   *result*: 1×1

Diagnostics

None.

Also see

[M-4] **stata** — Stata interface functions

Title

[M-5] **st_rclear( )** — Clear r( ), e( ), or s( )

Syntax

void `st_rclear()`

void `st_eclear()`

void `st_sclear()`

Description

`st_rclear()` clears Stata's `r()` saved results.

`st_eclear()` clears Stata's `e()` saved results.

`st_sclear()` clears Stata's `s()` saved results.

Remarks

Returning results in `r()`, `e()`, or `s()` is one way of communicating results calculated in Mata back to Stata; see [M-1] **ado**. See [R] **saved results** for a description of `e()`, `r()`, and `s()`.

Use `st_rclear()`, `st_eclear()`, or `st_sclear()` to clear results, and then use `st_global()` to define macros, `st_numscalar()` to define scalars, and `st_matrix()` to define Stata matrices in `r()`, `e()`, or `s()`. For example,

```
st_rclear()
st_global("r(name)", "tab")        see [M-5] st_global( )
st_numscalar("r(N)", n1+n2)        see [M-5] st_numscalar( )
st_matrix("r(table)", X+Y)         see [M-5] st_matrix( )
```

It is not necessary to clear before saving, but it is considered good style unless it is your intention to add to previously saved results.

If a saved result already exists, `st_global()`, `st_numscalar()`, and `st_matrix()` may be used to redefine it and even to redefine it to a different type. For instance, continuing with our example, later in the same code might appear

```
if (...) {
    st_matrix("r(name)", X)
}
```

Saved result `r(name)` was previously defined as a macro containing `"tab"`, and, even so, can now be redefined to become a matrix.

If you want to eliminate a particular saved result, use `st_global()` to change its contents to `""`:

```
st_global("r(name)", "")
```

Do this regardless of the type of the saved result. Here we use st_global() to clear saved result r(name), which might be a macro and might be a matrix.

Conformability

st_rclear(), st_eclear(), and st_sclear() take no arguments and return void.

Diagnostics

st_rclear(), st_eclear(), and st_sclear() cannot fail.

Also see

[M-5] **st_global( )** — Obtain strings from and put strings into global macros

[M-5] **st_numscalar( )** — Obtain values from and put values into Stata scalars

[M-5] **st_matrix( )** — Obtain and put Stata matrices

[M-4] **stata** — Stata interface functions

Title

[M-5] **st_store( )** — Modify values stored in current Stata dataset

Syntax

void	_st_store(*real scalar i*, *real scalar j*, *real scalar x*)
void	st_store(*real matrix i*, *rowvector j*, *real matrix X*) (1,2)
void	st_store(*real matrix i*, *rowvector j*, *scalar selectvar*, *real matrix X*) (1,2,3)
void	_st_sstore(*real scalar i*, *real scalar j*, *string scalar s*)
void	st_sstore(*real matrix i*, *rowvector j*, *string matrix X*) (1,2)
void	st_sstore(*real matrix i*, *rowvector j*, *scalar selectvar*, *string matrix X*) (1,2,3)

where

1. *i* may be specified in the same way as with st_data().

2. *j* may be specified in the same way as with st_data(), except that time-series operators may not be specified.

3. *selectvar* may be specified in the same way as with st_data().

See [M-5] **st_data( )**.

Description

These functions mirror _st_data(), st_data(), and st_sdata(). Rather than returning the contents from the Stata dataset, these commands change those contents to be as given by the last argument.

Remarks

See [M-5] **st_data( )**.

Conformability

_st_store(*i*, *j*, *x*), _st_sstore(*i*, *j*, *x*):

i:	1×1
j:	1×1
x:	1×1
result:	*void*

st_store(*i*, *j*, *X*), st_sstore(*i*, *j*, *X*):

i:	$n \times 1$ or $n_2 \times 2$
j:	$1 \times k$
X:	$n \times k$
result:	*void*

st_store(i, j, *selectvar*, X), st_sstore(i, j, *selectvar*, X):

i:	$n \times 1$	or	$n_2 \times 2$	
j:	$1 \times k$			
selectvar:	1×1			
X:	$(n - e) \times k$,	where e is number of observations excluded by *selectvar*		
result:	*void*			

Diagnostics

_st_store(i, j, x) and _st_sstore(i, j, s) do nothing if i or j is out of range; they do not abort with error.

st_store(i, j, X) and st_sstore(i, j, s) abort with error if any element of i or j is out of range. j may be specified as a vector of variable names or as a vector of variable indices. If names are specified, abbreviations are allowed. If you do not want this, use st_varindex() (see [M-5] **st_varindex**( )) to translate variable names into variable indices.

st_store() and st_sstore() abort with error if X is not p-conformable with the matrix that st_data() (st_sdata()) would return.

Also see

[M-5] **st_data**( ) — Load copy of current Stata dataset

[M-5] **st_addvar**( ) — Add variable to current Stata dataset

[M-4] **stata** — Stata interface functions

Title

[M-5] **st_subview( )** — Make view from view

Syntax

$$void \quad \texttt{st\_subview}(X, \textit{transmorphic matrix } V, \textit{real matrix } i, \textit{real matrix } j)$$

where

1. The type of X does not matter; it is replaced.

2. V is typically a view, but that is not required. V, however, must be real or string.

Description

`st_subview`(X, V, i, j) creates new view matrix X from existing view matrix V. V is to have been created from a previous call to `st_view()` (see [M-5] **st_view( )**) or `st_subview()`.

Although `st_subview()` is intended for use with view matrices, it may also be used when V is a regular matrix. Thus code may be written in such a way that it will work without regard to whether a matrix is or is not a view.

i may be specified as a 1×1 scalar, a 1×1 scalar containing missing, as a column vector of row numbers, as a row vector specifying a row-number range, or as a $k \times 2$ matrix specifying both:

  a. `st_subview`$(X,V, 1,2)$ makes X equal to the first row of the second column of V.

  b. `st_subview`$(X,V, .,2)$ makes X equal to all rows of the second column of V.

  c. `st_subview`$(X,V, (1\backslash 2\backslash 5),2)$ makes X equal to rows 1, 2, and 5 of the second column of V.

  d. `st_subview`$(X,V, (1,5),2)$ makes X equal to rows 1 through 5 of the second column of V.

  e. `st_subview`$(X,V, (1,5\backslash 7,9),2)$ makes X equal to rows 1 through 5 and 7 through 9 of the second column of V.

  f. When a range is specified, any element of the range (i_1,i_2) may be set to contribute zero observations if $i_2 = i_1 - 1$. For example, $(1,0)$ is not an error and neither is $(1,0\backslash 5,7)$.

j may be specified in the same way as i, except transposed, to specify the selected columns:

  a. `st_subview`$(X,V, 2,.)$ makes X equal to all columns of the second row of V.

  b. `st_subview`$(X,V, 2,(1,2,5))$ makes X equal to columns 1, 2, and 5 of the second row of V.

  c. `st_subview`$(X,V, 2,(1\backslash 5))$ makes X equal to columns 1 through 5 of the second row of V.

  d. `st_subview`$(X,V, 2,((1\backslash 5),(7\backslash 9)))$ makes X equal to columns 1 through 5 and 7 through 9 of the second row of V.

e. When a range is specified, any element of the range $(j_1 \backslash j_2)$ may be set to contribute zero columns if $j_2 = j_1 - 1$. For example, (1\0) is not an error and neither is ((1\0),(5\7)).

Obviously, notations for i and j can be specified simultaneously:

a. st_subview(X,V, .,.) makes X a duplicate of V.

b. st_subview(X,V, .,(1\5)) makes X equal to columns 1 through 5 of all rows of X.

c. st_subview(X,V, (10,25),(1\5)) makes X equal to columns 1 through 5 of rows 10 through 25 of X.

Also, st_subview() may be used to create views with duplicate variables or observations from V.

Remarks

Say that you need to make a calculation on matrices X and Y, which might be views. Perhaps the calculation is invsym($X'X$)*$X'Y$. Regardless, you start as follows:

```
st_view(X, ., "v2 v3 v4", 0)
st_view(Y, ., "v1 v7"   , 0)
```

You are already in trouble. You smartly coded fourth argument as 0, meaning exclude the missing values, but you do not know that the same observations were excluded in the manufacturing of X as in the manufacturing of Y.

If you had previously created a touse variable in your dataset marking the observations to be used in the calculation, one solution would be

```
st_view(X, ., "v2 v3 v4", "touse")
st_view(Y, ., "v1 v7"   , "touse")
```

That solution is recommended, but let's assume you did not do that. The other solution is

```
st_view(M, ., "v2 v3 v4 v1 v7", 0)
st_subview(X, M, ., (1,2,3))
st_subview(Y, M, ., (4,5))
```

The first call to st_view() will eliminate observations with missing values on any of the variables, and the second two st_subview() calls will create the matrices you wanted, obtaining them from the correctly formed M. Basically, the two st_subview() calls amount to the same thing as

```
X = M[., (1,2,3)]
Y = M[., (4,5)]
```

but you do not want to code that because then matrices X and Y would contain copies of the data, and you are worried that the dataset might be large.

For a second example, let's pretend that you are processing a panel dataset and making calculations from matrix X within panel. Your code looks something like

```
st_view(id, ., "panelid", 0)
for (i=1; i<=rows(id); i=j+1) {
        j = endobs(id, i)
        st_view(X, (i,j), "v1 v2 ...", 0)
        ...
}
```

where you have previously written function endobs() to be

```
scalar endobs(vector id, scalar i)
{
        scalar   j
        for (j=i+1; j<=rows(id); j++) {
                if (id[j]!=id[i]) return(j-1)
        }
        return(rows(id))
}
```

In any case, there could be a problem. Missing values of variable panelid might not align with missing values of variables v1, v2, etc. The result could be that observation and row numbers are not in accordance or that there appears to be a group that, in fact, has all missing data. The right way to handle the problem is

```
st_view(M, ., "panelid v1 v2 ...", 0)
st_subview(id, M, ., 1)
for (i=1; i<=rows(id); i=j+1) {
        j = endobs(id, i)
        st_subview(X, M, (i,j), (2\cols(M)))
        ...
}
```

Conformability

st_subview(X, V, i, j):

input:

V:	$r \times c$
i:	$1 \times 1, n \times 1$, or $n_2 \times 2$
j:	$1 \times 1, 1 \times k$, or $2 \times k_2$

output:

X:	$n \times k$

Diagnostics

st_subview(X, V, i, j) aborts with error if i or j are out of range. i and j refer to row and column numbers of V, not observation and variable numbers of the underlying Stata dataset.

Also see

[M-5] **st_view( )** — Make matrix that is a view onto current Stata dataset

[M-5] **select( )** — Select rows or columns

[M-4] **stata** — Stata interface functions

Title

[M-5] st_tempname( ) — Temporary Stata names

Syntax

> *string scalar*   st_tempname()
>
> *string rowvector*   st_tempname(*real scalar n*)
>
> *string scalar*   st_tempfilename()
>
> *string rowvector*   st_tempfilename(*real scalar n*)

Description

st_tempname() returns a Stata temporary name, the same as would be returned by Stata's tempvar and tempname commands; see [P] **macro**.

st_tempname(*n*) returns *n* temporary Stata names, $n \geq 0$.

st_tempfilename() returns a Stata temporary filename, the same as would be returned by Stata's tempfile command; see [P] **macro**.

st_tempfilename(*n*) returns *n* temporary filenames, $n \geq 0$.

Remarks

Remarks are presented under the following headings:

> *Creating temporary objects*
> *When temporary objects will be eliminated*

Creating temporary objects

st_tempname()s can be used to name Stata's variables, matrices, and scalars. Although in Stata a distinction is drawn between tempvars and tempnames, there is no real distinction, and so st_tempname() handles both in Mata. For instance, one can create a temporary variable by coding

```
idx = st_addvar("double", st_tempname())
```

See [M-5] **st_addvar( )**.

One creates a temporary file by coding

```
fh = fopen(st_tempfilename(), "w")
```

See [M-5] **fopen( )**.

When temporary objects will be eliminated

Temporary objects do not vanish when the Mata function ends, nor when Mata itself ends. They are removed when the ado-file (or do-file) calling Mata terminates.

Forget Mata for a minute. Stata eliminates temporary variables and files when the program that created them ends. That same rule applies to Mata: Stata eliminates them, not Mata, and that means that the ado-file or do-file that called Mata will eliminate them when that ado-file or do-file ends. Temporary variables and files are not eliminated by Mata when the Mata function ends. Thus Mata functions can create temporary objects for use by their ado-file callers, should that prove useful.

Conformability

st_tempname(), st_tempfilename():
> *result*: 1×1

st_tempname(n), st_tempfilename(n):
> n: 1×1
> *result*: $1 \times n$

Diagnostics

st_tempname(n) and st_tempfilename(n) abort with error if $n < 0$ and return J(1,0,"") if $n = 0$.

Also see

[M-5] **st_addvar( )** — Add variable to current Stata dataset

[M-4] **stata** — Stata interface functions

Title

> [M-5] **st_tsrevar( )** — Create time-series op.varname variables

Syntax

> *real rowvector*   st_tsrevar(*string rowvector s*)
>
> *real rowvector* _st_tsrevar(*string rowvector s*)

Description

st_tsrevar(*s*) is the equivalent of Stata's [TS] **tsrevar** programming command: it generates temporary variables containing the evaluation of any *op.varname* combinations appearing in *s*.

_st_tsrevar(*s*) does the same thing as st_tsrevar(). The two functions differ in how they respond to invalid elements of *s*. st_tsrevar() aborts with error, and _st_tsrevar() places missing in the appropriate element of the returned result.

Remarks

Both of these functions help achieve efficiency when using views and time-series variables. Assume that in *vars*, you have a list of Stata variable names, some of which might contain time-series *op.varname* combinations such as 1.gnp. For example, *vars* might contain

> *vars* = "gnp r l.gnp"

If you wanted to create in *V* a view on the data, you would usually code

> st_view(*V*, ., *vars*)

We are not going to do that, however, because we plan to do many calculations with *V* and, to speed execution, we want any *op.varname* combinations evaluated just once, as *V* is created. Of course, if efficiency were our only concern, we would code

> *V* = st_data(., *vars*)

Assume, however, that we have lots of data, so memory is an issue, and yet we still want as much efficiency as possible given the constraint of not copying the data. The solution is to code

> st_view(*V*, ., st_tsrevar(tokens(*vars*)))

st_tsrevar() will create temporary variables for each *op.varname* combination (1.gnp in our example), and then return the Stata variable indices of each of the variables, whether newly created or already existing. If gnp was the second variable in the dataset, r was the 23rd, and in total there were 54 variables, then returned by st_tsrevar() would be (2, 23, 55). Variable 55 is new, created by st_tsrevar(), and it contains the values of 1.gnp. The new variable is temporary and will be dropped automatically at the appropriate time.

Conformability

st_tsrevar(*s*), _st_tsrevar(*s*):
         s: $1 \times c$
    result: $1 \times c$

Diagnostics

st_tsrevar( ) aborts with error if any variable name is not found or any *op*.*varname* combination is invalid.

_st_tsrevar( ) puts missing in the appropriate element of the returned result for any variable name that is not found or any *op*.*varname* combination that is invalid.

Also see

[M-5] **st_varname( )** — Obtain variable names from variable indices

[M-4] **stata** — Stata interface functions

[M-5] **st_varindex( )** — Obtain variable indices from variable names

Title

> **[M-5] st_updata( )** — Determine or set data-have-changed flag

Syntax

> *real scalar*   st_updata( )
>
> *void*   st_updata(*real scalar value*)

Description

st_updata( ) returns 0 if the data in memory have not changed since they were last saved and returns 1 otherwise.

st_updata(*value*) sets the data-have-changed flag to 0 if *value* = 0 and 1 otherwise.

Remarks

Stata's describe command reports whether the data have changed since they were last saved. Stata's use command refuses to load a new dataset if the data currently in memory have not been saved since they were last changed. Other components of Stata also react to the data-have-changed flag.

st_updata( ) allows you to respect that same flag.

Also, as a Mata programmer, you must set the flag if your function changes the data in memory. Mata attempts to set the flag for you (for instance, when you add a new variable using st_addvar( ) [see [M-5] **st_addvar( )**]), but there are other places where the flag ought to be set, and you must do so. For instance, Mata does not set the flag every time you change a value in the dataset. Setting the flag what may be many thousands of times would reduce performance too much.

Moreover, even when Mata does set the flag, it might do so inappropriately, because the logic of your program fooled Mata. For instance, perhaps you added a variable and later dropped it. In such cases, the appropriate code is

```
priorupdatavalue = st_updata()
...
st_updata(priorupdatavalue)
```

Conformability

```
st_updata():
        result:      1 × 1

st_updata(value):
        value:       1 × 1
        result:      void
```

Diagnostics

None.

Also see

[M-4] **stata** — Stata interface functions

Title

> **[M-5] st_varformat( )** — Obtain/set format, etc., of Stata variable

Syntax

> *string scalar*   st_varformat(*scalar var*)
>
> *void*   st_varformat(*scalar var*, *string scalar fmt*)
>
> *string scalar*   st_varlabel(*scalar var*)
>
> *void*   st_varlabel(*scalar var*, *string scalar label*)
>
> *string scalar*   st_varvaluelabel(*scalar var*)
>
> *void*   st_varvaluelabel(*scalar var*, *string scalar labelname*)

where *var* contains a Stata variable name or a Stata variable index.

Description

st_varformat(*var*) returns the display format associated with *var*, such as "%9.0gc". st_varformat(*var*, *fmt*) changes *var*'s display format.

st_varlabel(*var*) returns the variable label associated with *var*, such as "Sex of Patient", or it returns "" if *var* has no variable label. st_varformat(*var*, *label*) changes *var*'s variable label.

st_varvaluelabel(*var*) returns the value-label name associated with *var*, such as "origin", or it returns "" if *var* has no value label. st_varvaluelabel(*var*, *labelname*) changes the value-label name associated with *var*.

Conformability

st_varformat(*var*), st_varlabel(*var*), st_varvaluelabel(*var*):

var:	1×1
result:	1×1

st_varformat(*var*, *fmt*), st_varlabel(*var*, *label*), st_varvaluelabel(*var*, *labelname*):

var:	1×1
value:	1×1
result:	*void*

Diagnostics

In all functions, if *var* is specified as a name, abbreviations are not allowed.

All functions abort with error if *var* is not a valid Stata variable.

st_varformat(*var*, *fmt*) aborts with error if *fmt* does not contain a valid display format for *var*.

st_varlabel(*var*, *label*) will truncate *label* if it is too long.

st_varvaluelabel(*var*, *labelname*) aborts with error if *var* is a Stata string variable or if *labelname* does not contain a valid name (assuming *labelname* is not ""). It is not required, however, that the label name exist.

Also see

[M-4] **stata** — Stata interface functions

Title

[M-5] **st_varindex( )** — Obtain variable indices from variable names

Syntax

> *real rowvector*   st_varindex(*string rowvector s*)
>
> *real rowvector*   st_varindex(*string rowvector s*, *real scalar abbrev*)
>
> *real rowvector*   _st_varindex(*string rowvector s*)
>
> *real rowvector*   _st_varindex(*string rowvector s*, *real scalar abbrev*)

Description

st_varindex(*s*) returns the variable index associated with each variable name recorded in *s*. st_varindex(*s*) does not allow variable-name abbreviations such as "pr" for "price".

st_varindex(*s*, *abbrev*) does the same thing but allows you to specify whether variable-name abbreviations are to be allowed. Abbreviations are allowed if *abbrev* $\neq$ 0. st_varindex(*s*) is equivalent to st_varindex(*s*, 0).

_st_varindex() does the same thing as st_varindex(). The two functions differ in how they respond when a name is not found. st_varindex() aborts with error, and _st_varindex() places missing in the appropriate element of the returned result.

Remarks

These functions require that each element of *s* contain a variable name, such as

> *s* = ("price", "mpg", "weight")

If you have one string containing multiple names

> *s* = ("price mpg weight")

then use tokens() to split it into the desired form, as in

> *k* = st_varindex(tokens(*s*))

See [M-5] **tokens( )**.

Conformability

st_varindex(*s*, *abbrev*), _st_varindex(*s*, *abbrev*):

s:	1 × *k*	
abbrev:	1 × 1	(optional)
result:	1 × *k*	

Diagnostics

st_varindex( ) aborts with error if any name is not found.

_st_varindex( ) puts missing in the appropriate element of the returned result for any name that is not found.

Also see

[M-5] **st_varname( )** — Obtain variable names from variable indices

[M-5] **tokens( )** — Obtain tokens from string

[M-4] **stata** — Stata interface functions

Title

> **[M-5] st_varname( )** — Obtain variable names from variable indices

Syntax

> *string rowvector*   st_varname(*real rowvector k*)

> *string rowvector*   st_varname(*real rowvector k, real scalar tsmap*)

Description

st_varname(*k*) returns the Stata variable names associated with the variable indices stored in *k*. For instance, with the automobile data in memory

> names = st_varname((1..3))

results in names being ("make", "price", "mpg").

st_varname(*k*, *tsmap*) does the same thing but allows you to specify whether you want the actual or logical variable names of any time-series–operated variables created by the Mata function st_tsrevar() (see [M-5] **st_tsrevar( )**) or by the Stata command tsrevar (see [TS] **tsrevar**).

st_varname(*k*) is equivalent to st_varname(*k*, 0); actual variable names are returned.

st_varname(*k*, 1) returns logical variable names.

Remarks

To understand the actions of st_varname(*k*, 1), pretend that variable 58 was created by st_tsrevar():

> k = st_tsrevar(("gnp", "r", "l.gnp"))

Pretend that k now contains (12, 5, 58). Variable 58 is a new, temporary variable, containing l.gnp values. Were you to ask for the actual names of the variables

> actualnames = st_varname(k)

actualnames would contain ("gnp", "r", "__00004a"), although the name of the last variable will vary because it is a temporary variable. Were you to ask for the logical names,

> logicalnames = st_varname(k, 1)

you would get back ("gnp", "r", "L.gnp").

Conformability

st_varname(*k*, *tsmap*)

k:	$1 \times c$	
tsmap:	1×1	(optional)
result:	$1 \times c$	

Diagnostics

st_varname(k) and st_varname(k, *tsmap*) abort with error if any element of k is less than 1 or greater than st_nvar( ); see [M-5] **st_nvar( )**.

Also see

[M-5] **st_varindex( )** — Obtain variable indices from variable names

[M-5] **st_tsrevar( )** — Create time-series op.varname variables

[M-4] **stata** — Stata interface functions

Title

[M-5] st_varrename( ) — Rename Stata variable

Syntax

void st_varrename(*scalar var*, *string scalar newname*)

where *var* contains a Stata variable name or a Stata variable index.

Description

st_varrename(*var*, *newname*) changes the name of *var* to *newname*.

If *var* is specified as a name, abbreviations are not allowed.

Conformability

st_varrename(*var*, *newname*):
    var: 1×1
    newname: 1×1
    result: *void*

Diagnostics

st_varrename(*var*, *newname*) aborts with error if *var* is not a valid Stata variable or if *newname* is not a valid name or if a variable named *newname* already exists.

Also see

[M-4] **stata** — Stata interface functions

Title

[M-5] **st_vartype( )** — Storage type of Stata variable

Syntax

>   *string scalar*   st_vartype(*scalar var*)
>
>   *real scalar*   st_isnumvar(*scalar var*)
>
>   *real scalar*   st_isstrvar(*scalar var*)

where *var* contains a Stata variable name or a Stata variable index.

Description

In all the functions, if *var* is specified as a name, abbreviations are not allowed.

st_vartype(*var*) returns the storage type of the *var*, such as float, double, or str18.

st_isnumvar(*var*) returns 1 if *var* is a numeric variable and 0 otherwise.

st_isstrvar(*var*) returns 1 if *var* is a string variable and 0 otherwise.

Remarks

st_isstrvar(*var*) and st_isnumvar(*var*) are antonyms. Both functions are provided merely for convenience; they tell you nothing that you cannot discover from st_vartype(*var*).

Conformability

st_vartype(*var*):
>   *var*: 1×1
>   *result*: 1×1

st_isnumvar(*var*), st_isstrvar(*var*):
>   *var*: 1×1
>   *result*: 1×1

Diagnostics

All functions abort with error if *var* is not a valid Stata variable.

Also see

[M-4] **stata** — Stata interface functions

Title

> **[M-5] st_view( )** — Make matrix that is a view onto current Stata dataset

Syntax

void   st_view(*V*, *real matrix i*, *j*)

void   st_view(*V*, *real matrix i*, *j*, *scalar selectvar*)

void   st_sview(*V*, *real matrix i*, *j*)

void   st_sview(*V*, *real matrix i*, *j*, *scalar selectvar*)

where

1. The type of *V* does not matter; it is replaced.

2. *i* may be specified in the same way as with st_data( ).

3. *j* may be specified in the same way as with st_data( ). Factor variables and time-series–operated variables may be specified.

4. *selectvar* may be specified in the same way as with st_data( ).

See [M-5] **st_data( )**.

Description

st_view( ) and st_sview( ) create a matrix that is a view onto the current Stata dataset.

Remarks

Remarks are presented under the following headings:

> *Overview*
> *Advantages and disadvantages of views*
> *When not to use views*
> *Cautions when using views 1: Conserving memory*
> *Cautions when using views 2: Assignment*
> *Efficiency*

Overview

st_view( ) serves the same purpose as st_data( )—and st_sview( ) serves the same purpose as st_sdata( )—except that, rather than returning a matrix that is a copy of the underlying values, st_view( ) and st_sview( ) create a matrix that is a view onto the Stata dataset itself.

To understand the distinction, consider

```
X = st_data(., "mpg displ weight")
```

and

```
st_view(X, ., "mpg displ weight")
```

Both commands fill in matrix X with the same data. However, were you to code

```
X[2,1] = 123
```

after the `st_data()` setup, you would change the value in the matrix X, but the Stata dataset would remain unchanged. After the `st_view()` setup, changing the value in the matrix would cause the value of mpg in the second observation to change to 123.

Advantages and disadvantages of views

Views make it easy to change the dataset, and that can be an advantage or a disadvantage, depending on your goals.

Putting that aside, views are in general better than copies because 1) they take less time to set up and 2) they consume less memory. The memory savings can be considerable. Consider a 100,000-observation dataset on 30 variables. Coding

```
X = st_data(., .)
```

creates a new matrix that is 24 MB in size. Meanwhile, the total storage requirement for

```
st_view(X, ., .)
```

is roughly 128 bytes!

There is a cost; when you use the matrix X, it takes longer to access the individual elements. You would have to do a lot of calculation with X, however, before that sum of the longer access times would equal the initial savings in setup time, and even then, the longer access time is probably worth the savings in memory.

When not to use views

Do not use views as a substitute for scalars. If you are going to loop through the data an observation at a time, and if every usage you will make of X is in scalar calculations, use `_st_data()`. There is nothing faster for that problem.

Putting aside that extreme, views become more efficient relative to copies the larger they are; i.e., it is more efficient to use `st_data()` for small amounts of data, especially if you are going to make computationally intensive calculations with it.

Cautions when using views 1: Conserving memory

If you are using views, it is probably because you are concerned about memory, and if you are, you want to be careful to avoid making copies of views. Copies of views are not views; they are copies. For instance,

```
st_view(V, ., .)
Y = V
```

That innocuous looking Y = V just made a copy of the entire dataset, meaning that if the dataset had 100,000 observations on 30 variables, Y now consumes 24 MB. Coding Y = V may be desirable in certain circumstances, but in general, it is better to set up another view.

Similarly, watch out for subscripts. Consider the following code fragment

```
st_view(V, ., .)
for (i=1; i<=cols(V); i++) {
    sum = colsum(V[,i])
    ...
}
```

The problem in the above code is the V[,i]. That creates a new column vector containing the values from the ith column of V. Given 100,000 observations, that new column vector needs 800k of memory. Better to code would be

```
for (i=1; i<=cols(V); i++) {
    st_view(v, ., i)
    sum = colsum(v)
    ...
}
```

If you need V and v, that is okay. You can have many views of the data setup simultaneously.

Similarly, be careful using views with operators. X'X makes a copy of X in the process of creating the transpose. Use functions such as cross() (see [M-5] **cross( )**) that are designed to minimize the use of memory.

Do not be overly concerned about this issue. Making a copy of a column of a view amounts to the same thing as introducing a temporary variable in a Stata program—something that is done all the time.

Cautions when using views 2: Assignment

The ability to assign to a view and so change the underlying data can be either convenient or dangerous, depending on your goals. When making such assignments, there are two things you need be aware of.

The first is more of a Stata issue than it is a Mata issue. Assignment does not cause promotion. Coding

```
V[1,2] = 4059.125
```

might store 4059.125 in the first observation of the second variable of the view. Or, if that second variable is an int, what will be stored is 4059, or if it is a byte, what will be stored is missing.

The second caution is a Mata issue. To reassign all the values of the view, code

V[.,.] = *matrix_expression*

Do not code

V = *matrix_expression*

The second expression does not assign to the underlying dataset, it redefines V to be a regular matrix.

Mata does not allow the use of views as the destination of assignment when the view contains factor variables or time-series–operated variables such as i.rep78 or l.gnp.

Efficiency

Whenever possible, specify argument i of $\texttt{st\_view}(V, i, j)$ and $\texttt{st\_sview}(V, i, j)$ as . (missing value) or as a row vector range (e.g., (i_1, i_2)) rather than as a column vector list.

Specify argument j as a real row vector rather than as a string whenever $\texttt{st\_view}()$ and $\texttt{st\_sview}()$ are used inside loops with the same variables (and the view does not contain factor variables nor time-series–operated variables). This prevents Mata from having to look up the same names over and over again.

Conformability

$\texttt{st\_view}(V, i, j)$, $\texttt{st\_sview}(V, i, j)$:

    input:

i:	$n \times 1$	or	$n_2 \times 2$
j:	$1 \times k$	or	1×2 containing k elements when expanded

    output:

V:	$n \times k$

$\texttt{st\_view}(V, i, j, \textit{selectvar})$, $\texttt{st\_sview}(V, i, j, \textit{selectvar})$:

    input:

i:	$n \times 1$	or	$n_2 \times 2$
j:	$1 \times k$	or	1×2 containing k elements when expanded
selectvar:	1×1		

    output:

V:	$(n - e) \times k$, where e is number of observations excluded by *selectvar*

Diagnostics

$\texttt{st\_view}(i, j[, \textit{selectvar}])$ and $\texttt{st\_sview}(i, j[, \textit{selectvar}])$ abort with error if any element of i is outside the range of observations or if a variable name or index recorded in j is not found. Variable-name abbreviations are allowed. If you do not want this and no factor variables nor time-series–operated variables are specified, use $\texttt{st\_varindex}()$ (see [M-5] **st_varindex( )**) to translate variable names into variable indices.

$\texttt{st\_view}()$ and $\texttt{st\_sview}()$ abort with error if any element of i is out of range as described under the heading *Details of observation subscripting using st_data( ) and st_sdata( )* in [M-5] **st_data( )**.

Some functions do not allow views as arguments. If $\texttt{example}(X)$ does not allow views, you can still use it by coding

    ... example(X=V) ...

because that will make a copy of view V in X. Most functions that do not allow views mention that in their *Diagnostics* section, but some do not because it was unexpected that anyone would want to use a view in that case. If a function does not allow a view, you will see in the traceback log:

```
    : myfunction(...)
            example():  3103  view found where array required
              mysub():     -  function returned error
         myfunction():     -  function returned error
             <istmt>:     -  function returned error
      r(3103);
```

The above means that function example() does not allow views.

Reference

Gould, W. W. 2005. Mata Matters: Using views onto the data. *Stata Journal* 5: 567–573.

Also see

[M-5] **st_subview( )** — Make view from view

[M-5] **select( )** — Select rows or columns

[M-5] **st_viewvars( )** — Variables and observations of view

[M-5] **st_data( )** — Load copy of current Stata dataset

[M-4] **stata** — Stata interface functions

Title

[M-5] st_viewvars( ) — Variables and observations of view

Syntax

> *real rowvector* st_viewvars(*matrix V*)
>
> *real vector* st_viewobs(*matrix V*)

where V is required to be a view.

Description

st_viewvars(V) returns the indices of the Stata variables corresponding to the columns of V.

st_viewobs(V) returns the Stata observation numbers corresponding to the rows of V. Returned is either a 1×2 row vector recording the observation range or an $N \times 1$ column vector recording the individual observation numbers.

Remarks

The results returned by these two functions are suitable for inclusion as arguments in subsequent calls to st_view() and st_sview(); see [M-5] **st_view( )**.

Conformability

```
st_viewvars(V):
        V:    N × k
    result:   1 × k

st_viewobs(V):
        V:    N × k
    result:   1 × 2  or  N × 1
```

Diagnostics

st_viewvars(V) and st_viewobs(V) abort with error if V is not a view.

Also see

[M-5] **st_view( )** — Make matrix that is a view onto current Stata dataset

[M-4] **stata** — Stata interface functions

Title

> **[M-5] st_vlexists( )** — Use and manipulate value labels

Syntax

real scalar	st_vlexists(*name*)
void	st_vldrop(*name*)
string matrix	st_vlmap(*name*, *real matrix values*)
real matrix	st_vlsearch(*name*, *string matrix text*)
void	st_vlload(*name*, *values*, *text*)
void	st_vlmodify(*name*, *real colvector values*, *real colvector text*)

where *name* is *string scalar* and where the types of *values* and *text* in st_vlload( ) are irrelevant because they are replaced.

Description

st_vlexists(*name*) returns 1 if value label *name* exists and returns 0 otherwise.

st_vldrop(*name*) drops value label *name* if it exists.

st_vlmap(*name*, *values*) maps *values* through value label *name* and returns the result.

st_vlsearch(*name*, *text*) does the reverse; it returns the value corresponding to the text.

st_vlload(*name*, *values*, *text*) places value label *name* into *values* and *text*.

st_vlmodify(*name*, *values*, *text*) creates a new value label or modifies an existing one.

Remarks

Value labels are named and record a mapping from numeric values to text. For instance, a value label named sexlbl might record that 1 corresponds to male and 2 to female. Values labels are attached to Stata numeric variables. If a Stata numeric variable had the value label sexlbl attached to it, then the 1s and 2s in the variable would display as male and female. How other values would appear—if there were other values—would not be affected.

Remarks are presented under the following headings:

> *Value-label mapping*
> *Value-label creation and editing*
> *Loading value labels*

Value-label mapping

Let us consider value label sexlbl mapping 1 to male and 2 to female.
st_vlmap("sexlbl", *values*) would map the $r \times c$ matrix values through sexlbl and return an
$r \times c$ string matrix containing the result. Any values for which there was no mapping would result
in "". Thus

```
: res = st_vlmap("sexlbl", 1)
: res
male
: res = st_vlmap("sexlbl", (2,3,1))
: res
            1       2       3

  1   female          male
```

st_vlsearch(*name*, *text*) performs the reverse mapping:

```
: txt = st_vlsearch("sexlbl", ("female","","male"))
: txt
       1   2   3

  1    2   .   1
```

Value-label creation and editing

st_vlmodify(*name*, *values*, *text*) creates new value labels and modifies existing ones.

If value label sexlbl did not exist, coding

```
: st_vlmodify("sexlbl", (1\2), ("male"\"female"))
```

would create it. If the value label did previously exist, the above would modify it so that 1 now
corresponds to male and 2 to female, regardless of what 1 or 2 previously corresponded to, if they
corresponded to anything. Other mappings that might have been included in the value label remain
unchanged. Thus

```
: st_vlmodify("sexlbl", 3, "unknown")
```

would add another mapping to the label. Values are deleted by specifying the text as "", so

```
: st_vlmodify("sexlbl", 3, "")
```

would remove the mapping for 3 (if there was a mapping). If you remove all the mappings, the value
label itself is automatically dropped:

```
: st_vlmodify("sexlbl", (1\2), (""\""))
```

results in value label sexlbl being dropped if 1 and 2 were the final mappings in it.

Loading value labels

st_vlload(*name*, *values*, *text*) returns the value label in *values* and *text*, where you can do with it as you please. Thus you could code

```
st_vlload("sexlbl", values, text)
    ...
st_vldrop("sexlbl")
st_vlmodify("sexlbl", values, text)
```

Conformability

st_vlexists(*name*):
>   *name*:   1 × 1
>   *result*:   1 × 1

st_vldrop(*name*):
>   *name*:   1 × 1
>   *result*:   void

st_vlmap(*name*, *values*):
>   *name*:   1 × 1
>   *values*:   r × c
>   *result*:   r × c

st_vlsearch(*name*, *text*):
>   *name*:   1 × 1
>   *text*:   r × c
>   *result*:   r × c

st_vlload(*name*, *values*, *text*):

>   *input*:
>   *name*:   1 × 1

>   *output*:
>   *values*:   k × 1
>   *text*:   k × 1

st_vlmodify(*name*, *values*, *text*):
>   *name*:   1 × 1
>   *values*:   m × 1
>   *text*:   m × 1
>   *result*:   void

Diagnostics

The only conditions under which the above functions abort with error is when *name* is malformed or Mata is out of memory. Functions tolerate all other problems.

st_vldrop(*name*) does nothing if value label *name* does not exist.

st_vlmap(*name*, *values*) returns J(rows(values), cols(values), "") if value label *name* does not exist. When the value label does exist, individual values for which there is no recorded mapping are returned as "".

st_vlsearch(*name*, *text*): returns J(rows(values), cols(values), .) if value label *name* does not exist. When the value label does exist, individual text values for which there is no corresponding value are returned as . (missing).

st_vlload(*name*, *values*, *text*): sets *values* and *text* to be 0 × 1 when value label *name* does not exist.

st_vlmodify(*name*, *values*, *text*): creates the value label if it does not already exist. Value labels may map only integers and .a, .b, ..., .z. Attempts to insert a mapping for . are ignored. Noninteger values are truncated to integer values. If an element of *text* is "", then the corresponding mapping is removed.

Also see

[M-4] **stata** — Stata interface functions

Title

Syntax

void	stata(*cmd*)
void	stata(*cmd*, *nooutput*)
void	stata(*cmd*, *nooutput*, *nomacroexpand*)
real scalar	_stata(*cmd*)
real scalar	_stata(*cmd*, *nooutput*)
real scalar	_stata(*cmd*, *nooutput*, *nomacroexpand*)

where

cmd:	*string scalar*
nooutput:	*real scalar*
nomacroexpand:	*real scalar*

Description

stata(*cmd*) executes the Stata command contained in the string scalar *cmd*. Output from the command appears at the terminal, and any macros contained in *cmd* are expanded.

stata(*cmd*, *nooutput*) does the same thing, but if *nooutput* $\neq$ 0, output produced by the execution is not displayed. stata(*cmd*, 0) is equivalent to stata(*cmd*).

stata(*cmd*, *nooutput*, *nomacroexpand*) does the same thing but, before execution, suppresses expansion of any macros contained in *cmd* if *nomacroexpand* $\neq$ 0. stata(*cmd*, 0, 0) is equivalent to stata(*cmd*).

_stata() repeats the syntaxes of stata(). The difference is that, whereas stata() aborts with error if the execution results in a nonzero return code, _stata() returns the resulting return code.

Remarks

The command you execute may invoke a process that causes another instance of Mata to be invoked. For instance, Stata program *A* calls Mata function *m1*(), which executes stata() to invoke Stata program *B*, which in turn calls Mata function *m2*(), which . . . .

stata(*cmd*) and _stata(*cmd*) execute *cmd* at the current run level. This means that any local macros refer to local macros in the caller's space. Consider the following:

```
program example
        ...
        local x = "value from A"
        mata: myfunc()
        display "'x'"
        ...
end

mata void myfunc()
{
        stata('"local x = "new value""')
}
```

After example executes mata: myfunc(), 'x' will be "new value".

That stata() and _stata() work that way was intentional: Mata functions can modify the caller's environment so that they may create temporary variables for the caller's use, etc., and you only have to exercise a little caution. Executing stata() functions to run other ado-files and programs will cause no problems because other ado-files and programs create their own new environment in which temporary variables, local macros, etc., are private.

Also, do not use stata() or _stata() to execute a multiline command or to execute the first line of what could be considered a multiline command. Once the first line is executed, Stata will fetch the remaining lines from the caller's environment. For instance, consider

── begin myfile.do ───────

```
mata void myfunc()
{
    stata("if (1==1) {")
}
mata: myfunc()
display "hello"
}
```

── end myfile.do ───────

In the example above, myfunc() will consume the display "hello" and } lines.

Conformability

stata(*cmd*, *nooutput*, *nomacroexpand*):
cmd:	1 × 1	
nooutput:	1 × 1	(optional)
nomacroexpand:	1 × 1	(optional)
result:	*void*	

_stata(*cmd*, *nooutput*, *nomacroexpand*):
cmd:	1 × 1	
nooutput:	1 × 1	(optional)
nomacroexpand:	1 × 1	(optional)
result:	1 × 1	

Diagnostics

stata( ) aborts with error if *cmd* is too long (exceedingly unlikely), if macro expansion fails, or if execution results in a nonzero return code.

_stata( ) aborts with error if *cmd* is too long.

Also see

[M-3] **mata stata** — Execute Stata command

[M-4] **stata** — Stata interface functions

Title

> **[M-5] stataversion( )** — Version of Stata being used

Syntax

> *real scalar*   staticversion( )

> *real scalar*   statasetversion( )

> *void*   statasetversion(*real scalar version*)

Note: the version number is multiplied by 100: Stata 2.0 is 200, Stata 5.1 is 510, and Stata 13.0 is 1300.

Description

stataversion( ) returns the version of Stata/Mata that is running, multiplied by 100. For instance, if you have Stata 13 installed on your computer, stataversion( ) returns 1300.

statasetversion( ) returns the version of Stata that has been set by the user—the version of Stata that Stata is currently emulating—multiplied by 100. Usually stataversion( ) == statasetversion( ). If the user has set a previous version—say, version 8 by typing version 8 in Stata—statasetversion( ) will return a number less than stataversion( ).

statasetversion(*version*) allows you to reset the version being emulated. Results are the same as using Stata's version command. *version*, however, is specified as an integer equal to 100 times the version you want.

Remarks

It is usually not necessary to reset statasetversion( ). If you do reset statasetversion( ), good form is to set it back when you are finished:

```
current_version = statasetversion()
statasetversion(desired_version)
. . .
statasetversion(current_version)
```

Conformability

```
stataversion():
    result:      1 × 1
statasetversion():
    result:      1 × 1
statasetversion(version):
    version:     1 × 1
    result:      void
```

Diagnostics

statasetversion(*version*) aborts with error if *version* is less than 100 or greater than stataversion().

Also see

[M-5] **bufio( )** — Buffered (binary) I/O

[M-5] **byteorder( )** — Byte order used by computer

[M-4] **programming** — Programming functions

Title

[M-5] strdup( ) — String duplication

Syntax

$n * s$

$s * n$

$n :* s$

$s :* n$

where n is real and s is string.

Description

There is no strdup() function. Instead, the multiplication operator is used:

```
3*"example" = "exampleexampleexample"
0*"this" = ""
```

Conformability

$n*s$, $s*n$:

n:	1×1
s:	$r \times c$
result:	$r \times c$

$n:*s$, $s:*n$:

n:	$r_1 \times c_1$
s:	$r_2 \times c_2$, n and s c-conformable
result:	$\max(r_1,r_2) \times \max(c_1,c_2)$

Diagnostics

If $n < 0$, the result is as if $n = 0$: "" is returned.

If n is not an integer, the result is as if $\mathrm{trunc}(n)$ were specified.

Also see

[M-4] **string** — String manipulation functions

Title

Syntax

> *real matrix*   strlen(*string matrix s*)

Description

strlen(*s*) returns the length of—the number of characters contained in—the string *s*.

When *s* is not a scalar, strlen() returns element-by-element results.

Remarks

Stata understands strlen() as a synonym for its own length() function, so you can use the function named strlen() in both your Stata and Mata code. Do not, however, use length() in Mata when you mean strlen(). Mata's length() function returns the length (number of elements) of a vector.

Conformability

strlen(*s*):
$$s: \quad r \times c$$
$$result: \quad r \times c$$

Diagnostics

strlen(*s*), when *s* is a binary string (a string containing binary 0), returns the overall length of the string, not the location of the binary 0. Use strpos(*s*, char(0)) if you want the location of the binary 0.

Also see

[M-5] **strpos( )** — Find substring in string

[M-5] **fmtwidth( )** — Width of %fmt

[M-4] **string** — String manipulation functions

Title

[M-5] strmatch( ) — Determine whether string matches pattern

Syntax

> *real matrix* strmatch(*string matrix s*, *string matrix pattern*)

Description

strmatch(*s*, *pattern*) returns 1 if *s* matches *pattern* and 0 otherwise.

When arguments are not scalars, strmatch() returns element-by-element results.

Remarks

In *pattern*, * means that 0 or more characters go here and ? means that exactly one character goes here. Thus *pattern*="*" matches anything and *pattern*="?p*x" matches all strings whose second character is p and whose last character is x.

Stata understands strmatch() as a synonym for its own match() function, so you can use the strmatch() function in both your Stata and Mata code.

Conformability

strmatch(*s*, *pattern*):

s:	$r_1 \times c_1$
pattern:	$r_2 \times c_2$, *s* and *pattern* r-conformable
result:	$\max(r_1,r_2) \times \max(c_1,c_2)$

Diagnostics

In strmatch(*s*, *pattern*), if *s* or *pattern* contain a binary 0 (they usually would not), the strings are considered to end at that point.

Also see

[M-4] **string** — String manipulation functions

Title

[M-5] **strofreal( )** — Convert real to string

Syntax

> *string matrix* strofreal(*real matrix R*)

> *string matrix* strofreal(*real matrix R*, *string matrix format*)

Description

strofreal(R) returns R as a string using Stata's %9.0g format. strofreal(R) is equivalent to strofreal(R, "%9.0g").

strofreal(R, *format*) returns R as a string formatted using *format*.

Leading blanks are trimmed from the result.

When arguments are not scalars, strofreal() returns element-by-element results.

Conformability

strofreal(R, *format*):

R:	$r_1 \times c_1$
format:	$r_2 \times c_2$, R and *format* r-conformable (optional)
result:	$\max(r_1, r_2) \times \max(c_1, c_2)$

Diagnostics

strofreal(R, *format*) returns "." if *format* is invalid.

Also see

[M-5] **strtoreal( )** — Convert string to real

[M-4] **string** — String manipulation functions

Title

[M-5] strpos( ) — Find substring in string

Syntax

> *real matrix*   strpos(*string matrix haystack, string matrix needle*)

Description

strpos(*haystack, needle*) returns the location of the first occurrence of *needle* in *haystack* or 0 if *needle* does not occur.

When arguments are not scalars, strpos( ) returns element-by-element results.

Remarks

When working with binary strings, one can find the location of the binary 0 using strpos(*s*, char(0)).

Conformability

strpos(*haystack, needle*):

haystack:	$r_1 \times c_1$
needle:	$r_2 \times c_2$, *haystack* and *needle* r-conformable
result:	$\max(r_1, r_2) \times \max(c_1, c_2)$

Diagnostics

strpos(*haystack, needle*) returns 0 if *needle* is not found in *haystack*.

Also see

[M-4] **string** — String manipulation functions

Title

[M-5] **strreverse( )** — Reverse string

Syntax

string matrix   strreverse(*string matrix s*)

Description

strreverse(*s*) returns *s* with its characters in reverse order.

When *s* is not a scalar, strreverse() returns element-by-element results.

Remarks

Stata understands strreverse() as a synonym for its own reverse() function, so you can use the strreverse() name in both your Stata and Mata code.

Conformability

strreverse(*s*)
$$\begin{array}{ll} s: & r \times c \\ result: & r \times c \end{array}$$

Diagnostics

None.

Also see

[M-4] **string** — String manipulation functions

Title

> **[M-5] strtoname( )** — Convert a string to a Stata name

Syntax

> *string matrix*   strtoname(*string matrix s*, *real scalar p*)
>
> *string matrix*   strtoname(*string matrix s*)

Description

strtoname(*s*, *p*) returns *s* translated into a Stata name. Each character in *s* that is not allowed in a Stata name is converted to an underscore character, _. If the first character in *s* is a numeric character and *p* is not 0, then the result is prefixed with an underscore. The result is truncated to 32 characters.

strtoname(*s*) is equivalent to strtoname(*s*, 1).

When arguments are not scalar, strtoname( ) returns element-by-element results.

Remarks

strtoname("StataName") returns "StataName".

strtoname("not a Stata name") returns "not_a_Stata_name".

strtoname("0 is off") returns "_0_is_off".

strtoname("0 is off", 0) returns "0_is_off".

Conformability

strtoname(*s*, *p*):
         s: $r \times c$
        p: 1×1
  result: $r \times c$

strtoname(*s*):
        s: $r \times c$
  result: $r \times c$

Diagnostics

None.

Also see

[M-4] **string** — String manipulation functions

Title

Syntax

> *real matrix*   strtoreal(*string matrix S*)
>
> *real scalar*   _strtoreal(*string matrix S*, *R*)

Description

strtoreal(*S*) returns *S* converted to real. Elements of *S* that cannot be converted are returned as . (missing value).

_strtoreal(*S*, *R*) does the same as above—it returns the converted values in *R*—and it returns the number of elements that could not be converted. In such cases, the corresponding value of *R* contains . (missing).

Remarks

strtoreal("1.5") returns (numeric) 1.5.

strtoreal("-2.5e+1") returns (numeric) −25.

strtoreal("not a number") returns (numeric) . (missing).

Typically, strtoreal(*S*) and _strtoreal(*S*, *R*) are used with scalars, but if applied to a vector or matrix *S*, element-by-element results are returned.

In performing the conversion, leading and trailing blanks are ignored: "1.5" and " 1.5 " both convert to (numeric) 1.5, but "1.5 kilometers" converts to . (missing). Use strtoreal(tokens(*S*)[1]) to convert just the first space-delimited part.

All Stata numeric formats are understood, such as 0, 1, −2, 1.5, 1.5e+2, and −1.0x+8, as well as the missing-value codes ., .a, .b, ..., .z.

Thus using strtoreal(*S*), if an element of *S* converts to . (missing), you cannot tell whether the element was valid and equal to "." or the element was invalid and so defaulted to . (missing), such as if *S* contained "cat" or "dog" or "1.5 kilometers".

When it is important to distinguish between these cases, use _strtoreal(*S*, *R*). The conversion is returned in *R* and the function returns the number of elements that were invalid. If _strtoreal() returns 0, then all values were valid.

Conformability

strtoreal(*S*):

    input:

                S: $r \times c$

    output:

          result: $r \times c$

_strtoreal(*S*, *R*):

    input:

                S: $r \times c$

    output:

                R: $r \times c$

          result: 1×1

Diagnostics

strtoreal(*S*) returns a missing value wherever an element of *S* cannot be converted to a number.

_strtoreal(*S*, *R*) does the same, but the result is returned in *R*.

Also see

[M-5] **strofreal( )** — Convert real to string

[M-4] **string** — String manipulation functions

Title

[M-5] strtrim( ) — Remove blanks

Syntax

> *string matrix* stritrim(*string matrix s*)
>
> *string matrix* strltrim(*string matrix s*)
>
> *string matrix* strrtrim(*string matrix s*)
>
> *string matrix* strtrim(*string matrix s*)

Description

stritrim(*s*) returns *s* with all consecutive, internal blanks collapsed to one blank.

strltrim(*s*) returns *s* with leading blanks removed.

strrtrim(*s*) returns *s* with trailing blanks removed.

strtrim(*s*) returns *s* with leading and trailing blanks removed.

When *s* is not a scalar, these functions return element-by-element results.

Remarks

Stata understands stritrim(), strltrim(), strrtrim(), and strtrim(), as synonyms for its own itrim(), ltrim(), rtrim(), and trim() functions, so you can use the str*() names in both your Stata and Mata code.

Conformability

stritrim(*s*), strltrim(*s*), strrtrim(*s*), strtrim(*s*):
s:	$r \times c$
result:	$r \times c$

Diagnostics

None.

Also see

[M-4] **string** — String manipulation functions

Title

[M-5] strupper( ) — Convert string to uppercase (lowercase)

Syntax

string matrix   strupper(*string matrix s*)

string matrix   strlower(*string matrix s*)

string matrix   strproper(*string matrix s*)

Description

strupper(*s*) returns *s*, converted to uppercase.

strlower(*s*) returns *s*, converted to lowercase.

strproper(*s*) returns a string with the first letter capitalized and any other letters capitalized that immediately follow characters that are not letters; all other letters are converted to lowercase.

When *s* is not a scalar, these functions return element-by-element results.

Remarks

strproper("mR. joHn a. sMitH") returns Mr. John A. Smith.

strproper("jack o'reilly") returns Jack O'Reilly.

strproper("2-cent's worth") returns 2-Cent'S Worth.

Also, Stata understands strupper(), strlower(), and strproper() as synonyms for its own upper(), lower(), and proper() functions, so you can use the str*() names in both your Stata and Mata code.

Conformability

strupper(*s*), strlower(*s*), strproper(*s*):

s:	$r \times c$
result:	$r \times c$

Diagnostics

None.

Also see

[M-4] **string** — String manipulation functions

Title

[M-5] **subinstr( )** — Substitute text

Syntax

string matrix subinstr(*string matrix s*, *string matrix old*, *string matrix new*)

string matrix subinstr(*string matrix s*, *string matrix old*, *string matrix new*,
  real matrix cnt)

string matrix subinword(*string matrix s*, *string matrix old*, *string matrix new*)

string matrix subinword(*string matrix s*, *string matrix old*, *string matrix new*,
  real matrix cnt)

Description

subinstr(*s*, *old*, *new*) returns *s* with all occurrences of *old* changed to *new*.

subinstr(*s*, *old*, *new*, *cnt*) returns *s* with the first *cnt* occurrences of *old* changed to *new*. All occurrences are changed if *cnt* contains missing.

subinword(*s*, *old*, *new*) returns *s* with all occurrences of *old* on word boundaries changed to *new*.

subinword(*s*, *old*, *new*, *cnt*) returns *s* with the first *cnt* occurrences of *old* on word boundaries changed to *new*. All occurrences are changed if *cnt* contains missing.

When arguments are not scalars, these functions return element-by-element results.

Remarks

subinstr("th thin man", "th", "the") returns "the thein man".

subinword("th thin man", "th", "the") returns "the thin man".

Conformability

subinstr(*s*, *old*, *new*, *cnt*), subinword(*s*, *old*, *new*, *cnt*):

s:	$r_1 \times c_1$
old:	$r_2 \times c_2$
new:	$r_3 \times c_3$
cnt:	$r_4 \times c_4$ (optional); *s*, *old*, *new*, *cnt* r-conformable
result:	$\max(r_1,r_2,r_3,r_4) \times \max(c_1,c_2,c_3,c_4)$

Diagnostics

subinstr(*s*, *old*, *new*, *cnt*) and subinword(*s*, *old*, *new*, *cnt*) treat *cnt* < 0 as if *cnt* = 0 was specified; the original string *s* is returned.

Also see

[M-4] **string** — String manipulation functions

Title

> **[M-5] sublowertriangle( )** — Return a matrix with zeros above a diagonal

Syntax

> *numeric matrix* sublowertriangle(*numeric matrix A* [, *numeric scalar p*])
>
> *void* _sublowertriangle(*numeric matrix A* [, *numeric scalar p*])

where argument *p* is optional.

Description

sublowertriangle(*A*, *p*) returns *A* with the elements above a diagonal set to zero. In the returned matrix, $A[i, j] = 0$ for all $i - j < p$. If it is not specified, *p* is set to zero.

_sublowertriangle() mirrors sublowertriangle() but modifies *A*. _sublowertriangle(*A*, *p*) sets $A[i, j] = 0$ for all $i - j < p$. If it is not specified, *p* is set to zero.

Remarks

Remarks are presented under the following headings:

> *Get lower triangular of a matrix*
> *Nonsquare matrices*

Get lower triangular of a matrix

If *A* is a square matrix, then sublowertriangle(*A*, 0) = lowertriangle(*A*). sublowertriangle() is a generalization of lowertriangle().

We begin by defining A

```
: A = (1, 2, 3 \ 4, 5, 6 \ 7, 8, 9)
```

sublowertriangle(A, 0) returns A with zeros above the main diagonal as does lowertriangle():

```
: sublowertriangle(A, 0)
         1   2   3

    1    1   0   0
    2    4   5   0
    3    7   8   9
```

sublowertriangle(A, 1) returns A with zeros in the main diagonal and above.

```
: sublowertriangle(A, 1)
    1   2   3
  ┌─────────────┐
1 │  0   0   0  │
2 │  4   0   0  │
3 │  7   8   0  │
  └─────────────┘
```

sublowertriangle(A, p) can take negative p. For example, setting p $= -1$ yields

```
: sublowertriangle(A, -1)
     1   2   3
  ┌─────────────┐
1 │  1   2   0  │
2 │  4   5   6  │
3 │  7   8   9  │
  └─────────────┘
```

Nonsquare matrices

sublowertriangle( ) and _sublowertriangle( ) may be used with nonsquare matrices.

For instance, we define a nonsquare matrix A

```
: A = (1, 2, 3, 4 \ 5, 6, 7,  8 \ 9, 10, 11, 12)
```

We use sublowertriangle( ) to obtain the lower triangular of A

```
: sublowertriangle(A, 0)
     1    2    3    4
  ┌──────────────────────┐
1 │  1    0    0    0     │
2 │  5    6    0    0     │
3 │  9   10   11    0     │
  └──────────────────────┘
```

Conformability

sublowertriangle(A, p):
   input:

                A: $r \times c$

                p: 1×1 (optional)

   output:

             result: $r \times c$

_sublowertriangle(A, p):
   input:

                A: $r \times c$

                p: 1×1 (optional)

   output:

                A: $r \times c$

Diagnostics

None.

Also see

[M-4] **manipulation** — Matrix manipulation

Title

[M-5] _substr( ) — Substitute into string

Syntax

> *void* _substr(*string scalar s*, *string scalar tosub*, *real scalar pos*)

Description

_substr(*s*, *tosub*, *pos*) substitutes *tosub* into *s* at position *pos*. The first position of *s* is *pos* = 1. _substr() may be used with text or binary strings.

Do not confuse _substr() with substr(), which extracts substrings; see [M-5] **substr( )**.

Remarks

If *s* contains "abcdef", then _substr(*s*, "XY", 2) changes *s* to contain "aXYdef".

Conformability

_substr(*s*, *tosub*, *pos*):

> *input*:
>
> | | |
> |---:|:---:|
> | *s*: | 1 × 1 |
> | *tosub*: | 1 × 1 |
> | *pos*: | 1 × 1 |
>
> *output*:
>
> | | |
> |---:|:---:|
> | *s*: | 1 × 1 |

Diagnostics

_substr(*s*, *tosub*, *pos*) does nothing if *tosub*=="".

_substr(*s*, *tosub*, *pos*) may not be used to extend *s*: _substr() aborts with error if substituting *tosub* into *s* would result in a string longer than the original *s*. _substr() also aborts with error if *pos* ≤ 0 or *pos* ≥ . unless *tosub* is "".

_substr(*s*, *tosub*, *pos*) aborts with error if *s* or *tosub* are views.

Also see

[M-4] **string** — String manipulation functions

Title

[M-5] substr( ) — Extract substring

Syntax

> *string matrix* substr(*string matrix s*, *real matrix b*, *real matrix l*)

> *string matrix* substr(*string matrix s*, *real matrix b*)

Description

substr(s, b, l) returns the substring of s starting at position b and continuing for a length of l, where

1. b specifies the starting position; the first character of the string is $b = 1$.

2. $b > 0$ is interpreted as distance from the start of the string; $b = 2$ means starting at the second character.

3. $b < 0$ is interpreted as distance from the end of string; $b = -1$ means starting at the last character; $b = -2$ means starting at the second from the last character.

4. l specifies the length; $l = 2$ means for two characters.

5. $l < 0$ is treated the same as $l = 0$: no characters are copied.

6. $l \geq$. is interpreted to mean to the end of the string.

substr(s, b) is equivalent to substr(s, b, .) for strings that do not contain binary 0. If there is a binary 0 to the right of b, the substring from b up to but not including the binary 0 is returned.

When arguments are not scalars, strmatch( ) returns element-by-element results.

Conformability

substr(s, b, l):

s:	$r_1 \times c_1$
b:	$r_2 \times c_2$
l:	$r_3 \times c_3$; s, b, and l r-conformable
result:	$\max(r_1, r_2, r_3) \times \max(c_1, c_2, c_3)$

substr(s, b):

s:	$r_1 \times c_1$
b:	$r_2 \times c_2$; s and b r-conformable
result:	$\max(r_1, r_2) \times \max(c_1, c_2)$

Diagnostics

In substr(s, b, l) and substr(s, b), if b describes a position before the beginning of the string or after the end, "" is returned. If $b + l$ describes a position to the right of the end of the string, results are as if a smaller value for l were specified.

Also see

[M-4] **string** — String manipulation functions

Title

Syntax

$\quad$ *numeric colvector* $\quad$ rowsum(*numeric matrix Z* $\left[\, , \, missing\right]$)

$\quad$ *numeric rowvector* $\quad$ colsum(*numeric matrix Z* $\left[\, , \, missing\right]$)

$\quad$ *numeric scalar* $\quad\quad$ sum(*numeric matrix Z* $\left[\, , \, missing\right]$)

$\quad$ *numeric colvector* $\quad$ quadrowsum(*numeric matrix Z* $\left[\, , \, missing\right]$)

$\quad$ *numeric rowvector* $\quad$ quadcolsum(*numeric matrix Z* $\left[\, , \, missing\right]$)

$\quad$ *numeric scalar* $\quad\quad$ quadsum(*numeric matrix Z* $\left[\, , \, missing\right]$)

where optional argument *missing* is a real scalar that determines how missing values in Z are treated:

1. Specifying *missing* as 0 is equivalent to not specifying the argument; missing values in Z are treated as contributing 0 to the sum.

2. Specifying *missing* as 1 (or nonzero) specifies that missing values in Z are to be treated as missing values and to turn the sum to missing.

Description

rowsum(Z) and rowsum(Z, *missing*) return a column vector containing the sum over the rows of Z.

colsum(Z) and colsum(Z, *missing*) return a row vector containing the sum over the columns of Z.

sum(Z) and sum(Z, *missing*) return a scalar containing the sum over the rows and columns of Z.

quadrowsum(), quadcolsum(), and quadsum() are quad-precision variants of the above functions. The sum is accumulated in quad precision and then rounded to double precision and returned.

Argument *missing* determines how missing values are treated. If *missing* is not specified, results are the same as if *missing* = 0 were specified: missing values are treated as zero. If *missing* = 1 is specified, missing values are treated as missing values.

These functions may be used with real or complex matrix Z.

Remarks

All functions return the same type as the argument, real if argument is real, complex if complex.

Conformability

rowsum(Z, *missing*), quadrowsum(Z, *missing*):

Z:	$r \times c$	
missing:	1×1	(optional)
result:	$r \times 1$	

colsum(Z, *missing*), quadcolsum(Z, *missing*):

Z:	$r \times c$	
missing:	1×1	(optional)
result:	$1 \times c$	

sum(Z, *missing*), quadsum(Z, *missing*):

Z:	$r \times c$	
missing:	1×1	(optional)
result:	1×1	

Diagnostics

If *missing* $= 0$, missing values are treated as contributing zero to the sum; they do not turn the sum to missing. Otherwise, missing values turn the sum to missing.

Also see

[M-5] **mean( )** — Means, variances, and correlations

[M-5] **runningsum( )** — Running sum of vector

[M-5] **cross( )** — Cross products

[M-4] **mathematical** — Important mathematical functions

[M-4] **utility** — Matrix utility functions

Title

[M-5] svd( ) — Singular value decomposition	

Syntax

void	svd(*numeric matrix A*, *U*, *s*, *Vt*)
real colvector	svdsv(*numeric matrix A*)
void	_svd(*numeric matrix A*, *s*, *Vt*)
real colvector	_svdsv(*numeric matrix A*)
real scalar	_svd_la(*numeric matrix A*, *s*, *Vt*)

Description

svd(*A*, *U*, *s*, *Vt*) calculates the singular value decomposition of A: $m \times n$, $m \geq n$, returning the result in U, s, and Vt. Singular values returned in s are sorted from largest to smallest.

svdsv(*A*) returns the singular values of A: $m \times n$, $m \geq n$ or $m < n$ (i.e., no restriction), in a column vector of length min(*m*,*n*). U and Vt are not calculated.

_svd(*A*, *s*, *Vt*) does the same as svd( ), except that it returns U in A. Use of _svd( ) conserves memory.

_svdsv(*A*) does the same as svdsv( ), except that, in the process, it destroys A. Use of _svdsv( ) conserves memory.

_svd_la( ) is the interface into the [M-1] **LAPACK** SVD routines and is used in the implementation of the previous functions. There is no reason you should want to use it.

Remarks

Remarks are presented under the following headings:

> *Introduction*
> *Possibility of convergence problems*

Documented here is the thin SVD, appropriate for use with A: $m \times n$, $m \geq n$. See [M-5] **fullsvd( )** for the full SVD, appropriate for use in all cases. The relationship between the two is discussed in *Relationship between the full and thin SVDs* in [M-5] **fullsvd( )**.

Use of the thin SVD—the functions documented here—is preferred when $m \geq n$.

Introduction

The SVD is used to compute accurate solutions to linear systems and least-squares problems, to compute the 2-norm, and to determine the numerical rank of a matrix.

The singular value decomposition (SVD) of A: $m \times n$, $m \geq n$, is given by

$$A = U * \mathrm{diag}(s) * V'$$

where

U:	$m \times n$ and $U'U = \mathrm{I}(n)$
s:	$n \times 1$
V:	$n \times n$ and orthogonal (unitary)

When A is complex, the transpose operator $'$ is understood to mean the conjugate transpose operator.

Vector s contains the singular values, and those values are real even when A is complex. s is ordered so that the largest singular value appears first, then the next largest, and so on.

Function $\mathrm{svd}(A, U, s, Vt)$ returns U, s, and $Vt = V'$.

Function $\mathrm{svdsv}(A)$ returns s, omitting the calculation of U and Vt. Also, whereas $\mathrm{svd}()$ is suitable for use only in the case $m \geq n$, $\mathrm{svdsv}()$ may be used in all cases.

Possibility of convergence problems

It is possible, although exceedingly unlikely, that the SVD routines could fail to converge. svd(), svdsv(), _svd(), and _svdsv() then return singular values in s equal to missing.

In coding, it is perfectly acceptable to ignore this possibility because (1) it is so unlikely and (2) even if the unlikely event occurs, the missing values will properly reflect the situation. If you do wish to check, in addition to checking missing(s)>0, you must also check missing(A)==0 because that is the other reason s could contain missing values. Convergence was not achieved if missing(s) > 0 & missing(A)==0. If you are calling one of the destructive-of-A versions of SVD, remember to check missing(A)==0 before extracting singular values.

Conformability

svd(A, U, s, Vt):

  input:

A:	$m \times n$, $m \geq n$

  output:

U:	$m \times n$
s:	$n \times 1$
Vt:	$n \times n$

svdsv(A):

A:	$m \times n$,	$m \geq n$ or $m < n$
result:	$\min(m, n) \times 1$	

_svd(A, s, Vt):

input:

A:	$m \times n$,	$m \geq n$

output:

A:	$m \times n$,	contains U
s:	$n \times 1$	
Vt:	$n \times n$	

_svdsv(A):

input:

A:	$m \times n$,	$m \geq n$ or $m < n$

output:

A:	0×0	
result:	$\min(m, n) \times 1$	

_svd_la(A, s, Vt):

input:

A:	$m \times n$,	$m \geq n$

output:

A:	$m \times n$,	contains U
s:	$n \times 1$	
Vt:	$n \times n$	
result:	1×1	

Diagnostics

svd(A, U, s, Vt) and _svd(A, s, Vt) return missing results if A contains missing. In all other cases, the routines should work, but there is the unlikely possibility of convergence problems, in which case missing results will also be returned; see *Possibility of convergence problems* above.

svdsv(A) and _svdsv(A) return missing results if A contains missing values or if there are convergence problems.

_svd( ) and _svdsv( ) abort with error if A is a view.

Direct use of _svd_la( ) is not recommended.

Also see

[M-5] **fullsvd( )** — Full singular value decomposition

[M-5] **svsolve( )** — Solve AX=B for X using singular value decomposition

[M-5] **pinv( )** — Moore–Penrose pseudoinverse

[M-5] **norm( )** — Matrix and vector norms

[M-5] **rank( )** — Rank of matrix

[M-4] **matrix** — Matrix functions

Title

[M-5] **svsolve( )** — Solve AX=B for X using singular value decomposition

Syntax

numeric matrix   svsolve(*A*, *B*)

numeric matrix   svsolve(*A*, *B*, *rank*)

numeric matrix   svsolve(*A*, *B*, *rank*, *tol*)

real scalar   _svsolve(*A*, *B*)

real scalar   _svsolve(*A*, *B*, *tol*)

where

A:	*numeric matrix*
B:	*numeric matrix*
rank:	irrelevant; *real scalar* returned
tol:	*real scalar*

Description

svsolve(*A*, *B*, ...), uses singular value decomposition to solve $AX = B$ and return X. When A is singular, svsolve() computes the minimum-norm least-squares generalized solution. When *rank* is specified, in it is placed the rank of A.

_svsolve(*A*, *B*, ...) does the same thing, except that it destroys the contents of A and it overwrites B with the solution. Returned is the rank of A.

In both cases, *tol* specifies the tolerance for determining whether A is of full rank. *tol* is interpreted in the standard way—as a multiplier for the default if *tol* > 0 is specified and as an absolute quantity to use in place of the default if *tol* $\leq$ 0 is specified.

Remarks

svsolve(*A*, *B*, ...) is suitable for use with square or nonsquare, full-rank or rank-deficient matrix A. When A is of full rank, qrsolve() returns the same solution as lusolve() (see [M-5] **lusolve( )**), ignoring roundoff error. When A is singular, svsolve() returns the minimum-norm least-squares generalized solution. qrsolve() (see [M-5] **qrsolve( )**), an alternative, returns a generalized least-squares solution that amounts to dropping rows of A.

Remarks are presented under the following headings:

> *Derivation*
> *Relationship to inversion*
> *Tolerance*

Derivation

We wish to solve for X

$$AX = B \tag{1}$$

Perform singular value decomposition on A so that we have $A = USV'$. Then (1) can be rewritten

$$USV'X = B$$

Premultiplying by U' and remembering that $U'U = I$, we have

$$SV'X = U'B$$

Matrix S is diagonal and thus its inverse is easily calculated, and we have

$$V'X = S^{-1}U'B$$

When we premultiply by V, remembering that $VV' = I$, the solution is

$$X = VS^{-1}U'B \tag{2}$$

See [M-5] **svd( )** for more information on the SVD.

Relationship to inversion

For a general discussion, see *Relationship to inversion* in [M-5] **lusolve( )**.

For an inverse based on the SVD, see [M-5] **pinv( )**. `pinv(A)` amounts to `svsolve(A, I(rows(A)))`, although `pinv()` has separate code that uses less memory.

Tolerance

In (2) above, we are required to calculate the inverse of diagonal matrix S. The generalized solution is obtained by substituting zero for the ith diagonal element of S^{-1}, where the ith diagonal element of S is less than or equal to *eta* in absolute value. The default value of *eta* is

$$eta = \text{epsilon}(1) * \text{rows}(A) * \max(S)$$

If you specify *tol* > 0, the value you specify is used to multiply *eta*. You may instead specify *tol* $\leq$ 0 and then the negative of the value you specify is used in place of *eta*; see [M-1] **tolerance**.

(*Continued on next page*)

Conformability

svsolve(*A*, *B*, *rank*, *tol*):

  input:

	A:	$m \times n$	
	B:	$m \times k$	
	tol:	1×1	(optional)

  output:

	rank:	1×1	(optional)
	result:	$n \times k$	

_svsolve(*A*, *B*, *tol*):

  input:

	A:	$m \times n$	
	B:	$m \times k$	
	tol:	1×1	(optional)

  output:

	A:	0×0	
	B:	$m \times k$	
	result:	1×1	

Diagnostics

svsolve(*A*, *B*, ...) and _svsolve(*A*, *B*, ...) return missing results if *A* or *B* contain missing.

_svsolve(*A*, *B*, ...) aborts with error if *A* (but not *B*) is a view.

Also see

[M-5] **solvelower( )** — Solve AX=B for X, A triangular

[M-5] **cholsolve( )** — Solve AX=B for X using Cholesky decomposition

[M-5] **lusolve( )** — Solve AX=B for X using LU decomposition

[M-5] **qrsolve( )** — Solve AX=B for X using QR decomposition

[M-4] **matrix** — Matrix functions

[M-4] **solvers** — Functions to solve AX=B and to obtain A inverse

Title

Syntax

$void$ swap(*transmorphic matrix A*, *transmorphic matrix B*)

Description

swap(A, B) interchanges the contents of A and B. A and B are not required to be of the same type or dimension.

Remarks

There is no faster way than swap(A, B) to assign $A=B$ when you do not care about the contents of B after the assignment. For instance, you have the code

        A = B
        B = ...(matrix expression)...

Faster is

        swap(A, B)
        B = ...(matrix expression)...

The execution time of swap() is independent of the size of A and B, and swap() conserves memory to boot. Pretend that B is a 900 × 900 matrix. After $A=B$ is executed, but before B is reassigned, two copies of the 900 × 900 matrix exist. That does not happen with swap().

Conformability

swap(A, B):

input:

A:	$r_1 \times c_1$
B:	$r_2 \times c_2$

output:

A:	$r_2 \times c_2$
B:	$r_1 \times c_1$

Diagnostics

swap(A, B) works only with variables. Do not code, for instance, swap($A[i,j]$, $A[j,i]$). It is not an error, but it will have no effect.

Also see

[M-4] **programming** — Programming functions

Title

Syntax

numeric matrix   Toeplitz(*numeric colvector c1*, *numeric rowvector r1*)

Description

Toeplitz(*c1*, *r1*) returns the Toeplitz matrix defined by *c1* being its first column and *r1* being its first row. A Toeplitz matrix T is characterized by $T[i,j] = T[i-1, j-1]$, $i, j > 1$. In a Toeplitz matrix, each diagonal is constant.

Vectors *c1* and *r1* specify the first column and first row of T.

Remarks

$c1[1]$ is used to fill $T[1,1]$ and $r1[1]$ is not used.

To obtain the symmetric (Hermitian) Toeplitz matrix, code Toeplitz(v, v') (if v is a column vector), or Toeplitz(v', v) if v is a row vector.

Conformability

Toeplitz(*c1*, *r1*):

c1:	$r \times 1$
r1:	$1 \times c$
result:	$r \times c$

Diagnostics

None. The top left element is defined by $c1[1]$, and $r1[1]$ is not used.

Otto Toeplitz (1881–1940) was born in Breslau, Germany (now Wrocław, Poland), and educated there in mathematics. He researched and taught at universities in Göttingen, Kiel, and Bonn, making many contributions to algebra and analysis, but he was dismissed in 1935 for being a Jew. Toeplitz emigrated to Palestine in 1939 but died a few months later in Jerusalem. He was fascinated by the history of mathematics and wrote a popular work with Hans Rademacher, *The Enjoyment of Mathematics*.

Reference

Robinson, A. 1976. Toeplitz, Otto. In Vol. 13 of *Dictionary of Scientific Biography*, ed. C. C. Gillispie. New York: Scribner's.

Also see

[M-4] **standard** — Functions to create standard matrices

Title

Syntax

t = tokeninit([*wchars* [, *pchars* [, *qchars* [, *allownum* [, *allowhex*]]]]])

t = tokeninitstata( )

void	tokenset(*t*, *real scalar s*)
string rowvector	tokengetall(*t*)
string scalar	tokenget(*t*)
string scalar	tokenpeek(*t*)
string scalar	tokenrest(*t*)
real scalar	tokenoffset(*t*)
void	tokenoffset(*t*, *real scalar offset*)
string scalar	tokenwchars(*t*)
void	tokenwchars(*t*, *string scalar wchars*)
string rowvector	tokenpchars(*t*)
void	tokenpchars(*t*, *string rowvector pchars*)
string rowvector	tokenqchars(*t*)
void	tokenqchars(*t*, *string rowvector qchars*)
real scalar	tokenallownum(*t*)
void	tokenallownum(*t*, *real scalar allownum*)
real scalar	tokenallowhex(*t*)
void	tokenallowhex(*t*, *real scalar allowhex*)

where

t is *transmorphic* and contains the parsing environment information. You obtain a t from
tokeninit( ) or tokeninitstata( ) and then pass t to the other functions.

wchars is a *string scalar* containing the characters to be treated as white space, such as " ",
(" "+char(9)), or "".

pchars is a *string rowvector* containing the strings to be treated as parsing characters, such as
"" and (">", "<", ">=", "<="). "" and J(1,0,"") are given the same interpretation:
there are no parsing characters.

qchars is a *string rowvector* containing the character pairs to be treated as quote characters. "" (i.e., empty string) is given the same interpretation as J(1,0,""); there are no quote characters. *qchars* = (‘""""’) (i.e., the two-character string quote indicates that " is to be treated as open quote and " is to be treated as close quote. *qchars* = (‘""""’, ‘"‘"’"’) indicates that, in addition, ‘" is to be treated as open quote and "’ as close quote. In a syntax that did not use < and > as parsing characters, *qchars* = ("<>") would indicate that < is to be treated as open quote and > as close quote.

allownum is a *string scalar* containing 0 or 1. *allownum* = 1 indicates that numbers such as 12.23 and 1.52e+02 are to be returned as single tokens even in violation of other parsing rules.

allowhex is a *string scalar* containing 0 or 1. *allowhex* = 1 indicates that numbers such as 1.921fb54442d18X+001 and 1.0x+a are to be returned as single tokens even in violation of other parsing rules.

Description

These functions provide advanced parsing. If you simply wish to convert strings into row vectors by separating on blanks, converting "mpg weight displ" into ("mpg", "weight", "displ"), see [M-5] **tokens( )**.

Remarks

Remarks are presented under the following headings:

> *Concepts*
>> *White-space characters*
>> *Parsing characters*
>> *Quote characters*
>> *Overrides*
>> *Setting the environment to parse on blanks with quote binding*
>> *Setting the environment to parse full Stata syntax*
>> *Setting the environment to parse tab-delimited files*
>
> *Function overview*
>> *tokeninit() and tokeninitstata()*
>> *tokenset()*
>> *tokengetall()*
>> *tokenget(), tokenpeek(), and tokenrest()*
>> *tokenoffset()*
>> *tokenwchars(), tokenpchars(), and tokenqchars()*
>> *tokenallownum and tokenallowhex()*

Concepts

Parsing refers to splitting a string into pieces, which we will call tokens. Parsing as implemented by the token*() functions is defined by (1) the white-space characters *wchars*, (2) the parsing characters *pchars*, and (3) the quote characters *qchars*.

White-space characters

Consider the string "this that what". If there are no white-space characters, no parsing characters, and no quote characters, i.e., if *wchars* = *pchars* = *qchars* = "", then the result of parsing "this that what" would be one token that would be the string just as it is: "this that what".

If *wchars* were instead " ", then parsing "this that what" results in ("this", "that", "what"). Parsing "this that   what" (note the multiple blanks) would result in the same thing. White-space characters separate one token from the next but are not otherwise significant.

Parsing characters

If we instead left *wchars* = "" and set *pchars* = " ", "this that what" parses into ("this", " ", "that", " ", "what") and parsing "this that   what" results in ("this", " ", "that", " ", " ", " ", "what").

pchars are like *wchars* except that they are themselves significant.

pchars do not usually contain space. A more reasonable definition of *pchars* is ("+", "-"). Then parsing "x+y" results in ("x", "+", "y"). Also, the parsing characters can be character combinations. If *pchars* = ("+", "-", "++", "--"), then parsing "x+y++" results in ("x", "+", "y", "++") and parsing "x+++y" results in ("x", "++", "+", "y"). Longer *pchars* are matched before shorter ones regardless of the order in which they appear in the *pchars* vector.

Quote characters

qchars specifies the quote characters. Pieces of the string being parsed that are surrounded by quotes are returned as one token, ignoring the separation that would usually occur because of the *wchars* and *pchars* definitions. Consider the string

        mystr= "x = y"

Let *wchars* = " " and *pchars* include "=". That by itself would result in the above string parsing into the five tokens

mystr	=	"x	=	y"

Now let *qchars* = ('""'); i.e., *qchars* is the two-character string "". Parsing then results in the three tokens

mystr	=	"x = y"

Each element of *qchars* contains a character pair: the open character followed by the close character. We defined those two characters as " and " above, that is, as being the same. The two characters can differ. We might define the first as ' and the second as '. When the characters are different, quotations can nest. The quotation "he said "hello"" makes no sense because that parses into ("he said ", hello, ""). The quotation 'he said 'hello'', however, makes perfect sense and results in the single token 'he said 'hello''.

The quote characters can themselves be multiple characters. You can define open quote as '" and close as "': *qchars* = ('"'""'"'). Or you can define multiple sets of quotation characters, such as *qchars* = ('""''', '"'""'"').

The quote characters do not even have to be quotes at all. In some context you might find it convenient to specify them as ("()"). With that definition, "$(2 \times (3 + 2))$" would parse into one token. Specifying them like this can be useful, but in general we recommend against it. It is usually better to write your code so that quote characters really are quote characters and to push the work of handling other kinds of nested expressions back onto the caller.

Overrides

The `token*()` functions provide two overrides: *allownum* and *allowhex*. These have to do with parsing numbers. First, consider life without overrides. You have set *wchars* = " " and *pchars* = ("=", "+", "-", "*", "/"). You attempt to parse

The result is

when what you wanted was

y | = | x | + | 1e+13

Setting *allownum* = 1 will achieve the desired result. *allownum* specifies that, when a token could be interpreted as a number, the number interpretation is to be taken even in violation of the other parsing rules.

Setting *allownum* = 1 will not find numbers buried in the middle of strings, such as the 1e+3 in "x is 1e+3", but if the number occurs at the beginning of the token according to the parsing rules set by *wchars* and *pchars*, *allownum* = 1 will continue the token in violation of those rules if that results in a valid number.

The override *allowhex* is similar and Stata specific. Stata (and Mata) provide a unique and useful way of writing hexadecimal floating-point numbers in a printable, short, and precise way: π can be written 1.921fb54442d18X+001. Setting *allowhex* = 1 allows such numbers.

Setting the environment to parse on blanks with quote binding

Stata's default rule for parsing do-file arguments is "parse on blanks and bind on quotes". The settings for duplicating that behavior are

> *wchars* = " "
>
> *pchars* = ("")
>
> *qchars* = (`""""'` , `"`"""`"'`)
>
> *allownum* = 0
>
> *allowhex* = 0

This behavior can be obtained by coding

> *t* = tokeninit(" ", "", (`""""'`, `"`"""`"'`), 0, 0)

or by coding

> *t* = tokeninit()

because in `tokeninit()` the arguments are optional and "parse on blank with quote binding" is the default.

With those settings, parsing ‘"first second "third fourth" fifth"’ results in
("first", "second", ‘""third fourth""’, "fifth").

This result is a little different from that of Stata because the third token includes the quote binding
characters. Assume that the parsed string was obtained by coding

```
res = tokengetall(t)
```

The following code will remove the open and close quotes, should that be desirable.

```
for (i=1; i<=cols(res); i++) {
        if (res[i]==‘"""’) {
                res[i] = substr(res[i], 2, strlen(res[i])-2)
        }
        else if (substr(res[i], 1, 2)=="‘" + ‘"""’) {
                res[i] = substr(res[i], 3, strlen(res[i])-4)
        }
}
```

Setting the environment to parse full Stata syntax

To parse full Stata syntax, the settings are

$wchars =$ " "

$pchars = ($ "\",   "~",   "!",   "=",   ":",   ";",   ",",
            "?",   "!",   "@",   "#",   "==",   "!=",   ">=",
            "<=",   "<",   ">",   "&",   "|",   "&&",   "||",
            "+",   "-",   "++",   "--",   "*",   "/",   "^",
            "(",   ")",   "[",   "]",   "{",   "}" $)$

$qchars = ($ ‘"""’, ‘"‘""’"’, char(96)+char(39)$)$

$allownum = 1$

$allowhex = 1$

The above is a slight oversimplification. Stata is an interpretive language and Stata does not require
users to type filenames in quotes, although Stata does allow it. Thus "\" is sometimes a parsing
character and sometimes not, and the same is true of "/". As Stata parses a line from left to right,
it will change *pchars* between two tokenget() calls when the next token could be or is known to
be a filename. Sometimes Stata peeks ahead to decide which way to parse. You can do the same by
using the tokenpchars() and tokenpeek() functions.

To obtain the above environment, code

```
t = tokeninitstata()
```

Setting the environment to parse tab-delimited files

The token*() functions can be used to parse lines from tab-delimited files. A tab-delimited file
contains lines of the form

⟨*field1*⟩⟨*tab*⟩⟨*field2*⟩⟨*tab*⟩⟨*field3*⟩

The parsing environment variables are

$$wchars = \verb|""|$$

$$pchars = (\ \texttt{char(9)}\ )\quad (\text{i.e., } tab)$$

$$qchars = (\ \verb|""|\ )$$

$$allownum = 0$$

$$allowhex = 0$$

To set this environment, code

t = `tokeninit("", char(9), "", 0, 0)`

Say that you then parse the line

Farber, William⟨*tab*⟩   2201.00⟨*tab*⟩12

The results will be

`("Farber, William", char(9), " 2201.00", char(9), "12")`

If the line were

Farber, William⟨*tab*⟩⟨*tab*⟩12

the result would be

`("Farber, William", char(9), char(9), "12")`

The tab-delimited format is not well defined when the missing fields occur at the end of the line. A line with the last field missing might be recorded

Farber, William⟨*tab*⟩   2201.00⟨*tab*⟩

or

Farber, William⟨*tab*⟩   2201.00

A line with the last two fields missing might be recorded

Farber, William⟨*tab*⟩⟨*tab*⟩

or

Farber, William⟨*tab*⟩

or

Farber, William

The following program would correctly parse lines with missing fields regardless of how they are recorded:

```
real rowvector readtabbed(transmorphic t, real scalar n)
{
        real scalar       i
        string rowvector  res
        string scalar     token

        res = J(1, n, "")
        i = 1
        while ((token = tokenget(t))!="") {
                if (token==char(9)) i++
                else res[i] = token
        }
        return(res)
}
```

Function overview

The basic way to proceed is to initialize the parsing environment and store it in a variable,

```
t = tokeninit(...)
```

and then set the string s to be parsed,

```
tokenset(t, s)
```

and finally use tokenget() to obtain the tokens one at a time (tokenget() returns "" when the end of the line is reached), or obtain all the tokens at once using tokengetall(t). That is, either

```
while((token = tokenget(t)) != "") {
        ... process token ...
}
```

or

```
tokens = tokengetall(t)
for (i=1; i<=cols(tokens); i++) {
        ... process tokens[i] ...
}
```

After that, set the next string to be parsed,

```
tokenset(t, nextstring)
```

and repeat.

tokeninit() and tokeninitstata()

tokeninit() and tokeninitstata() are alternatives. tokeninitstata() is generally unnecessary unless you are writing a fairly complicated function.

Whichever function you use, code

> *t* = tokeninit(...)

or

> *t* = tokeninitstata()

If you declare *t*, declare it transmorphic. *t* is in fact a structure containing all the details of your parsing environment, but that is purposely hidden from you so that you cannot accidentally modify the environment.

tokeninit() allows up to five arguments:

> *t* = tokeninit(*wchars*, *pchars*, *qchars*, *allownum*, *allowhex*)

You may omit arguments from the end. If omitted, the default values of the arguments are

> *allowhex* = 0
>
> *allownum* = 0
>
> *qchars* = (‘"""’ , ‘"‘""›"’)
>
> *pchars* = ("")
>
> *wchars* = " "

Notes

1. Concerning *wchars*:

   a. *wchars* is a *string scalar*. The white-space characters appear one after the other in the string. The order in which the characters appear is irrelevant.

   b. Specify *wchars* as " " to treat blank as white space.

   c. Specify *wchars* as " "+char(9) to treat blank and *tab* as white space. Including *tab* is necessary only when strings to be parsed are obtained from a file; strings obtained from Stata already have the *tab* characters removed.

   d. Any character can be treated as a white-space character, including letters.

   e. Specify *wchars* as "" to specify that there are no white-space characters.

2. Concerning *pchars*:

   a. *pchars* is a *string rowvector*. Each element of the vector is a separate parse character. The order in which the parse characters are specified is irrelevant.

   b. Specify *pchars* as ("+", "-") to make + and − parse characters.

   c. Parse characters may be character combinations such as ++ or >=. Character combinations may be up to four characters long.

   d. Specify *pchars* as "" or J(1,0,"") to specify that there are no parse characters. It makes no difference which you specify, but you will realize that J(1,0,"") is more logically consistent if you think about it.

3. Concerning *qchars*:

   a. *qchars* is a `string rowvector`. Each element of the vector contains the open followed by the close characters. The order in which sets of quote characters are specified is irrelevant.

   b. Specify *qchars* as (`'""""'`) to make " an open and close character.

   c. Specify *qchars* as (`'""""'`, `'"‘"’"'`) to make "" and `'""'` quote characters.

   d. Individual quote characters can be up to two characters long.

   e. Specify *qchars* as `""` or `J(1,0,"")` to specify that there are no quote characters.

tokenset()

After `tokeninit()` or `tokeninitstata()`, you are not yet through with initialization. You must `tokenset(s)` to specify the string scalar you wish to parse. You `tokenset()` one line, parse it, and if you have more lines, you `tokenset()` again and repeat the process. Often you will need to parse only one line. Perhaps you wish to write a program to parse the argument of a complicated option in a Stata ado-file. The structure is

```
program ...
        ...
        syntax ... [, ... MYoption(string) ...]
        mata: parseoption('"'myoption'"')
        ...
end

mata:
void parseoption(string scalar option)
{
        transmorphic    t

        t = tokeninit(...)
        tokenset(t, option)
        ...
}
end
```

Notes

1. When you `tokenset(s)`, the contents of *s* are not stored. Instead, a pointer to *s* is stored. This approach saves memory and time, but it means that if you change *s* after setting it, you will change the subsequent behavior of the `token*()` functions.

2. Simply changing *s* is not sufficient to restart parsing. If you change *s*, you must `tokenset(s)` again.

tokengetall()

You have two alternatives in how to process the tokens. You can parse the entire line into a row vector containing all the individual tokens by using `tokengetall()`,

```
tokens = tokengetall(t)
```

or you can use `tokenget()` to process the tokens one at a time, which is discussed in the next section.

Using `tokengetall()`, `tokens[1]` will be the first token, `tokens[2]` the second, and so on. There are, in total, `cols(tokens)` tokens. If the line was empty or contained only white-space characters, `cols(tokens)` will be 0.

tokenget(), tokenpeek(), and tokenrest()

`tokenget()` returns the tokens one at a time and returns `""` when the end of the line is reached. The basic loop for processing all the tokens in a line is

```
while ( (token = tokenget( t )) != "") {
        . . .
}
```

`tokenpeek()` allows you to peek ahead at the next token without actually getting it, so whatever is returned will be returned again by the next call to `tokenget()`. `tokenpeek()` is suitable only for obtaining the next token after `tokenget()`. Calling `tokenpeek()` twice in a row will not return the next two tokens; it will return the next token twice. To obtain the next two tokens, code

```
        . . .
current = tokenget(t)             // get the current token
        . . .
t2 = t                            // copy parse environment
next_1 = tokenget(t2)             // peek at next token
next_1 = tokenget(t2)             // peek at token after that
        . . .
current = tokenget(t)             // get next token
```

If you declare *t2*, declare it `transmorphic`.

`tokenrest()` returns the unparsed portion of the `tokenset()` string. Assume that you have just gotten the first token by using `tokenget()`. `tokenrest()` would return the rest of the original string, following the first token, unparsed. `tokenrest(t)` returns `substr(`*original_string*`, tokenoffset(t), .)`.

tokenoffset()

`tokenoffset()` is useful only when you are using the `tokenget()` rather than `tokengetall()` style of programming. Let the original string you `tokenset()` be "this is an example". Right after you have `tokenset()` this string, `tokenoffset()` is 1:

```
        this is an example
        |
tokenoffset() = 1
```

After getting the first token (say it is `"this"`), `tokenoffset()` is 5:

```
        this is an example
            |
tokenoffset() = 5
```

`tokenoffset()` is always located on the first character following the last character parsed.

The syntax of `tokenoffset()` is

        `tokenoffset(`*t*`)`

and

        `tokenoffset(` *t, newoffset*`)`

The first returns the current offset value. The second resets the parser's location within the string.

tokenwchars(), tokenpchars(), and tokenqchars()

`tokenwchars()`, `tokenpchars()`, and `tokenqchars()` allow resetting the current *wchars*, *pchars*, and *qchars*. As with `tokenoffset()`, they come in two syntaxes.

With one argument, *t*, they return the current value of the setting. With two arguments, *t* and *newvalue*, they reset the value.

Resetting in the midst of parsing is an advanced issue. The most useful of these functions is `tokenpchars()`, since for interactive grammars, it is sometimes necessary to switch on and off a certain parsing character such as /, which in one context means division and in another is a file separator.

tokenallownum and tokenallowhex()

These two functions allow obtaining the current values of *allownum* and *allowhex* and resetting them.

Conformability

`tokeninit(`*wchars, pchars, qchars, allownum, allowhex*`)`:

wchars:	1×1	(optional)
pchars:	$1 \times c_p$	(optional)
qchars:	$1 \times c_q$	(optional)
allownum:	1×1	(optional)
allowhex:	1×1	(optional)
result:	*transmorphic*	

`tokeninitstata()`:

result:	*transmorphic*

`tokenset(`*t, s*`)`:

t:	*transmorphic*
s:	1×1
result:	*void*

`tokengetall(`*t*`)`:

t:	*transmorphic*
result:	$1 \times k$

`tokenget(`*t*`)`, `tokenpeek(`*t*`)`, `tokenrest(`*t*`)`:

t:	*transmorphic*
result:	1×1

tokenoffset(t), tokenwchars(t), tokenallownum(t), tokenallowhex(t):

t: *transmorphic*
result: 1×1

tokenoffset(t, *newvalue*), tokenwchars(t, *newvalue*),
    tokenallownum(t, *newvalue*), tokenallowhex(t, *newvalue*):

t: *transmorphic*
newvalue: 1×1
result: *void*

tokenpchars(t), tokenqchars(t):

t: *transmorphic*
result: $1 \times c$

tokenpchars(t, *newvalue*), tokenqchars(t, *newvalue*):

t: *transmorphic*
newvalue: $1 \times c$
result: *void*

Diagnostics

None.

Also see

[M-5] **tokens( )** — Obtain tokens from string

[M-4] **programming** — Programming functions

[M-4] **string** — String manipulation functions

Title

Syntax

> *string rowvector*   tokens(*string scalar s*)
>
> *string rowvector*   tokens(*string scalar s*, *string scalar parsechars*)

Description

tokens(*s*) returns the contents of *s*, split into words.

tokens(*s*, *parsechars*) returns the contents of *s* split into tokens based on *parsechars*.

tokens(*s*) is equivalent to tokens(*s*, " ").

If you need more advanced parsing, see [M-5] **tokenget( )**.

Remarks

tokens() is commonly used to split a string containing a sequence of variable names into a row vector, each element of which contains one variable name:

> tokens("mpg weight displacement") = ("mpg", "weight", "displacement")

Some Stata interface functions require that variable names be specified in this form. This is required, for instance, by st_varindex(); see [M-5] **st_varindex( )**. If you had a string scalar vars containing one or more variable names, you could obtain their variable indices by coding

> indices = st_varindex(tokens(vars))

Conformability

tokens(*s*, *parsechars*)

s:	1×1	
parsechars:	1×1	(optional)
result:	$1 \times w$,	w = number of words (tokens) in *s*

Diagnostics

If *s* contains "", tokens() returns J(1,0,"").

If *s* contains double-quoted or compound-double-quoted material, the quotes are stripped and that material is returned as one token. For example,

> tokens(`"this "is an" example"') = ("this", "is an", "example")

If *s* contains quoted material and the quotes do not match, results are as if the appropriate number of close quotes were added to the end of *s*. For example,

```
tokens('"this "is an example"') = ("this", "is an example")
```

Also see

[M-5] **invtokens( )** — Concatenate string rowvector into string scalar

[M-5] **tokenget( )** — Advanced parsing

[M-4] **string** — String manipulation functions

Title

Syntax

numeric scalar trace(*numeric matrix A*)

numeric scalar trace(*numeric matrix A, numeric matrix B*)

numeric scalar trace(*numeric matrix A, numeric matrix B, real scalar t*)

Description

trace(*A*) returns the sum of the diagonal elements of *A*. Returned result is real if *A* is real, complex if *A* is complex.

trace(*A, B*) returns trace(*AB*), the calculation being made without calculating or storing the off-diagonal elements of *AB*. Returned result is real if *A* and *B* are real and is complex otherwise.

trace(*A, B, t*) returns trace(*AB*) if $t = 0$ and returns trace(*A'B*) otherwise, where, if either *A* or *B* is complex, transpose is understood to mean conjugate transpose. Returned result is real if *A* and *B* are real and is complex otherwise.

Remarks

trace(*A, B*) returns the same result as trace(*A*B*) but is more efficient if you do not otherwise need to calculate *A*B*.

trace(*A, B, 1*) returns the same result as trace(*A'B*) but is more efficient.

For real matrices *A* and *B*,

$$\text{trace}(A') = \text{trace}(A)$$

$$\text{trace}(AB) = \text{trace}(BA)$$

and for complex matrices,

$$\text{trace}(A') = \text{conj}(\text{trace}(A))$$

$$\text{trace}(AB) = \text{trace}(BA)$$

where, for complex matrices, transpose is understood to mean conjugate transpose.

Thus for real matrices,

to calculate	code
trace(AB)	trace(A, B)
trace($A'B$)	trace(A, B, 1)
trace(AB')	trace(A, B, 1)
trace($A'B'$)	trace(A, B)

and for complex matrices,

to calculate	code
trace(AB)	trace(A, B)
trace($A'B$)	trace(A, B, 1)
trace(AB')	conj(trace(A, B, 1))
trace($A'B'$)	conj(trace(A, B))

Transpose in the first column means conjugate transpose.

Conformability

trace(A):

             A: $n \times n$

    result: 1×1

trace(A, B):

             A: $n \times m$

             B: $m \times n$

    result: 1×1

trace(A, B, t)

             A: $n \times m$ if $t = 0$, $m \times n$ otherwise

             B: $m \times n$

             t: 1×1

    result: 1×1

Diagnostics

trace(A) aborts with error if A is not square.

trace(A, B) and trace(A, B, t) abort with error if the matrices are not conformable or their product is not square.

The trace of a 0×0 matrix is 0.

Also see

[M-4] **matrix** — Matrix functions

Title

[M-5] _transpose( ) — Transposition in place

Syntax

> *void* _transpose(*numeric matrix A*)

Description

_transpose(*A*) replaces *A* with *A′*. Coding _transpose(*A*) is equivalent to coding $A = A′$, except that execution can take a little longer and less memory is used. When *A* is complex, *A* is replaced with its conjugate transpose; see [M-5] **transposeonly( )** if transposition without conjugation is desired.

Remarks

In some calculation, you need *A′*

> $X = \ldots$ *calculation using A′* $\ldots$

If *A* is large, you can save considerable memory by coding

```
_transpose(A)
X = ... calculation using A ...
_transpose(A)
```

Conformability

_transpose(*A*):

> *input*:
>> *A*: $r \times c$
>
> *output*:
>> *A*: $c \times r$

Diagnostics

_transpose(*A*) aborts with error if *A* is a view.

Also see

[M-2] **op_transpose** — Conjugate transpose operator

[M-5] **transposeonly( )** — Transposition without conjugation

[M-5] **conj( )** — Complex conjugate

[M-4] **manipulation** — Matrix manipulation

Title

Syntax

numeric matrix	transposeonly(*numeric matrix A*)
void	_transposeonly(*numeric matrix A*)

Description

transposeonly(*A*) returns *A* with its rows and columns interchanged. When *A* is real, the actions of transposeonly(*A*) are indistinguishable from coding *A'*; see [M-2] **op_transpose**. The returned result is the same, and the execution time is the same, too. When *A* is complex, however, transposeonly(*A*) is equivalent to coding conj(*A'*), but transposeonly() obtains the result more quickly.

_transposeonly(*A*) interchanges the rows and columns of *A* in place—without use of additional memory—and returns the transposed (but not conjugated) result in *A*.

Remarks

transposeonly() is useful when you are coding in the programming, rather than the mathematical, sense. Say that you have two row vectors, a and b, and you want to place the two vectors together in a matrix R, and you want to turn them into column vectors. If a and b were certain to be real, you could just code

```
R = (a', b')
```

The above line, however, would result in not just the organization but also the values recorded in R changing if a or b were complex. The solution is to code

```
R = (transposeonly(a), transposeonly(b))
```

The above line will work for real or complex a and b. If you were concerned about memory consumption, you could instead code

```
R = (a \ b)
_transposeonly(R)
```

Conformability

transposeonly(*A*):

A:	$r \times c$
result:	$c \times r$

`_transposeonly(A):`

    input:

           A: $r \times c$

    output:

           A: $c \times r$

Diagnostics

`_transposeonly(A)` aborts with error if A is a view.

Also see

[M-2] **op_transpose** — Conjugate transpose operator

[M-5] **_transpose( )** — Transposition in place

[M-4] **manipulation** — Matrix manipulation

Title

Syntax

> *real matrix* trunc(*real matrix R*)
>
> *real matrix* floor(*real matrix R*)
>
> *real matrix* ceil(*real matrix R*)
>
> *real matrix* round(*real matrix R*)
>
> *real matrix* round(*real matrix R*, *real matrix U*)

Description

These functions convert noninteger values to integers by moving toward 0, moving down, moving up, or rounding. These functions are typically used with scalar arguments, and they return a scalar in that case. When used with vectors or matrices, the operation is performed element by element.

trunc(R) returns the integer part of R.

floor(R) returns the largest integer i such that $i \le R$.

ceil(R) returns the smallest integer i such that $i \ge R$.

round(R) returns the integer closest to R.

round(R, U) returns the values of R rounded in units of U and is equivalent to round((R:/U)):*U. For instance, round(R, 2) returns R rounded to the closest even number. round(R, .5) returns R rounded to the closest multiple of one half. round(R, 1) returns R rounded to the closest integer and so is equivalent to round(R).

Remarks

Remarks are presented under the following headings:

> *Relationship to Stata's functions*
> *Examples of rounding*

Relationship to Stata's functions

trunc() is equivalent to Stata's int() function.

floor(), ceil(), and round() are equivalent to Stata's functions of the same name.

Examples of rounding

x	$\texttt{trunc}(x)$	$\texttt{floor}(x)$	$\texttt{ceil}(x)$	$\texttt{round}(x)$
1	1	1	1	1
1.3	1	1	2	1
1.6	1	1	2	2
−1	−1	−1	−1	−1
−1.3	−1	−2	−1	−1
−1.6	−1	−2	−1	−2

Conformability

$\texttt{trunc}(R)$, $\texttt{floor}(R)$, $\texttt{ceil}(R)$:

        R: $r \times c$

   result: $r \times c$

$\texttt{round}(R)$:

        R: $r \times c$

   result: $r \times c$

$\texttt{round}(R, U)$:

        R: $r_1 \times c_1$

        U: $r_2 \times c_2$,   *R* and *U* r-conformable

   result: $\max(r_1, r_2) \times \max(c_1, c_2)$

Diagnostics

Most Stata and Mata functions return missing when arguments contain missing, and in particular, return . whether the argument is ., .a, .b, ..., .z. The logic is that performing the operation on a missing value always results in the same missing-value result. For example, $\texttt{sqrt}(.a){=}{=}.$.

These functions, however, when passed a missing value, return the particular missing value. Thus $\texttt{trunc}(.a){=}{=}.a$, $\texttt{floor}(.b){=}{=}.b$, $\texttt{ceil}(.c){=}{=}.c$, and $\texttt{round}(.d){=}{=}.d$.

For $\texttt{round}()$ with two arguments, this applies to the first argument and only when the second argument is not missing. If the second argument is missing (whether ., .a, ..., or .z), then . is returned.

Also see

[M-4] **scalar** — Scalar mathematical functions

Title

[M-5] uniqrows( ) — Obtain sorted, unique values

Syntax

> *transmorphic matrix* uniqrows(*transmorphic matrix P*)

Description

uniqrows(*P*) returns a sorted matrix containing the unique rows of *P*.

Remarks

```
: x
          1    2    3

    1     4    5    7
    2     4    5    6
    3     1    2    3
    4     4    5    6

: uniqrows(x)
          1    2    3

    1     1    2    3
    2     4    5    6
    3     4    5    7
```

Conformability

uniqrows(*P*)
- *P*: $r_1 \times c_1$
- *result*: $r_2 \times c_1, \quad r_2 \le r_1$

Diagnostics

In uniqrows(*P*), if rows(*P*)==0, J(0, cols(*P*), missingof(*P*)) is returned.

If rows(*P*)>0 and cols(*P*)==0, J(1, 0, missingof(*P*)) is returned.

Also see

[M-5] **sort( )** — Reorder rows of matrix

[M-4] **manipulation** — Matrix manipulation

Title

> **[M-5] unitcircle( )** — Complex vector containing unit circle

Syntax

>   *complex colvector*   unitcircle(*real scalar n*)

Description

unitcircle(*n*) returns a column vector containing $C(\cos(\theta), \sin(\theta))$ for $0 \leq \theta \leq 2\pi$ in *n* points.

Conformability

```
unitcircle(n):
        n:      1 × 1
   result:      n × 1
```

Diagnostics

None.

Also see

[M-4] **standard** — Functions to create standard matrices

Title

[M-5] **unlink( )** — Erase file

Syntax

$void$ unlink(*real scalar filename*)

$real\ scalar$ _unlink(*real scalar filename*)

Description

unlink(*filename*) erases *filename* if it exists, does nothing if *filename* does not exist, and aborts with error if *filename* exists but cannot be erased.

_unlink(*filename*) does the same, except that, if *filename* cannot be erased, rather than aborting with error, _unlink() returns a negative error code. _unlink() returns 0 if *filename* was erased or *filename* did not exist.

Remarks

To remove directories, see rmdir() in [M-5] **chdir( )**.

Conformability

unlink(*filename*)
    filename: 1×1
    result: *void*

_unlink(*filename*)
    filename: 1×1
    result: 1×1

Diagnostics

unlink(*filename*) aborts with error when _unlink() would give a negative result.

_unlink(*filename*) returns a negative result if the file cannot be erased and returns 0 otherwise. If the file did not exist, 0 is returned. When there is an error, most commonly returned are -3602 (filename invalid) or -3621 (file is read-only).

Also see

[M-4] **io** — I/O functions

Title

[M-5] **valofexternal( )** — Obtain value of external global

Syntax

transmorphic matrix `valofexternal(`*string scalar name*`)`

Description

`valofexternal(`*name*`)` returns the contents of the external global matrix, vector, or scalar whose name is specified by *name*; it returns `J(0,0,.)` if the external global is not found.

Also see *Linking to external globals* in [M-2] **declarations**.

Remarks

Also see [M-5] **findexternal( )**. Rather than returning a pointer to the external global, as does `findexternal()`, `valofexternal()` returns the contents of the external global. This is useful when the external global contains a scalar:

```
tol = valofexternal("tolerance")
if (tol==J(0,0,.)) tol = 1e-6
```

Using `findexternal()`, one alternative would be

```
if ((p = findexternal("tolerance"))==NULL) tol = 1e-6
else tol = *p
```

For efficiency reasons, use of `valofexternal()` should be avoided with nonscalar objects; see [M-5] **findexternal( )**.

Conformability

`valofexternal(`*name*`)`:
> *name*: 1×1
> *result*: $r \times c$ or 0×0 if not found

Diagnostics

`valofexternal()` aborts with error if *name* contains an invalid name.

`valofexternal(`*name*`)` returns `J(0,0,.)` if *name* does not exist.

Also see

[M-5] **findexternal( )** — Find, create, and remove external globals

[M-4] **programming** — Programming functions

Title

[M-5] **Vandermonde( )** — Vandermonde matrices

Syntax

numeric matrix Vandermonde(*numeric colvector x*)

Description

Vandermonde(*x*) returns the Vandermonde matrix containing the geometric progression of x in each row

$$\begin{bmatrix} 1 & x_1 & x_1^2 & x_1^3 & \cdots & x_1^{n-1} \\ 1 & x_2 & x_2^2 & x_2^3 & \cdots & x_2^{n-1} \\ \vdots & \vdots & \vdots & \vdots & \ddots & \vdots \\ 1 & x_n & x_n^2 & x_n^3 & \cdots & x_n^{n-1} \end{bmatrix}$$

where $n = $ rows(*x*). Some authors use the transpose of the above matrix.

Remarks

Vandermonde matrices are useful in polynomial interpolation.

Conformability

Vandermonde(*x*):
$$x: \quad n \times 1$$
$$result: \quad n \times n$$

Diagnostics

None.

Alexandre-Théophile Vandermonde (1735–1796) was born in Paris. His first passion was music (particularly the violin) and he turned to mathematics only at the age of 35. Four papers dated 1771 and 1772 are his entire mathematical output, although all contain good work. He also worked in experimental science and the theory of music, arguing that musicians should ignore all theory and trust their trained ears, and was busy with various committees and other administration. Vandermonde was a strong supporter of the French Revolution. He is now best known for the Vandermonde determinant, even though it does not appear in any of his papers, and for the associated matrix. Lebesgue later conjectured that the attribution arises from a misreading of Vandermonde's notation.

Reference

Jones, P. S. 1976. Vandermonde, Alexandre-Théophile. In Vol. 13 of *Dictionary of Scientific Biography*, ed. C. C. Gillispie, 571–572. New York: Scribner's.

Also see

[M-4] **standard** — Functions to create standard matrices

Title

[M-5] **vec( )** — Stack matrix columns

Syntax

transmorphic colvector vec(*transmorphic matrix T*)

transmorphic colvector vech(*transmorphic matrix T*)

transmorphic matrix invvech(*transmorphic colvector v*)

Description

vec(*T*) returns *T* transformed into a column vector with one column stacked onto the next.

vech(*T*) returns square and typically symmetric matrix *T* transformed into a column vector; only the lower half of the matrix is recorded.

invvech(*v*) returns vech()-style column vector *v* transformed into a symmetric (Hermitian) matrix.

Remarks

Remarks are presented under the following headings:

> *Example of vec( )*
> *Example of vech( ) and invvech( )*

Example of vec( )

```
: x
          1   2   3
      ┌─────────────┐
    1 │  1   2   3  │
    2 │  4   5   6  │
      └─────────────┘

: vec(x)
          1
      ┌─────┐
    1 │  1  │
    2 │  4  │
    3 │  2  │
    4 │  5  │
    5 │  3  │
    6 │  6  │
      └─────┘
```

Example of vech( ) and invvech( )

```
: x
[symmetric]
        1    2    3
    1 │ 1
    2 │ 2    4
    3 │ 3    6    9

: v = vech(x)
: v
        1
    1 │ 1
    2 │ 2
    3 │ 3
    4 │ 4
    5 │ 6
    6 │ 9

: invvech(v)
[symmetric]
        1    2    3
    1 │ 1
    2 │ 2    4
    3 │ 3    6    9
```

Conformability

vec(T):

$$T: \quad r \times c$$
$$result: \quad r * c \times 1$$

vech(T):

$$T: \quad n \times n$$
$$result: \quad (n(n+1)/2 \times 1)$$

invvech(v):

$$v: \quad (n(n+1)/2 \times 1)$$
$$result: \quad n \times n$$

Diagnostics

vec(T) cannot fail.

vech(T) aborts with error if T is not square. vech( ) records only the lower triangle of T; it does not require T be symmetric.

invvech(v) aborts with error if v does not have 0, 1, 3, 6, 10, ... rows.

Also see

[M-4] **manipulation** — Matrix manipulation

[M-6] Mata glossary of common terms

Title

[M-6] Glossary

Description

Commonly used terms are defined here.

Mata glossary

arguments

The values a function receives are called the function's arguments. For instance, in lud(A, L, U), A, L, and U are the arguments.

broad type

Two matrices are said to be of the same broad type if the elements in each are numeric, are string, or are pointers. Mata provides two numeric types, real and complex. The term *broad type* is used to mask the distinction within numeric and is often used when discussing operators or functions. One might say, "The comma operator can be used to join the rows of two matrices of the same broad type," and the implication of that is that one could join a real to a complex. The result would be complex. Also see *type, eltype, and orgtype.*

c-conformability

Matrix, vector, or scalar A is said to be c-conformable with matrix, vector, or scalar B if they have the same number of rows and columns (they are *p-conformable*), or if they have the same number of rows and one is a vector, or if they have the same number of columns and one is a vector, or if one or the other is a scalar. c stands for colon; c-conformable matrices are suitable for being used with Mata's :*op* operators. A and B are c-conformable if and only if

A	B
$r \times c$	$r \times c$
$r \times 1$	$r \times c$
$1 \times c$	$r \times c$
1×1	$r \times c$
$r \times c$	$r \times 1$
$r \times c$	$1 \times c$
$r \times c$	1×1

The idea behind c-conformability is generalized elementwise operation. Consider $C=A:*B$. If A and B have the same number of rows and have the same number of columns, then $||C_{ij}|| = ||A_{ij}*B_{ij}||$. Now say that A is a column vector and B is a matrix. Then $||C_{ij}|| = ||A_i*B_{ij}||$: each element of A is applied to the entire row of B. If A is a row vector, each column of A is applied to the entire column of B. If A is a scalar, A is applied to every element of B. And then all the rules repeat, with the roles of A and B interchanged. See [M-2] **op_colon** for a complete definition.

class programming

See *object-oriented programming*.

colon operators

Colon operators are operators preceded by a colon, and the colon indicates that the operator is to be performed elementwise. $A:*B$ indicates element-by-element multiplication, whereas $A*B$ indicates matrix multiplication. Colons may be placed in front of any operator. Usually one thinks of elementwise as meaning $c_{ij} = a_{ij} <op> b_{ij}$, but in Mata, elementwise is also generalized to include c-conformability. See [M-2] **op_colon**.

column-major order

Matrices are stored as vectors. Column-major order specifies that the vector form of a matrix is created by stacking the columns. For instance,

```
: A
        1   2
   1 ┌ 1   4 ┐
   2 │ 2   5 │
   3 └ 3   6 ┘
```

is stored as

```
        1   2   3   4   5   6
   1 ┌ 1   2   3   4   5   6 ┐
```

in column-major order. The LAPACK functions use column-major order. Mata uses row-major order. See *row-major order*.

colvector

See *vector, colvector, and rowvector*.

complex

A matrix is said to be complex if its elements are complex numbers. Complex is one of two numeric types in Stata, the other being real. Complex is generally used to describe how a matrix is stored and not the kind of numbers that happen to be in it: complex matrix Z might happen to contain real numbers. Also see *type, eltype, and orgtype*.

condition number

The condition number associated with a numerical problem is a measure of that quantity's amenability to digital computation. A problem with a low condition number is said to be well conditioned, whereas a problem with a high condition number is said to be ill conditioned.

Sometimes reciprocals of condition numbers are reported and yet authors will still refer to them sloppily as condition numbers. Reciprocal condition numbers are often scaled between 0 and 1, with values near epsilon(1) indicating problems.

conformability

Conformability refers to row-and-column matching between two or more matrices. For instance, to multiply $A*B$, A must have the same number of columns as B has rows. If that is not true, then the matrices are said to be nonconformable (for multiplication).

Three kinds of conformability are often mentioned in the Mata documentation: *p-conformability*, *c-conformability*, and *r-conformability*.

conjugate

If $z = a + b$i, the conjugate of z is conj$(z) = a - b$i. The conjugate is obtained by reversing the sign of the imaginary part. The conjugate of a real number is the number itself.

conjugate transpose

See *transpose*.

data matrix

A dataset containing n observations on k variables in often stored in an $n \times k$ matrix. An observation refers to a row of that matrix; a variable refers to a column. When the rows are observations and the columns are variables, the matrix is called a data matrix.

declarations

Declarations state the *eltype* and *orgtype* of functions, arguments, and variables. In

```
real matrix myfunc(real vector A, complex scalar B)
{
        real scalar i
        ...
}
```

the real matrix is a function declaration, the real vector and complex scalar are argument declarations, and real scalar i is a variable declaration. The real matrix states the function returns a real matrix. The real vector and complex scalar state the kind of arguments myfunc() expects and requires. The real scalar i helps Mata to produce more efficient compiled code.

Declarations are optional, so the above could just as well have read

```
function myfunc(A, B)
{
    ...
}
```

When you omit the function declaration, you must substitute the word function.

When you omit the other declarations, transmorphic matrix is assumed, which is fancy jargon for a matrix that can hold anything. The advantages of explicit declarations are that they reduce the chances you make a mistake either in coding or in using the function, and they assist Mata in producing more efficient code. Working interactively, most people omit the declarations.

See [M-2] **declarations** for more information.

defective matrix

An $n \times n$ matrix is defective if it does not have n linearly independent eigenvectors.

diagonal matrix

A matrix is diagonal if its off-diagonal elements are zero; A is diagonal if $A[i,j]==0$ for $i!=j$. Usually, diagonal matrices are also *square*. Some definitions require that a diagonal matrix also be a square matrix.

diagonal of a matrix

The diagonal of a matrix is the set of elements $A[i,j]$.

eigenvalues and eigenvectors

A scalar, λ, is said to be an eigenvalue of square matrix $\mathbf{A}$: $n \times n$ if there is a nonzero column vector $\mathbf{x}$: $n \times 1$ (called an eigenvector) such that

$$\mathbf{Ax} = \lambda\mathbf{x} \tag{1}$$

Equation (1) can also be written

$$(\mathbf{A} - \lambda\mathbf{I})\mathbf{x} = 0$$

where $\mathbf{I}$ is the $n \times n$ identity matrix. A nontrivial solution to this system of n linear homogeneous equations exists if and only if

$$\det(\mathbf{A} - \lambda\mathbf{I}) = 0 \tag{2}$$

This nth-degree polynomial in λ is called the characteristic polynomial or characteristic equation of A, and the eigenvalues λ are its roots, also known as the characteristic roots.

The eigenvector defined by (1) is also known as the right eigenvector, because matrix $\mathbf{A}$ is postmultiplied by eigenvector $\mathbf{x}$. See [M-5] **eigensystem( )** and *left eigenvectors*.

eltype

See *type, eltype, and orgtype.*

epsilon(1), etc.

epsilon(1) refers to the unit roundoff error associated with a computer, also informally called machine precision. It is the smallest amount by which a number may differ from 1. For IEEE double-precision variables, epsilon(1) is approximately 2.22045e–16.

epsilon(x) is the smallest amount by which a real number can differ from x, or an approximation thereof; see [M-5] **epsilon( )**.

exp

exp is used in syntax diagrams to mean "any valid expression may appear here"; see [M-2] **exp**.

external variable

See *global variable.*

function

The words *program* and *function* are used interchangeably. The programs that you write in Mata are in fact functions. Functions receive arguments and optionally return results.

Examples of functions that are included with Mata are sqrt(), ttail(), and substr(). Such functions are often referred to as the built-in functions or the library functions. Built-in functions refer to functions implemented in the C code that implements Mata, and library functions refer to functions written in the Mata programming language, but many users use the words interchangeably because how functions are implemented is of little importance. If you have a choice between using a built-in function and a library function, however, the built-in function will usually execute more quickly and the library function will be easier to use. Mostly, however, features are implemented one way or the other and you have no choice.

Also see *underscore functions.*

For a list of the functions that Mata provides, see [M-4] **intro**.

generalized eigenvalues

A scalar, λ, is said to be a generalized eigenvalue of a pair of $n \times n$ square numeric matrices $\mathbf{A}$, $\mathbf{B}$ if there is a nonzero column vector $\mathbf{x}$: $n \times 1$ (called a generalized eigenvector) such that

$$\mathbf{A}\mathbf{x} = \lambda\mathbf{B}\mathbf{x} \tag{1}$$

Equation (1) can also be written

$$(\mathbf{A} - \lambda\mathbf{B})\mathbf{x} = 0$$

A nontrivial solution to this system of n linear homogeneous equations exists if and only if

$$\det(\mathbf{A} - \lambda\mathbf{B}) = 0 \qquad (2)$$

In practice, the generalized eigenvalue problem for the matrix pair $(\mathbf{A}, \mathbf{B})$ is usually formulated as finding a pair of scalars (w, b) and a nonzero column vector $\mathbf{x}$ such that

$$w\mathbf{A}\mathbf{x} = b\mathbf{B}\mathbf{x}$$

The scalar w/b is a generalized eigenvalue if b is not zero.

Infinity is a generalized eigenvalue if b is zero or numerically close to zero. This situation may arise if $\mathbf{B}$ is singular.

The Mata functions that compute generalized eigenvalues return them in two complex vectors, $\mathbf{w}$ and $\mathbf{b}$ of length n. If $\mathbf{b}[i] = 0$, the ith generalized eigenvalue is infinite, otherwise the ith generalized eigenvalue is $\mathbf{w}[i]/\mathbf{b}[i]$.

global variable

Global variables, also known as external variables and as global external variables, refer to variables that are common across programs and which programs may access without the variable being passed as an argument.

The variables you create interactively are global variables. Even so, programs cannot access those variables without engaging in another step, and global variables can be created without your creating them interactively.

To access (and create if necessary) global external variables, you declare the variable in the body of your program:

```
function myfunction(...)
{
        external real scalar globalvar
        ...
}
```

See *Linking to external globals* in [M-2] **declarations**.

There are other ways of creating and accessing global variables, but the declaration method is recommended. The alternatives are crexternal(), findexternal(), and rmexternal() documented in [M-5] **findexternal( )** and valofexternal() documented in [M-5] **valofexternal( )**.

Hermitian matrix

Matrix A is Hermitian if it is equal to its conjugate transpose; $A = A'$; see *transpose*. This means that each off-diagonal element a_{ij} must equal the conjugate of a_{ji}, and that the diagonal elements must be real. The following matrix is Hermitian:

$$\begin{bmatrix} 2 & 4+5i \\ 4-5i & 6 \end{bmatrix}$$

The definition $A = A'$ is the same as the definition for a symmetric matrix, although usually the word *symmetric* is reserved for real matrices and Hermitian, for complex matrices. In this manual, we use the word *symmetric* for both; see *symmetric matrices*.

Hessenberg decomposition

The Hessenberg decomposition of a matrix, **A**, can be written as

$$\mathbf{Q'AQ = H}$$

where **H** is in upper Hessenberg form and **Q** is orthogonal if **A** is real or unitary if **A** is complex. See [M-5] **hessenbergd( )**.

Hessenberg form

A matrix, **A**, is in upper Hessenberg form if all entries below the first subdiagonal are zero: $A_{ij} = 0$ for all $i > j + 1$.

A matrix, **A**, is in lower Hessenberg form if all entries above the first superdiagonal are zero: $A_{ij} = 0$ for all $j > i + 1$.

istmt

An *istmt* is an interactive statement, a statement typed at Mata's colon prompt.

J(r, c, value)

J( ) is the function that returns an $r \times c$ matrix with all elements set to *value*; see [M-5] **J( )**. Also, J( ) is often used in the documentation to describe the various types of *void* matrices; see *void matrix*. Thus the documentation might say that such-and-such returns J(0, 0, .) under certain conditions. That is another way of saying that such-and-such returns a 0×0 real matrix.

When r or c is 0, there are no elements to be filled in with *value*, but even so, *value* is used to determine the type of the matrix. Thus J(0, 0, 1i) refers to a 0×0 complex matrix, J(0, 0, "") refers to a 0×0 string matrix, and J(0, 0, NULL) refers to a 0×0 *pointer* matrix.

In the documentation, J( ) is used for more than describing 0×0 matrices. Sometimes, the matrices being described are $r \times 0$ or are $0 \times c$. Say that a function example(X) is supposed to return a column vector; perhaps it returns the last column of X. Now say that X is 0×0. Function example( ) still should return a column vector, and so it returns a 0×1 matrix. This would be documented by noting that example( ) returns J(0, 1, .) when X is 0×0.

LAPACK

LAPACK stands for Linear Algebra PACKage and forms the basis for many of Mata's linear algebra capabilities; see [M-1] **LAPACK**.

left eigenvectors

A vector $\mathbf{x}$: $n \times 1$ is said to be a left eigenvector of square matrix $\mathbf{A}$: $n \times n$ if there is a nonzero scalar, λ, such that

$$\mathbf{x}\mathbf{A} = \lambda\mathbf{x}$$

lval

lval stands for left-hand-side value and is defined as the property of being able to appear on the left-hand side of an equal-assignment operator. Matrices are *lvals* in Mata, and thus

```
X = ...
```

is valid. Functions are not *lvals*; thus, you cannot code

```
substr(mystr,1,3) = "abc"
```

lvals would be easy to describe except that *pointers* can also be lvals. Few people ever use pointers. See [M-2] **op_assignment** for a complete definition.

machine precision

See *epsilon(1)*, etc.

.mata file

By convention, we store the Mata source code for function *function()* in file *function*.mata; see [M-1] **source**.

matrix

The most general organization of data, containing *r* rows and *c* columns. Vectors, column vectors, row vectors, and scalars are special cases of matrices.

.mlib library

The object code of functions can be collected and stored in a library. Most Mata functions, in fact, are located in the official libraries provided with Stata. You can create your own libraries. See [M-3] **mata mlib**.

.mo file

The object code of a function can be stored in a .mo file, where it can be later reused. See [M-1] **how** and [M-3] **mata mosave**.

norm

A norm is a real-valued function $f(x)$ satisfying

$$
\begin{aligned}
f(0) \;&=\; 0 \\
f(x) \;&>\; 0 \qquad\qquad \text{for all } x \neq 0 \\
f(cx) \;&=\; |c| f(x) \\
f(x+y) \;&\leq\; f(x)+f(y)
\end{aligned}
$$

The word *norm* applied to a vector x usually refers to its Euclidean norm, $p = 2$ norm, or length: the square root of the sum of its squared elements. The are other norms, the popular ones being $p = 1$ (the sum of the absolute values of its elements) and $p = $ infinity (the maximum element). Norms can also be generalized to deal with matrices. See [M-5] **norm( )**.

NULL

A special value for a *pointer* that means "points to nothing". If you list the contents of a pointer variable that contains NULL, the address will show as 0x0. See *pointer*.

numeric

A matrix is said to be numeric if its elements are real or complex; see *type, eltype, and orgtype*.

object code

Object code refers to the binary code that Mata produces from the source code you type as input. See [M-1] **how**.

object-oriented programming

Object-oriented programming is a programming concept that treats programming elements as objects and concentrates on actions affecting those objects rather than merely on lists of instructions. Object-oriented programming uses classes to describe objects. Classes are much like structures with a primary difference being that classes can contain functions (known as methods) as well as variables. Unlike structures, however, classes may inherit variables and functions from other classes, which in theory makes object-oriented programs easier to extend and modify than non–object-oriented programs.

observations and variables

A dataset containing n observations on k variables in often stored in an $n \times k$ matrix. An observation refers to a row of that matrix; a variable refers to a column.

operator

An operator is +, −, and the like. Most operators are binary (or dyadic), such as + in *A+B* and *
in *C*D*. Binary operators also include logical operators such as & and | ("and" and "or") in *E&F*
and *G|H*. Other operators are unary (or monadic), such as ! (not) in !*J*, or both unary and binary,
such as − in −*K* and in *L−M*. When we say "operator" without specifying which, we mean binary
operator. Thus colon operators are in fact colon binary operators. See [M-2] **exp**.

optimization

Mata compiles the code that you write. After compilation, Mata performs an *optimization* step, the
purpose of which is to make the compiled code execute more quickly. You can turn off the optimization
step—see [M-3] **mata set**—but doing so is not recommended.

orgtype

See *type, eltype,* and *orgtype.*

orthogonal matrix and unitary matrix

A is orthogonal if *A* is *square* and *A′A==I*. The word orthogonal is usually reserved for real matrices;
if the matrix is complex, it is said to be unitary (and then transpose means conjugate-transpose). We
use the word orthogonal for both real and complex matrices.

If *A* is orthogonal, then $\det(A) = \pm 1$.

p-conformability

Matrix, vector, or scalar *A* is said to be p-conformable with matrix, vector, or scalar *B* if
rows(*A*)==rows(*B*) and cols(*A*)==cols(*B*). *p* stands for plus; p-conformability is one of the
properties necessary to be able to add matrices together. p-conformability, however, does not imply
that the matrices are of the same type. Thus (1,2,3) is p-conformable with (4,5,6) and with
("this","that","what") but not with (4\5\6).

permutation matrix and permutation vector

A *permutation matrix* is an $n \times n$ matrix that is a row (or column) permutation of the identity matrix.
If *P* is a permutation matrix, then *P*A* permutes the rows of *A* and *A*P* permutes the columns of *A*.
Permutation matrices also have the property that $P^{-1} = P'$.

A *permutation vector* is a $1 \times n$ or $n \times 1$ vector that contains a permutation of the integers 1,
2, ..., *n*. Permutation vectors can be used with subscripting to reorder the rows or columns of a
matrix. Permutation vectors are a memory-conserving way of recording permutation matrices; see
[M-1] **permutation**.

pointer

A matrix is said to be a pointer matrix if its elements are pointers.

A pointer is the address of a *variable*. Say that variable X contains a matrix. Another variable p might contain 137,799,016 and, if 137,799,016 were the address at which X were stored, then p would be said to point to X. Addresses are seldom written in base 10, and so rather than saying p contains 137,799,016, we would be more likely to say that p contains 0x836a568, which is the way we write numbers in base 16. Regardless of how we write addresses, however, p contains a number and that number corresponds to the address of another variable.

In our program, if we refer to p, we are referring to p's contents, the number 0x836a568. The monadic operator * is defined as "refer to the address" or "dereference": *p means X. We could code Y = *p or Y = X, and either way, we would obtain the same result. In our program, we could refer to $X[i,j]$ or $(*p)[i,j]$, and either way, we would obtain the i, j element of X.

The monadic operator & is how we put addresses into p. To load p with the address of X, we code $p = \&X$.

The special address 0 (zero, written in hexadecimal as 0x0), also known as NULL, is how we record that a pointer variable points to nothing. A pointer variable contains NULL or it contains a valid address of another variable.

See [M-2] **pointers** for a complete description of pointers and their use.

pragma

"(Pragmatic information) A standardised form of comment which has meaning to a compiler. It may use a special syntax or a specific form within the normal comment syntax. A pragma usually conveys non-essential information, often intended to help the compiler to optimise the program." See *The Free On-line Dictionary of Computing*, http://www.foldoc.org/, Editor Denis Howe. For Mata, see [M-2] **pragma**.

rank

Terms in common use are rank, row rank, and column rank. The row rank of a matrix A: $m \times n$ is the number of rows of A that are linearly independent. The column rank is defined similarly, as the number of columns that are linearly independent. The terms *row rank* and *column rank*, however, are used merely for emphasis; the ranks are equal and the result is simply called the rank of A.

For a square matrix A (where $m==n$), the matrix is invertible if and only if rank(A)==n. One often hears that A is of full rank in this case and rank deficient in the other. See [M-5] **rank( )**.

r-conformability

A set of two or more matrices, vectors, or scalars A, B, ..., are said to be r-conformable if each is *c-conformable* with a matrix of max(rows(A), rows(B), ...) rows and max(cols(A), cols(B), ...) columns.

r-conformability is a more relaxed form of *c-conformability* in that, if two matrices are c-conformable, they are r-conformable, but not vice versa. For instance, A: 1×3 and B: 3×1 are r-conformable but not c-conformable. Also, c-conformability is defined with respect to a pair of matrices only; r-conformability can be applied to a set of matrices.

r-conformability is often required of the arguments for functions that would otherwise naturally be expected to require scalars. See *r-conformability* in [M-5] **normal( )** for an example.

real

A matrix is said to be a real matrix if its elements are all reals and it is stored in a `real` matrix. Real is one of the two numeric types in Mata, the other being complex. Also see *type, eltype, and orgtype.*

row-major order

Matrices are stored as vectors. Row-major order specifies that the vector form of a matrix is created by stacking the rows. For instance,

```
: A
        1   2   3
   1    1   2   3
   2    4   5   6
```

is stored as

```
        1   2   3   4   5   6
   1    1   2   3   4   5   6
```

in row-major order. Mata uses row-major order. The LAPACK functions use column-major order. See *column-major order.*

rowvector

See *vector, colvector, and rowvector.*

scalar

A special case of a *matrix* with one row and one column. A scalar may be substituted anywhere a matrix, vector, column vector, or row vector is required, but not vice versa.

Schur decomposition

The Schur decomposition of a matrix, **A**, can be written as

$$Q'AQ = T$$

where **T** is in Schur form and **Q**, the matrix of Schur vectors, is orthogonal if **A** is real or unitary if **A** is complex. See [M-5] **schurd( )**.

Schur form

There are two Schur forms: real Schur form and complex Schur form.

A real matrix is in Schur form if it is block upper triangular with 1×1 and 2×2 diagonal blocks. Each 2×2 diagonal block has equal diagonal elements and opposite sign off-diagonal elements. The real eigenvalues are on the diagonal and complex eigenvalues can be obtained from the 2×24 diagonal blocks.

A complex square matrix is in Schur form if it is upper triangular with the eigenvalues on the diagonal.

source code

Source code refers to the human-readable code that you type into Mata to define a function. Source code is compiled into object code, which is binary. See [M-1] **how**.

square matrix

A matrix is square if it has the same number of rows and columns. A 3×3 matrix is square; a 3×4 matrix is not.

string

A matrix is said to be a string matrix if its elements are strings (text); see *type, eltype, and orgtype*. In Mata, a string may be text or binary and may be up to 2,147,483,647 characters (bytes) long.

structure

A structure is an *eltype*, indicating a set of variables tied together under one name. `struct mystruct` might be

```
struct mystruct {
        real scalar     n1, n2
        real matrix     X
}
```

If variable a was declared a `struct mystruct scalar`, then the scalar a would contain three pieces: two real scalars and one real matrix. The pieces would be referred to as `a.n1`, `a.n2`, and `a.X`. If variable b were also declared a `struct mystruct scalar`, it too would contain three pieces, `b.n1`, `b.n2`, and `b.X`. The advantage of structures is that they can be referred to as a whole. You can code `a.n1=b.n1` to copy one piece, or you can code `a=b` if you wanted to copy all three pieces. In all ways, a and b are variables. You may pass a to a subroutine, for instance, which amounts to passing all three values.

Structures variables are usually scalar, but they are not limited to being so. If A were a `struct mystruct matrix`, then each element of A would contain three pieces, and one could refer, for instance, to `A[2,3].n1`, `A[2,3].n2`, and `A[2,3].X`, and even to `A[2,3].X[3,2]`.

See [M-2] **struct**.

subscripts

Subscripts are how you refer to an element or even a submatrix of a matrix.

Mata provides two kinds of subscripts, known as list subscripts and range subscripts.

In list subscripts, $A[2,3]$ refers to the $(2,3)$ element of A. $A[(2\backslash3),(4,6)]$ refers to the submatrix made up of the second and third rows, fourth and sixth columns, of A.

In range subscripts, $A[|2,3|]$ also refers to the $(2,3)$ element of A. $A[|2,3\backslash4,6|]$ refers to the submatrix beginning at the $(2,3)$ element and ending at the $(4,6)$ element.

See [M-2] **subscripts** for more information.

symmetric matrices

Matrix A is symmetric if $A = A'$. The word *symmetric* is usually reserved for real matrices, and in that case, a symmetric matrix is a square matrix with a_{ij}==a_{ji}.

Matrix A is said to be Hermitian if $A = A'$, where the transpose operator is understood to mean the conjugate-transpose operator; see *Hermitian matrix*. In Mata, the $'$ operator is the conjugate-transpose operator, and thus, in this manual, we will use the word *symmetric* both to refer to real, symmetric matrices and to refer to complex, Hermitian matrices.

Sometimes, you will see us follow the word *symmetric* with a parenthesized Hermitian, as in, "the resulting matrix is symmetric (Hermitian)". That is done only for emphasis.

The inverse of a symmetric (Hermitian) matrix is symmetric (Hermitian).

symmetriconly

Symmetriconly is a word we have coined to refer to a square matrix whose corresponding off-diagonal elements are equal to each other, whether the matrix is real or complex. Symmetriconly matrices have no mathematical significance, but sometimes, in data-processing and memory-management routines, it is useful to be able to distinguish such matrices.

time-series–operated variable

Time-series–operated variables are a Stata concept. The term refers to *op*. *varname* combinations such as L.gnp to mean the lagged value of variable gnp. Mata's [M-5] **st_data( )** function works with time-series–operated variables just as it works with other variables, but many other Stata-interface functions do not allow *op*. *varname* combinations. In those cases, you must use [M-5] **st_tsrevar( )**.

traceback log

When a function fails—either because of a programming error or because it was used incorrectly—it produces a traceback log:

```
: myfunction(2,3)
              solve():  3200  conformability error
              mysub():     -  function returned error
         myfunction():     -  function returned error
             <istmt>:      -  function returned error
r(3200);
```

The log says that solve() detected the problem—arguments are not conformable—and that solve() was called by mysub() was called by myfunction() was called by what you typed at the keyboard. See [M-2] **errors** for more information.

transmorphic

Transmorphic is an *eltype*. A scalar, vector, or matrix can be transmorphic, which indicates that its elements may be real, complex, string, pointer, or even a structure. The elements are all the same type; you are just not saying which they are. Variables that are not declared are assumed to be transmorphic, or a variable can be explicitly declared to be transmorphic. Transmorphic is just fancy jargon for saying that the elements of the scalar, vector, or matrix can be anything and that, from one instant to the next, the scalar, vector, or matrix might change from holding elements of one type to elements of another.

See [M-2] **declarations**.

transpose

The transpose operator is written different ways in different books, including $'$, superscript $*$, superscript T, and superscript H. Here we use the $'$ notation: A' means the transpose of A, A with its rows and columns interchanged.

In complex analysis, the transpose operator, however it is written, is usually defined to mean the conjugate transpose; that is, one interchanges the rows and columns of the matrix and then one takes the conjugate of each element, or one does it in the opposite order—it makes no difference. Conjugation simply means reversing the sign of the imaginary part of a complex number: the conjugate of 1+2i is 1-2i. The conjugate of a real is the number itself; the conjugate of 2 is 2.

In Mata, $'$ is defined to mean conjugate transpose. Since the conjugate of a real is the number itself, A' is regular transposition when A is real. Similarly, we have defined $'$ so that it performs regular transposition for string and pointer matrices. For complex matrices, however, $'$ also performs conjugation.

If you have a complex matrix and simply want to transpose it without taking the conjugate of its elements, see [M-5] **transposeonly( )**. Or code conj(A'). The extra conj() will undo the undesired conjugation performed by the transpose operator.

Usually, however, you want transposition and conjugation to go hand in hand. Most mathematical formulas, generalized to complex values, work that way.

triangular matrix

A triangular matrix is a matrix with all elements equal to zero above the diagonal or all elements equal to zero below the diagonal.

A matrix A is *lower triangular* if all elements are zero above the diagonal, i.e., if $A[i,j]==0, j > i$.

A matrix A is *upper triangular* if all elements are zero below the diagonal, i.e., if $A[i,j]==0, j < i$.

A *diagonal matrix* is both lower and upper triangular. That is worth mentioning because any function suitable for use with triangular matrices is suitable for use with diagonal matrices.

A triangular matrix is usually *square*.

The inverse of a triangular matrix is a triangular matrix. The determinant of a triangular matrix is the product of the diagonal elements. The eigenvalues of a triangular matrix are the diagonal elements.

type, eltype, and orgtype

The *type* of a matrix (or vector or scalar) is formally defined as the matrix's *eltype* and *orgtype*, listed one after the other—such as real vector—but it can also mean just one or the other—such as the *eltype* real or the *orgtype* vector.

(Continued on next page)

eltype refers to the type of the elements. The *eltypes* are

real	numbers such as 1, 2, 3.4
complex	numbers such as 1+2i, 3+0i
string	strings such as "bill"
pointer	pointers such as &*varname*
struct	structures
numeric	meaning real or complex
transmorphic	meaning any of the above

orgtype refers to the organizational type. *orgtype* specifies how the elements are organized. The *orgtypes* are

matrix	two-dimensional arrays
vector	one-dimensional arrays
colvector	one-dimensional column arrays
rowvector	one-dimensional row arrays
scalar	single items

The fully specified type is the element and organization types combined, as in real vector.

underscore functions

Functions whose names start with an underscore are called underscore functions, and when an underscore function exists, usually a function without the underscore prefix also exists. In those cases, the function is usually implemented in terms of the underscore function, and the underscore function is harder to use but is faster or provides greater control. Usually, the difference is in the handling of errors.

For instance, function fopen() opens a file. If the file does not exist, execution of your program is aborted. Function _fopen() does the same thing, but if the file cannot be opened, it returns a special value indicating failure, and it is the responsibility of your program to check the indicator and to take the appropriate action. This can be useful when the file might not exist, and if it does not, you wish to take a different action. Usually, however, if the file does not exist, you will wish to abort, and use of fopen() will allow you to write less code.

unitary matrix

See *orthogonal matrix*.

variable

In a program, the entities that store values (*a*, *b*, *c*, . . . , *x*, *y*, *z*) are called variables. Variables are given names of 1 to 32 characters long. To be terribly formal about it: a variable is a container; it contains a matrix, vector, or scalar and is referred to by its variable name or by another variable containing a *pointer* to it.

Also, *variable* is sometimes used to refer to columns of data matrices; see *data matrix*.

vector, colvector, and rowvector

A special case of a matrix with either one row or one column. A vector may be substituted anywhere a matrix is required. A matrix, however, may not be substituted for a vector.

A `colvector` is a vector with one column.

A `rowvector` is a vector with one row.

A `vector` is either a `rowvector` or `colvector`, without saying which.

view

A view is a special type of matrix that appears to be an ordinary matrix, but in fact the values in the matrix are the values of certain or all variables and observations in the Stata dataset that is currently in memory. Its values are not just equal to the dataset's values; they are the dataset's values: if an element of the matrix is changed, the corresponding variable and observation in the Stata dataset also changes. Views are obtained by `st_view()` and are efficient; see [M-5] **st_view( )**.

void function

A function is said to be void if it returns nothing. For instance, the function [M-5] **printf( )** is a void function; it prints results, but it does not return anything in the sense that, say, [M-5] **sqrt( )** does. It would not make any sense to code x = `printf("hi there")`, but coding x = `sqrt(2)` is perfectly logical.

void matrix

A matrix is said to be void if it is 0×0, $r \times 0$, or $0 \times c$; see [M-2] **void**.

Also see

[M-0] **intro** — Introduction to the Mata manual

[M-1] **intro** — Introduction and advice

Subject and author index

This is the subject and author index for the *Mata Reference Manual*. Readers interested in topics other than Mata should see the combined subject index (and the combined author index) in the *Quick Reference and Index*. The combined index indexes the *Getting Started* manuals, the *User's Guide*, and all the reference manuals except this one.

N